AF865230

Excel

Get the Results You Want!

SmartStudy 10

Advanced-level Mathematics

Lyn Baker

PASCAL PRESS

© 2015 Lyn Baker and Pascal Press
Reprinted 2017, 2019, 2021, 2023, 2025

ISBN 978 1 74125 477 8

Pascal Press
PO Box 250
Glebe NSW 2037
www.pascalpress.com.au

Publisher: Vivienne Joannou
Project editor: Rosemary Peers
Answers checked by Peter Little
Typeset by Precision Typesetting (Barbara Nilsson)
Cover design by DiZign Pty Ltd
Printed by Vivar Printing/Green Giant Press

Reproduction and communication for educational purposes

The *Australian Copyright Act* 1968 (the Act) allows a maximum of one chapter or 10% of this book, whichever is the greater, to be copied by any educational institution for its educational purposes provided that the educational institution (or the body that administers it) has given a remuneration notice to Copyright Agency under the Act.

For details of the Copyright Agency licence for educational institutions contact:

Copyright Agency
Level 12, 66 Goulburn Street
Sydney NSW 2000
Telephone: (02) 9394 7600
Facsimile: (02) 9394 7601
E-mail: memberservices@copyright.com.au

Reproduction and communication for other purposes

Except as permitted under the Act (for example, a fair dealing for the purposes of study, research, criticism or review) no part of this book may be reproduced, stored in a retrieval system, or transmitted in any form or by any means without prior written permission. All inquiries should be made to the publisher at the address above.

Students
All care has been taken in compiling this book, but please check with your teacher about the exact requirements of the course as these can change from year to year.

TABLE OF CONTENTS

Real Numbers 1

Rational and Irrational Numbers 1

Surds 5

Logarithms 9

Functions and Other Graphs 13

Functions 13

Polynomials 17

Polynomial Expressions 17

Further Polynomials 21

Linear and Non-linear Relationships 25

Parabolas 25

Further Graphs 29

Curve-sketching Techniques 33

The Equation of a Straight Line 37

Equations 41

Algebraic Techniques 45

Further Factorisation 45

Quadratic Equations 49

Surface Area and Volume 53

Volume of Pyramids, Cones and Spheres 53

Surface Area of Pyramids, Cones and Spheres 57

Geometry 61

Geometry Review 61

Circle Geometry 65

Parts of Circles and Chord Properties 65

Angle Properties, Tangents and Secants 69

Pythagoras and Trigonometry 73

Problems in Two and Three Dimensions 73

Complementary Ratios, Exact Values and the Unit Circle 77

Sine Rule, Cosine Rule and Area of a Triangle 81

Mean and Standard Deviation 85

Investigations of Data Sets and Reports 89

Tips for the Sample Exam Papers **94**

Sample Exam Papers **95**

Paper 1: Level of Difficulty—Average 95

Paper 2: Level of Difficulty—Above Average 99

Paper 3: Level of Difficulty—Difficult 103

Worked Solutions **107**

Tests 107

Sample Exam Papers 173

Test & Exam Results **182**

Feedback Checklist **183**

TAKE THESE REVISION STEPS TO SUCCESS!

Step 1 Study Notes

- This section contains valuable **suggestions, tips** and **essential points** about the topic.
- When reading this section, **highlight** the points that are new to you or that you consistently forget. You could also **rewrite** these points so that you are making a summary of the Study Notes. Keep this summary in a prominent place at home—you could stick it on a mirror, wall or door.
- Work carefully through the **checklist** at the end of the section to ensure that you have mastered each of the listed skills.
- If there is something that you are **not sure about**, take time out to read through your class notes, textbooks or another relevant ***Excel*** book. (Check our website at www.pascalpress.com.au for more titles.)

Step 2 Skills Check

- Check that you know the **basic skills** you need to successfully complete the topic.
- Once you have completed the check, **mark your work** quickly by looking at the answers at the bottom of the page. This is **instant feedback** for you. The **worked solutions** are located at the back of the book if you want to check the working of any answer.
- If you get **several questions** wrong in this section then you should revise your basic skills.

Step 3 Intermediate Test

- This test has very similar questions to the ones you will get in your **class test** or **exam**.
- Look carefully for the **Hints** which are provided for the trickiest questions—these appear at the bottom of the page.
- **Marks** are allocated for each question. These are similar to the marks in your tests.
- Time yourself—**check** how much **time** you have got to complete the test. Also look at the **total marks** of the test to calculate approximately how much time you should spend on each question. For example, if there are twenty marks in total and twenty minutes have been allocated for completion of the test, then spend about one minute on each mark. If you cannot complete all the questions within the suggested time, you may need to revise the topic.
- Fill in the **Your Feedback** panel once you have marked your work in order to calculate your percentage mark. Then complete the **Test & Exam Results** on page 182 to keep a running total of all your test marks.

Step 4 Advanced Test

- This test features only **extension questions** such as problem-solving questions.
- This test is not like a class test, as all the questions in it are challenging. There are no easy questions in it. Mastery of questions in this test, however, will ensure you gain top marks in your class tests and exams, and will also prepare you for the mathematics you will learn next year.
- **Marks** are allocated for each question. These are similar to the marks in your tests.
- Time yourself—**check** how much **time** you have to complete the test. Also look at the **total marks** of the test to calculate approximately how much time you should spend on

each question. For example, if there are forty marks in total and forty minutes have been allocated for completion of the test, then spend about one minute on each mark.

- Fill in the **Your Feedback** panel once you have marked your work in order to calculate your percentage mark. Then complete the **Test & Exam Results** on page 182 to keep a running total of all your test marks.

Step 5 Check Your Solutions

- **Worked solutions** to all questions are found at the back of the book. Work through the solutions to any questions that you got wrong.
- **Longer questions** are usually worth two or more marks and will involve some working. You should set out all your working, because in Maths you may get some marks for your working even if your answer is wrong.
- The **ticks** that appear in the worked solutions indicate those parts of the working which receive marks. Therefore, even if your answer is wrong, you may be entitled to some marks for what you have written. Compare the worked solution to your own working to find out whether you are entitled to any marks for the question.
- If you still **cannot understand** how the correct answer was obtained, revise that part of the topic and, if necessary, refer to your class notes or ask your teacher for help. It is important to learn from your mistakes.

Step 6 Test & Exam Results

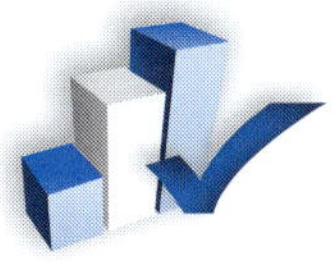

- Go to the **Test & Exam Results** section on page 182 to record your test score as a percentage. When you have completed all topics you will be able to determine your areas of weakness and your areas of strength.
- It is important that you know which **areas need further work**—to 'know what you don't know'. The more you prepare for the topic tests, the more successful you will be and the more you will remember when you sit your end-of-term/semester/year test or exam.

Step 7 Tips for the Sample Exam Papers

- These **useful tips** appear on page 94. Read them before you start one of the Sample Exams (see Step 8 below).

Step 8 Sample Exam Papers

- Three **Sample Exam Papers** are provided at the end of the book. These are of three levels of difficulty: Average, Above Average and Difficult. The Above Average paper will be very similar to your final examination.
- Before attempting the Sample Exam Papers, make sure that you have completed all of the **Tests** and have worked through the solutions to all questions that you answered incorrectly.
- Set aside the **time allowed** for the paper and complete it under **exam conditions**—no sneaking a look at your notes or textbooks! That way you will be better prepared for your final exam.
- **Worked solutions** to the Sample Exam Papers are found at the back of the book. Work through the solutions to any questions that you got wrong. Carefully note the **ticks** in the worked solutions and remember to give yourself marks for correct working. Write down your total marks for each section in the **Your Feedback** boxes at the end of each part of the paper, then add them up to get a total percentage for each test.

HOW TO USE THIS BOOK TO STUDY FOR A CLASS TEST, HALF-YEARLY OR END-OF-YEAR EXAM

Depending on your teacher or school, you will be given a variety of tests and exams each year. There may be a single-topic test, a test that covers a number of topics, a semester test or exam, or even a half-yearly or yearly exam.

Step 1

Find out which topics will be covered in the class test.

- To do this, look at your class workbook/textbook, laptop/tablet or online study program, and ask your teacher.
- For example, your class test may be on Rational and Irrational Numbers, Surds and Logarithms.

Step 2

Match the topics that your test is on to the topics in this book.

- For example, the material on pages 1 to 12 covers Rational and Irrational Numbers, Surds and Logarithms.

Step 3

Use this book to study the topics being tested.

- For example, the first three units in this book contain the topics you will study for your class test on Rational and Irrational Numbers, Surds and Logarithms!

Note:

- When you are using this book to study for a **half-yearly** test, follow the same steps as above—the only difference being that you will have more topics to revise of course!
- When you are using this book to study for an **end-of-year** test, you will more than likely need to study the whole book!

RATIONAL AND IRRATIONAL NUMBERS

Real Numbers

STUDY NOTES

1 An **integer** is a whole number (positive, negative or zero). A **rational number** is one that can be expressed as a fraction ($\frac{a}{b}$, where a and b are intgers). All integers, fractions and terminating or recurring decimals are rational. An irrational number cannot be expressed in the form $\frac{a}{b}$. Examples of irrational numbers are π, $\sqrt{2}$, $\sqrt{3}$ and $4\sqrt{5}$ ($4\sqrt{5}$ means $4 \times \sqrt{5}$).

2 **Real numbers** are numbers that can be represented by points on the number line. Real numbers include both rational and irrational numbers.

3 Decimals may either terminate or recur. **Terminating decimals** come to an end (e.g. 0.2, 0.125, 0.76543). **Recurring decimals** have digits that repeat. There may be only one digit that repeats (e.g. $0.\dot{1} = 0.111111\ldots$, $0.45\dot{6} = 0.45666666\ldots$) or a sequence of several digits (e.g. $0.\dot{2}8571\dot{4} = 0.28571428571428\ldots$).

4 Taking the **square root** is the opposite operation to **squaring**. So one operation undoes what the other does. For example, if a is a positive number $(\sqrt{a})^2 = a$ and $(\sqrt{a^2}) = a$.

5 Fractional indices are used to express roots. $x^{\frac{1}{n}} = \sqrt[n]{x}$ and $x^{\frac{m}{n}} = \sqrt[n]{x^m} = (\sqrt[n]{x})^m$. In particular, raising a number or algebraic term to the power of one half is the same as taking the **square root** **(** $a^{\frac{1}{2}} = \sqrt{a}$ **)**. This is because $\sqrt{a} \times \sqrt{a} = a$ and also $a^{\frac{1}{2}} \times a^{\frac{1}{2}} = a$ (by the index laws) so $a^{\frac{1}{2}}$ must be the same as $\sqrt{a}$. For example, $\sqrt{a^{16}} = (a^{16})^{\frac{1}{2}} = a^{16\times\frac{1}{2}} = a^8$. Also raising a number or algebraic term to the power of one third is the same as taking the **cube root**. For example, $64^{\frac{1}{3}} = \sqrt[3]{64} = 4$ and $\sqrt[3]{x^{12}} = x^4$ because $\sqrt[3]{x^{12}} = (x^{12})^{\frac{1}{3}} = x^{12\times\frac{1}{3}}$.

6 Make sure you can use **your calculator** to find powers and roots. Either the $\boxed{x^{\frac{1}{y}}}$ button or the $\boxed{\sqrt[\square]{\square}}$ button allows you to find roots, e.g. $\sqrt[5]{7776} = 7776^{\frac{1}{5}} = 6$.

7 Know how to use your calculator to **convert fractions to decimals** and vice versa. Some calculators round off recurring decimals in the display (e.g. $0.\dot{6}$ will be shown as 0.666666667).

8 There is a method to **convert recurring decimals to fractions**. Let x be the recurring decimal. Multiply by 10 if one digit repeats, 100 if 2 digits repeat, 1000 if 3 digits repeat and so on. Then subtract the first equation from the second and all repeating digits will cancel out.

For example: Let $x = 0.45454545454545\ldots$ (i)

$100x = 45.45454545454545\ldots$ (ii)

(ii) – (i) $99x = 45$

So $x = \frac{45}{99} = \frac{5}{11}$ $\quad\therefore\ 0.\dot{4}\dot{5} = \frac{5}{11}$

Checklist

Can you:

1 *Understand what is meant by rational, irrational and real numbers?* ☐
2 *Understand that squaring and taking the square root are opposite operations?* ☐
3 *Understand and use fractional indices?* ☐
4 *Convert recurring fractions to decimals with and without a calculator?* ☐
5 *Use you calculator to find powers and roots?* ☐

RATIONAL AND IRRATIONAL NUMBERS

Real Numbers

SKILLS CHECK

1 State whether each number is rational or irrational:

a $\frac{2}{3}$ **b** -7 **c** $\sqrt{3}$ **d** 0 **e** π

2 Which number is represented by x on the number line below?

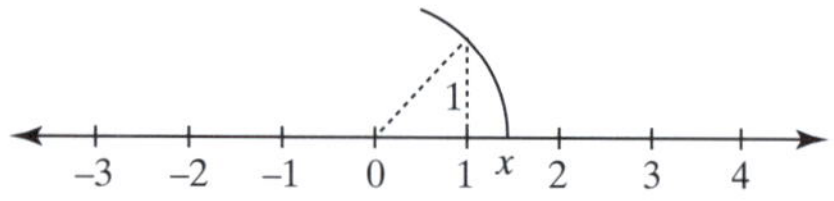

3 Find:

a $(\sqrt{4})^2$ **b** $(\sqrt{9})^2$ **c** $(\sqrt{36})^2$ **d** $(\sqrt{7})^2$

e $(\sqrt{5})^2$ **f** $(2\sqrt{2})^2$

4 Find the value of:

a $4^{\frac{1}{2}}$ **b** $8^{\frac{1}{3}}$ **c** $25^{\frac{1}{2}}$ **d** $1000^{\frac{1}{3}}$

e $32^{\frac{2}{5}}$ **f** $9^{\frac{3}{2}}$ **g** $\sqrt[8]{6561}$ **h** $\sqrt[6]{11\,390\,625}$

5 Write in index form:

a $\sqrt{x}$ **b** $\sqrt[3]{x}$ **c** $\frac{\sqrt{a}}{2}$

6 Write without indices:

a $m^{\frac{1}{2}}$ **b** $t^{\frac{1}{3}}$ **c** $4x^{\frac{1}{2}}$

d $a^{\frac{2}{3}}$ **e** $x^{\frac{3}{2}}$ **f** $(9n)^{\frac{1}{2}}$

7 Write each fraction as a recurring decimal:

a $\frac{2}{9}$ **b** $\frac{5}{12}$ **c** $1\frac{1}{6}$ **d** $\frac{4}{7}$

8 State whether each fraction is a terminating or recurring decimal:

a $\frac{2}{5}$ **b** $\frac{2}{3}$ **c** $\frac{3}{7}$ **d** $\frac{7}{8}$

9 Determine whether the following statements are true or false:

a $\sqrt{2a} = 2a^{\frac{1}{2}}$ **b** $0.\dot{3} = \frac{1}{3}$ **c** $0.\dot{5} = \frac{1}{5}$

10 Simplify:

a $2x^{\frac{1}{2}} \times 4x^{\frac{1}{2}}$ **b** $20p^{\frac{1}{2}} \div 2p^{\frac{1}{2}}$ **c** $6y^{\frac{1}{3}} \times 3y^{\frac{1}{3}}$

PAGE 107

Answers **1 a** rational **b** rational **c** irrational **d** rational **e** irrational **2** $\sqrt{2}$ **3 a** 4 **b** 9 **c** 36 **d** 7 **e** 5 **f** 8 **4 a** 2 **b** 2 **c** 5 **d** 10 **e** 4 **f** 27 **g** 3 **h** 15 **5 a** $x^{\frac{1}{2}}$ **b** $x^{\frac{1}{3}}$ **c** $\frac{1}{2}a^{\frac{1}{2}}$ **6 a** $\sqrt{m}$ **b** $\sqrt[3]{t}$ **c** $4\sqrt{x}$ **d** $\sqrt[3]{a^2}$ **e** $\sqrt{x^3}$ **f** $3\sqrt{n}$ **7 a** $0.\dot{2}$ **b** $0.41\dot{6}$ **c** $1.1\dot{6}$ **d** $0.\dot{5}7142\dot{8}$ **8 a** terminating **b** recurring **c** recurring **d** terminating **9 a** false **b** true **c** false **10 a** $8x$ **b** 10 **c** $18y^{\frac{2}{3}}$

RATIONAL AND IRRATIONAL NUMBERS

Real Numbers

20 MINUTES

INTERMEDIATE TEST

Part A Multiple Choice

1 Which is an irrational number?

A $2\frac{3}{5}$ B $0.\dot{7}$ C $\sqrt{6}$ D -1.8 (1 mark)

2 $8^{\frac{4}{3}} = ?$

A 10   B $10\frac{2}{3}$ C 12   D 16   (1 mark)

3 $\sqrt{64a^{64}} = ?$ *Hint 1*

A $8a^8$ B $8a^{32}$ C $32a^8$ D $32a^{32}$ (1 mark)

4 Which of these fractions is **not** a recurring decimal?

A $\frac{1}{3}$ B $\frac{5}{6}$ C $\frac{7}{8}$ D $\frac{7}{9}$ (1 mark)

5 $8n^{\frac{1}{3}} = ?$

A $2\sqrt[3]{n}$ B $8\sqrt[3]{n}$ C $\frac{2}{n^3}$ D $\frac{8}{n^3}$ (1 mark)

6 When written as a decimal, correct to one decimal place, $\frac{2}{3} = ?$

A 0.2   B 0.3   C 0.6   D 0.7   (1 mark)

7 $a^{\frac{m}{n}} = ?$

A $\sqrt[n]{a^m}$ B $\sqrt[m]{a^n}$ C $a^m \div a^n$ D none of these   (1 mark)

Part B Written Response

8 Without using a calculator, evaluate $\sqrt[3]{1000^5}$ *Hint 2*   (1 mark)

9 Simplify $\left(2x^{\frac{1}{3}}\right)^6 \div 8x^4$ (2 marks)

10 Simplify $(9x^2y^4)^{\frac{1}{2}}$ (1 mark)

11 Find $\sqrt{9x^{100}y^{16}}$ *Hint 3*   (1 mark)

12 Find $\sqrt[3]{8n^{27}}$ (1 mark)

13 a Express $\sqrt[5]{32^7}$ in index form.   (1 mark)

b Evaluate $\sqrt[5]{32^7}$ *Hint 4*   (1 mark)

14 Without using a calculator, write as a fraction in simplest form:

a $0.\dot{8}$ (2 marks)

b $0.1\dot{6}$ *Hint 5*   (3 marks)

Hint 1: What fractional power gives the square root?
Hint 2: Find the cube root of 1000 first.
Hint 3: Raise to the power of one half.
Hint 4: Use the index key on your calculator.
Hint 5: Only one number repeats. Multiply by ten.

Your Feedback

$\frac{\square}{20} \times 100\% = \square\%$

PAGES 107–108
PAGE 182

RATIONAL AND IRRATIONAL NUMBERS

Real Numbers

40 MINUTES

ADVANCED TEST

NOTE: This is a NON-CALCULATOR test.

1 Write as a terminating or repeating decimal:

a $\frac{7}{12}$ **b** $\frac{37}{40}$ **c** $\frac{11}{14}$ (1 mark each)

2 Change to a fraction in simplest form:

a $0.34\dot{5}$ **b** $0.\overline{345}$ **c** $0.3\dot{4}\dot{5}$ (2 marks each)

3 Is the number $3.2\dot{1}\dot{4}$ rational or irrational? Justify your answer. (1 mark)

4 If $x = \frac{2}{9}$, $y = \frac{3}{4}$ and $z = \frac{1}{6}$, evaluate $\frac{x^2y^3}{z^4}$ (2 marks)

5 The number e is irrational. A calculator gives the value of e as 2.718 281 828.

a Is the number e real? Justify your answer. (1 mark)

b George said that e is not irrational because it repeats. Explain why George is wrong. (1 mark)

6 Write in index form:

a $\frac{a}{b^4}$ **b** $\frac{1}{\sqrt{a^3}}$ **c** $\sqrt[3]{a^2}$ (1 mark each)

7 Write in surd form:

a $x^{-\frac{1}{4}}$ **b** $x^{\frac{5}{3}}$ **c** $x^{-2}y^{\frac{1}{2}}$ (1 mark each)

8 Find the value of:

a $16^{\frac{3}{4}}$ **b** $64^{\frac{2}{3}}$ **c** $100^{\frac{3}{2}}$

d $\left(\frac{1}{2}\right)^{-3}$ **e** $\left(1\frac{1}{2}\right)^{-1}$ **f** $\left(\frac{1}{4}\right)^{-\frac{1}{2}}$

g $25^{-\frac{1}{2}} \div 8^{\frac{2}{3}}$ **h** $64^{-\frac{3}{2}} \times \left(\frac{1}{2}\right)^{-5}$ **i** $\sqrt{\frac{3^4}{4^3}}$ (1 mark each)

9 Write $\sqrt[3]{a}$ in the form a^m and show why that value of m is correct. (2 marks)

10 Solve:

a $4^x = 2^3$ (2 marks)

b $5^{2x-1} = \frac{1}{125}$ (2 marks)

c $9^{4-x} = 27$ (2 marks)

11 Construct a number line and accurately show the position of $\sqrt{6}$ (3 marks)

Your Feedback

$\frac{\square}{40} \times 100\% = \square\%$

PAGES 108–109

PAGE 182

SURDS

Real Numbers

STUDY NOTES

1 A **surd** is a number involving the square root of a number that is not a perfect square. $\sqrt{2}$, $\sqrt{3}$ and $4\sqrt{5}$ are surds but $\sqrt{4}$ is not a surd because $\sqrt{4} = 2$. The number $\sqrt{3}$ means the positive number which when squared gives three. We say 'the square root of three' or 'root three'. We cannot take the square root of a negative number.

2 **Like surds** involve the square root of the same number. Only like surds can be **added** together or **subtracted**. For example, $3\sqrt{2} + 4\sqrt{2} = 7\sqrt{2}$ but $3\sqrt{2} + 4\sqrt{3} = 3\sqrt{2} + 4\sqrt{3}$.

3 Surds can be **simplified** if the number beneath the square root has a factor that is a perfect square. For example, $\sqrt{8} = \sqrt{4} \times \sqrt{2} = 2\sqrt{2}$.

4 Surds can be **multiplied** or **divided** using the following rules:

$$(\sqrt{x})^2 = \sqrt{x^2} = x \qquad \sqrt{x} \times \sqrt{y} = \sqrt{xy} \qquad \frac{\sqrt{x}}{\sqrt{y}} = \sqrt{\frac{x}{y}}$$

For example, $2\sqrt{3} \times 6\sqrt{5} = 12\sqrt{15}$, $\quad 8\sqrt{14} \div 4\sqrt{2} = 2\sqrt{7}$, $\quad \sqrt{3} \times \sqrt{6} = \sqrt{18} = \sqrt{9} \times \sqrt{2} = 3\sqrt{2}$.

5 **Rationalising the denominator** of a surd means changing the denominator so that it is a rational number. To rationalise the denominator of a surd such as $\dfrac{\sqrt{2}}{\sqrt{3}}$ we use the result that $(\sqrt{x})^2 = x$. So if we multiply the denominator by $\sqrt{3}$ it will be rational. If we multiply the denominator by $\sqrt{3}$ we must also multiply the numerator by $\sqrt{3}$. i.e. $\dfrac{\sqrt{2}}{\sqrt{3}} = \dfrac{\sqrt{2}}{\sqrt{3}} \times \dfrac{\sqrt{3}}{\sqrt{3}} = \dfrac{\sqrt{6}}{3}$.

6 **Binomials** involving surds can be multiplied using the rules for any binomial products.

$$(\sqrt{a} + \sqrt{b})(\sqrt{c} + \sqrt{d}) = \sqrt{ac} + \sqrt{ad} + \sqrt{bc} + \sqrt{bd}$$

7 Whenever a binomial product involves the **sum of two terms multiplied by the difference of the same two terms** the result is the difference of two squares. So if a binomial product involves the sum of two surds multiplied by the difference of the same two surds then the product will be rational.
For example, $(\sqrt{3} + \sqrt{2})(\sqrt{3} - \sqrt{2}) = (\sqrt{3})^2 - (\sqrt{2})^2 = 3 - 2 = 1$

8 The **conjugate** of a binomial surd is the binomial involving the same two surds but with a different sign. For example, the conjugate of $\sqrt{3} + \sqrt{2}$ is $\sqrt{3} - \sqrt{2}$. Whenever a binomial surd is multiplied by its conjugate the result is a rational number. So to rationalise the denominator involving a binomial multiply both the numerator and denominator by the conjugate.

For example, $\dfrac{5}{\sqrt{3}+\sqrt{2}} = \dfrac{5}{\sqrt{3}+\sqrt{2}} \times \dfrac{\sqrt{3}-\sqrt{2}}{\sqrt{3}-\sqrt{2}} = \dfrac{5\sqrt{3}-5\sqrt{2}}{3-2} = 5\sqrt{3} - 5\sqrt{2}$

Checklist

Can you:

1 *Understand what is meant by a surd?* ☐
2 *Add, subtract, multiply, divide and simplify surds?* ☐
3 *Understand that squaring a surd gives a whole number?* ☐
4 *Expand expressions involving surds?* ☐
5 *Find the product of binomial surds?* ☐
6 *Understand what is meant by a conjugate?* ☐
7 *Understand that the product of a binomial surd with its conjugate will be rational?* ☐
8 *Rationalise the denominator?* ☐

SURDS
Real Numbers

SKILLS CHECK

1 Find:

a $3\sqrt{7} + 5\sqrt{7}$ **b** $5\sqrt{2} - \sqrt{2}$

c $\sqrt{6} + \sqrt{6}$ **d** $7 - \sqrt{6} + 5\sqrt{6} - 2 + \sqrt{7}$

e $8\sqrt{3} - 2\sqrt{5}$ **f** $3\sqrt{10} + 2\sqrt{3} + 5\sqrt{3} - 2\sqrt{10}$

2 Simplify:

a $\sqrt{12}$ **b** $\sqrt{45}$ **c** $\sqrt{75}$ **d** $\sqrt{600}$

e $2\sqrt{20}$ **f** $3\sqrt{8}$ **g** $6\sqrt{300}$ **h** $5\sqrt{98}$

3 Find (after first simplifying where possible):

a $\sqrt{8} + 5\sqrt{2}$ **b** $\sqrt{18} - \sqrt{98}$ **c** $2\sqrt{300} + 3\sqrt{200} - 4\sqrt{27}$

4 Find these products. (Simplify where necessary.)

a $\sqrt{3} \times \sqrt{5}$ **b** $2\sqrt{6} \times \sqrt{7}$ **c** $3\sqrt{2} \times 5\sqrt{3}$

d $7\sqrt{2} \times \sqrt{2}$ **e** $3\sqrt{6} \times 4\sqrt{3}$ **f** $-2\sqrt{3} \times 5\sqrt{2} \times 7\sqrt{10}$

5 Find:

a $\sqrt{10} \div \sqrt{2}$ **b** $6\sqrt{15} \div 2\sqrt{3}$ **c** $8\sqrt{5} \div 2$ **d** $4\sqrt{7} \div \sqrt{7}$

6 Expand:

a $2(3\sqrt{2} - 5)$ **b** $\sqrt{3}(\sqrt{5} + 2)$ **c** $3\sqrt{7}(\sqrt{5} - 2\sqrt{7})$

7 Find these binomial products:

a $(\sqrt{3} + 2)(\sqrt{5} - 3)$ **b** $(\sqrt{7} + \sqrt{2})(\sqrt{6} + \sqrt{5})$

c $(\sqrt{2} - 1)(\sqrt{3} - 1)$ **d** $(\sqrt{5} + \sqrt{3})(\sqrt{5} - \sqrt{3})$

e $(2 - \sqrt{10})(2 + \sqrt{10})$ **f** $(2\sqrt{3} - \sqrt{5})(2\sqrt{3} + \sqrt{5})$

8 Find:

a $\frac{2}{\sqrt{5}} \times \frac{\sqrt{5}}{\sqrt{5}}$ **b** $\frac{3}{\sqrt{3}} \times \frac{\sqrt{3}}{\sqrt{3}}$ **c** $\frac{\sqrt{5}}{\sqrt{3}+2} \times \frac{\sqrt{3}-2}{\sqrt{3}-2}$

PAGES 109–110

Answers **1 a** $8\sqrt{7}$ **b** $4\sqrt{2}$ **c** $2\sqrt{6}$ **d** $5 + 4\sqrt{6} + \sqrt{7}$ **e** $8\sqrt{3} - 2\sqrt{5}$ **f** $\sqrt{10} + 7\sqrt{3}$ **2 a** $2\sqrt{3}$ **b** $3\sqrt{5}$ **c** $5\sqrt{3}$ **d** $10\sqrt{6}$ **e** $4\sqrt{5}$ **f** $6\sqrt{2}$ **g** $60\sqrt{3}$ **h** $35\sqrt{2}$ **3 a** $7\sqrt{2}$ **b** $-4\sqrt{2}$ **c** $8\sqrt{3} + 30\sqrt{2}$ **4 a** $\sqrt{15}$ **b** $2\sqrt{42}$ **c** $15\sqrt{6}$ **d** 14 **e** $36\sqrt{2}$ **f** $-140\sqrt{15}$ **5 a** $\sqrt{5}$ **b** $3\sqrt{5}$ **c** $4\sqrt{5}$ **d** 4 **6 a** $6\sqrt{2} - 10$ **b** $\sqrt{15} + 2\sqrt{3}$ **c** $3\sqrt{35} - 42$ **7 a** $\sqrt{15} - 3\sqrt{3} + 2\sqrt{5} - 6$ **b** $\sqrt{42} + \sqrt{35} + 2\sqrt{3} + \sqrt{10}$ **c** $\sqrt{6} - \sqrt{2} - \sqrt{3} + 1$ **d** 2 **e** -6 **f** 7 **8 a** $\frac{2\sqrt{5}}{5}$ **b** $\sqrt{3}$ **c** $2\sqrt{5} - \sqrt{15}$

SURDS

Real Numbers

25 MINUTES

INTERMEDIATE TEST

Part A Multiple Choice

1 $5\sqrt{5} - 4\sqrt{3} + 3\sqrt{5} - \sqrt{3} = ?$

A $8\sqrt{5} - 5\sqrt{3}$ **B** $8\sqrt{5} - 3\sqrt{3}$ **C** $8\sqrt{5} + 3\sqrt{3}$ **D** $8\sqrt{5} + 5\sqrt{3}$ (1 mark)

2 In simplest form $\sqrt{32} + \sqrt{8} = ?$ *Hint 1*

A $2\sqrt{10}$ **B** $3\sqrt{8}$ **C** $6\sqrt{2}$ **D** cannot be simplified (1 mark)

3 $\sqrt{100x^{100}} = ?$ *Hint 2*

A $10x^{10}$ **B** $10x^{50}$ **C** $50x^{10}$ **D** $50x^{50}$ (1 mark)

4 $\sqrt{3}(\sqrt{2} + \sqrt{3}) = ?$

A $\sqrt{6} + 3$ **B** $\sqrt{6} + \sqrt{3}$ **C** $\sqrt{15}$ **D** $3\sqrt{2}$ (1 mark)

5 $\frac{14}{\sqrt{7}} = ?$

A $\sqrt{2}$ **B** $\sqrt{7}$ **C** $2\sqrt{7}$ **D** $7\sqrt{2}$ (1 mark)

6 Between which two consecutive integers does $3\sqrt{2}$ lie?

A 2 and 3 **B** 3 and 4 **C** 4 and 5 **D** 5 and 6 (1 mark)

7 $\sqrt{63} - \sqrt{28} = ?$

A $7\sqrt{3} - 7\sqrt{2}$ **B** $7\sqrt{5}$ **C** $\sqrt{35}$ **D** $\sqrt{7}$ (1 mark)

Part B Written Response

8 Simplify $\frac{4\sqrt{12}}{2\sqrt{2}}$ (1 mark)

9 Express $3\sqrt{5}$ as an entire surd. *Hint 3* (1 mark)

10 Write with a rational denominator:

a $\frac{6}{\sqrt{2}}$ **b** $\frac{\sqrt{2}}{5\sqrt{10}}$ (2 marks each)

11 Expand $\sqrt{6}(\sqrt{2} + \sqrt{3})$ giving the answer in simplest surd form. (2 marks)

12 $\sqrt{75} + 2\sqrt{50} - 8\sqrt{27} = ?$ *Hint 4* (2 marks)

13 Expand and simplify:

a $(5\sqrt{2} - 2\sqrt{3})(3\sqrt{2} - 4\sqrt{3})$ **b** $(2\sqrt{7} - 3)(2\sqrt{7} + 3)$ (2 marks each)

14 Find x given that $\sqrt{48} + \sqrt{108} = x\sqrt{3}$ (2 marks)

15 Find $(3\sqrt{5} + 10\sqrt{6})^2$ giving the answer in simplest form. (2 marks)

Hint 1: Simplify before adding.
Hint 2: The square root is the same as a power of one half.
Hint 3: Write the surd in the form $\sqrt{x}$ with the entire number under the square root.
Hint 4: Simplify all the surds and then add the like surds.

Your Feedback

$\frac{\square}{25} \times 100\% = \square\%$

PAGE 110
PAGE 182

SURDS

Real Numbers

40 MINUTES

ADVANCED TEST

1 Simplify:

a $(6\sqrt{2})^2$ b $\dfrac{2\sqrt{24}}{8\sqrt{6}}$ c $\dfrac{\sqrt{56}}{\sqrt{7}}$ (1 mark each)

2 Find, in simplest surd form:

a $\sqrt{700} + \sqrt{200} - \sqrt{98}$ (2 marks)

b $\sqrt{45} - \sqrt{12} + \sqrt{20} - \sqrt{3}$ (2 marks)

c $-\sqrt{32} + 17 - 5\sqrt{288} - 36$ (2 marks)

3 Simplify:

a $3\sqrt{6} \times 5\sqrt{2}$ (2 marks)

b $\sqrt{15} - 2\sqrt{3} \times 3\sqrt{5}$ (2 marks)

4 a Find the exact value of h, in simplest surd form. (2 marks)

b Find the exact area of the triangle. (1 mark)

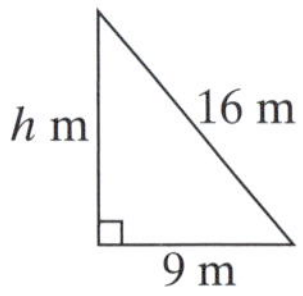

5 If $a < 0$, what is the value of $\sqrt{a^2}$? Justify your answer. (2 marks)

6 Simplify:

a $\sqrt{3456}$ b $\sqrt{2^5 \times 3^3}$ (2 marks each)

7 Show that $(2\sqrt{3} - 5\sqrt{2})^2 - (10 - \sqrt{6})^2$ is rational. (2 marks)

8 Expand and simplify:

a $\sqrt{6}(3\sqrt{2} - 2\sqrt{3} - \sqrt{6})$ (2 marks)

b $(\sqrt{5} - 1)(\sqrt{2} - 3\sqrt{10})$ (2 marks)

c $(5\sqrt{3} + \sqrt{7})(5\sqrt{3} - \sqrt{7}) - (4 - \sqrt{2})(4 + \sqrt{2})$ (2 marks)

9 Express in simplest form:

a $\dfrac{6\sqrt{2}}{\sqrt{3}} + \dfrac{\sqrt{12}}{\sqrt{2}}$ b $\dfrac{\sqrt{5} - 2}{\sqrt{5} + 2}$ c $\dfrac{6}{\sqrt{7} + \sqrt{5}}$ (2 marks each)

10 Find the values of a and b if $\dfrac{5 - \sqrt{7}}{3 + \sqrt{7}} = a + b\sqrt{7}$ (2 marks)

11 Show that $\dfrac{1}{5 - 2\sqrt{6}} + \dfrac{1}{5 + 2\sqrt{6}}$ is rational. (2 marks)

Your Feedback

$\dfrac{\square}{40} \times 100\% = \square\%$

PAGES 110–111

PAGE 182

LOGARITHMS

Real Numbers

STUDY NOTES

1 The **notation $\log_a b$** is used for logarithms. We say 'log to the base a of b'.

2 The **logarithm of a number to any (positive) base** is the index when the number is expressed as a power of the base. For example, because $2^3 = 8$, the logarithm to the base 2 of 8 is 3, i.e. $\log_2 8 = 3$.

3 $a^x = y$ and $\log_a y = x$ are **equivalent statements** ($a > 0$, $y > 0$). For example, $5^2 = 25$, $\therefore \log_5 25 = 2$. If $\log_m 81 = 4$, then $m^4 = 81$.

4 There are certain **laws of logarithms** that must be known. These log laws follow from the index laws.

$$\log_a x + \log_a y = \log_a xy \qquad \log_a x - \log_a y = \log_a \frac{x}{y} \qquad \log_a x^n = n\log_a x$$

5 **Other results** follow directly from the definition of logarithms.

$$\log_a a^x = x$$
$$\log_a a = 1 \qquad [\text{because } a^1 = a]$$
$$\log_a 1 = 0 \qquad [\text{because } a^0 = 1]$$
$$\log_a \frac{1}{x} = -\log_a x \qquad [\text{because } \frac{1}{x} = x^{-1}]$$

6 The log laws can be used to **evaluate** simple expressions or **simplify** expressions.
For example,
$$\begin{aligned}\log_2 36 - 2\log_2 3 &= \log_2 36 - \log_2 3^2\\ &= \log_2 36 - \log_2 9\\ &= \log_2 4 \qquad [36 \div 9]\\ &= 2 \qquad [2^2 = 4]\end{aligned}$$

7 Simple **equations** that contain indices or logarithms can be **solved**. For example, if $\log_3 \frac{1}{9} = x$ then $3^x = \frac{1}{9}$. Now $3^{-2} = \frac{1}{9}$ so the solution to $\log_3 \frac{1}{9} = x$ is $x = -2$.

8 $y = a^x$ and $y = \log_a x$ are **inverse functions**. The **graph of $y = \log_a x$** is the reflection of the graph of $y = a^x$ in the line $y = x$.

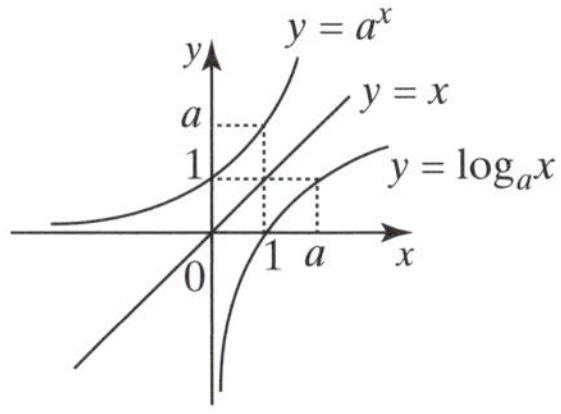

Checklist

Can you:

1. *Understand what is meant by a logarithm?* ☐
2. *Write equivalent statements involving indices and logarithms?* ☐
3. *Understand and use the log laws?* ☐
4. *Evaluate simple expressions and simplify expressions involving logarithms?* ☐
5. *Solve simple equations involving logarithms or indices?* ☐
6. *Recognise and draw the graph of $y = \log_a x$?* ☐

LOGARITHMS

Real Numbers

SKILLS CHECK

1 Complete:

a $16 = 4^2, \therefore \log_4 16 =$ ______

b $2^3 = 8, \therefore \log_2$ ______ $= 3$

c $2^4 = 16, \therefore \log\_\_ 16 =$ ______

d $9^{\frac{1}{2}} = 3, \therefore \log\_\_ \_\_ =$ ______

2 Write equivalent statements using logarithms:

a $9 = 3^2$ **b** $4^3 = 64$ **c** $5^{-1} = \frac{1}{5}$ **d** $36^{\frac{1}{2}} = 6$

3 Write equivalent statements using indices:

a $\log_2 4 = 2$ **b** $\log_5 125 = 3$ **c** $\log_2 \frac{1}{2} = -1$ **d** $\log_{25} 5 = \frac{1}{2}$

4 Evaluate:

a $\log_2 32$ **b** $\log_3 27$ **c** $\log_5 25$ **d** $\log_7 \frac{1}{7}$

e $\log_{16} 4$ **f** $\log_{10} 10000$ **g** $\log_9 \frac{1}{3}$ **h** $\log_{1000} 10$

i $\log_2 1$ **j** $\log_3 3$ **k** $\log_7 7$ **l** $\log_5 1$

5 Complete:

a $\log_a 3 + \log_a 5 = \log_a$ ______

b $\log_b 8 - \log_b 4 = \log_b$ ______

c $2\log_c 5 = \log_c$ ____

6 Evaluate:

a $\log_6 4 + \log_6 9$ **b** $\log_2 6 - \log_2 3$ **c** $2\log_{10} 2 + \log_{10} 25$

7 Simplify:

a $\log_x x^3$ **b** $7\log_a x + \log_a x^2$ **c** $3\log_a 4 - 5\log_a 2$

8 On the same set of axes sketch the graphs of $y = 2^x$ and $y = \log_2 x$.

9 Solve:

a $2^{x+1} = 16$ **b** $3^x = \frac{1}{3\sqrt{3}}$ **c** $\log_7 x = -2$ **d** $\log_8 2 = x$

PAGES 111–112

Answers **1 a** 2 **b** 8 **c** 2, 4 **d** $9, 3, \frac{1}{2}$ **2 a** $\log_3 9 = 2$ **b** $\log_4 64 = 3$ **c** $\log_5 \frac{1}{5} = -1$ **d** $\log_{36} 6 = \frac{1}{2}$ **3 a** $2^2 = 4$ **b** $5^3 = 125$ **c** $2^{-1} = \frac{1}{2}$ **d** $25^{\frac{1}{2}} = 5$ **4 a** 5 **b** 3 **c** 2 **d** -1 **e** $\frac{1}{2}$ **f** 4 **g** $-\frac{1}{2}$ **h** $\frac{1}{3}$ **i** 0 **j** 1 **k** 1 **l** 0 **5 a** $\log_a 15$ **b** $\log_b 2$ **c** $\log_c 25$ **6 a** 2 **b** 1 **c** 2 **7 a** 3 **b** $9\log_a x$ **c** $\log_a 2$ **8** see solutions **9 a** $x = 3$ **b** $x = -1.5$ **c** $x = \frac{1}{49}$ **d** $x = \frac{1}{3}$

LOGARITHMS

Real Numbers

25 MINUTES

INTERMEDIATE TEST

Part A Multiple Choice

1 Which is equivalent to $\dfrac{\log_m 9}{\log_m 3}$?

A 2 **B** 3 **C** $\log_m 3$ **D** $\log_m 6$ (1 mark)

2 Which statement is correct?

A $\log_a 8 + \log_a 4 = \log_a 12$ **B** $\log_a 8 \times \log_a 4 = \log_a 12$

C $\log_a 8 + \log_a 4 = \log_a 32$ **D** $\log_a 8 \times \log_a 4 = \log_a 32$ (1 mark)

3 If $a^x = y$, (where $a > 0$ and $y > 0$), then which is correct?

A $\log_x y = a$ **B** $\log_a y = x$ **C** $\log_y a = x$ **D** $\log_y x = a$ (1 mark)

4 $\log_8 4 = ?$

A $\dfrac{1}{2}$ **B** $\dfrac{2}{3}$ **C** $\dfrac{3}{2}$ **D** 2 (1 mark)

5 Which is equivalent to $\log_a b - \log_a c$?

A $\log_a \dfrac{b}{c}$ **B** $\log_a \dfrac{c}{b}$ **C** $\dfrac{\log_a b}{\log_a c}$ **D** $\dfrac{\log_a c}{\log_a b}$ (1 mark)

Part B Written Response

6 Simplify $\log_2 10 + \log_2 6 - \log_2 15$ (2 marks)

7 Evaluate $\log_3 12 - 2\log_3 2$ (2 marks)

8 Simplify $\log_a a^4 - \log_a a^3$ (2 marks)

9 Solve $9^x = \dfrac{1}{27}$ *Hint 1* (2 marks)

10 On the same set of axes sketch the graph of:

a $y = 3^x$ (2 marks)

b $y = \log_3 x$ (2 marks)

11 Show that $\log_a \dfrac{1}{5} = -\log_a 5$ *Hint 2* (2 marks)

12 Find the value of x if $\log_3 x - \log_3 2 = 7$ (2 marks)

13 If $\log_5 x = 0.683$ and $\log_5 y = 1.209$ find the value of $\log_5\left(\dfrac{x^2}{y}\right)$ *Hint 3* (2 marks)

14 If $\log_2 a = 3.17$ find the value of $\log_2 4a$. (2 marks)

Hint 1: Express both sides of the equation as powers of 3.

Hint 2: $\dfrac{1}{5}$ is what power of 5?

Hint 3: Use the log laws to find another expression for $\log_5\left(\dfrac{x^2}{y}\right)$.

Your Feedback

$\dfrac{\square}{25} \times 100\% = \square\%$

PAGE 112

PAGE 182

LOGARITHMS

Real Numbers

40 MINUTES

ADVANCED TEST

1 Evaluate:

a $\log_{32} 2$ b $\log_8 \left(\frac{1}{2}\right)$ c $\log_{100} 1000$ (1 mark each)

2 If $a^{3.48} = 11$ find:

a $\log_a 11$ b $\log_a \left(\frac{1}{11}\right)$ c $\log_a 121$ (1 mark each)

3 If $\log_3 2 = 0.630\,929\,75\ldots$, find to four decimal places, the value of:

a $\log_3 16$ (1 mark)

b $\log_3 36$ (2 marks)

4 If $\log_a 5 = x$, $\log_a 3 = y$ and $\log_a 2 = z$, find expressions for:

a $\log_a 50$ b $\log_a \left(\frac{5}{6}\right)$ (2 marks each)

5 Simplify:

a $\log_4 8 + \log_9 27$

b $\log_3 48 - 4\log_3 2$

c $\log_2 81 \div \log_2 \left(\sqrt[3]{3}\right)$

d $\log_6 \left(\frac{1}{9}\right) + \log_6 \left(\frac{1}{4}\right)$

e $-\log_a \left(\frac{1}{a}\right) - \log_a a^6$

f $\dfrac{\log_a 32}{\log_a 8}$ (2 marks each)

6 If $\log_a b = c$ find the value of $\log_a ab$. (2 marks)

7 If $\log_4 x = 0.862$ find the value of $\log_2 x$. (2 marks)

8 Find the shaded area: (2 marks)

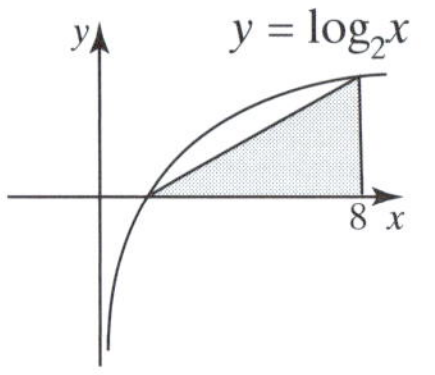

9 If $\log_a b = 1.209$ and $\log_a c = 0.683$, find the value of $\log_a \sqrt{\frac{b}{c}}$ (2 marks)

10 Find the value of x for which $\log_a x + \log_a 3 = 3\log_a 6$. (2 marks)

11 Solve $4^x = \dfrac{1}{2\sqrt{2}}$ (2 marks)

12 Solve $2\log_a x = \log_a (3x + 10)$ (3 marks)

Your Feedback

$\frac{\square}{40} \times 100\% = \square\%$

PAGE 113

PAGE 182

FUNCTIONS

Functions and Other Graphs

STUDY NOTES

1 A **function** is a **rule** or relationship where for **each input** there is **exactly one output**. Every member of one set is associated with exactly one member of a second set. The **notation** $\boldsymbol{f(x)}$ is used for functions. We say 'f of x'. For example, $f(x) = 5x$ means that the relationship is that for every x-value the value of the function is five times that x-value.

2 The **vertical line test** is an easy way to determine whether a sketched relationship is a function. If any vertical line intersects with the graph in only one point then the graph is that of a function. For example, in diagram I it doesn't matter where a vertical line is drawn it will only cut the graph in one place so the graph in I is that of a function. In diagram II, the line shown cuts the graph in two places so that graph is not that of a function, (even though it is possible to draw a line that cuts the graph in just one place).

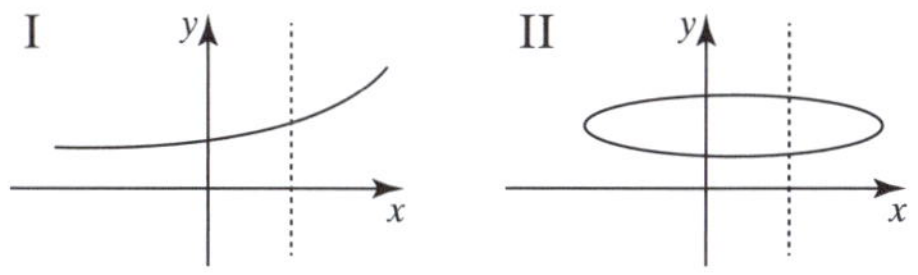

3 $f(x)$ is a function of x. In other words, the **variable** is x. We can have functions of other variables. $f(y)$ is a function where the variable is y. $f(t)$ is a function of t. $g(x)$ is also a function of x.

4 We can **substitute into a function**. $f(2)$ is the value of the function when $x = 2$. $f(-3)$ is the value of the function when $x = -3$. For example, if $f(x) = 3x - 7$ then $f(2) = 3 \times 2 - 7 = -1$ and $f(-3) = 3 \times -3 - 7 = -16$. When substituting we must take care to replace every x with the given value. For example, if $f(x) = x^2 - x$ then $f(3) = 3^2 - 3 = 6$.

5 The **permissible** $\boldsymbol{x}$**-values** of a function are those values of x for which the function exists. For example, if $f(x) = \sqrt{x}$ then $x \geq 0$. If $f(x) = \frac{3}{x}$ then x cannot be zero so the permissible x-values are all real values of x except $x = 0$.

6 The **permissible** $\boldsymbol{y}$**-values** are those values of y for which $f(x) = y$. For example, if $f(x) = 2x$ then the permissible y-values are all real numbers because every number is twice some other number. If $f(x) = x^2$ then $y \geq 0$ because no number squared will give a negative number.

7 An **inverse function** is found when the x and y values are switched. For example, the inverse function of $y = \frac{1}{2}x$ is $x = \frac{1}{2}y$. Rearranging gives $y = 2x$ so $y = \frac{1}{2}x$ and $y = 2x$ are inverse functions.

8 The **graph of** $\boldsymbol{y = f(x) + k}$ is the graph of $y = f(x)$ moved up k units. The graph of $y = f(x - a)$ is the graph of $y = f(x)$ moved across a units (to the right). The graphs of a function and its inverse are reflections in the line $y = x$ (if the number plane was folded along the line $y = x$ the two graphs would coincide).

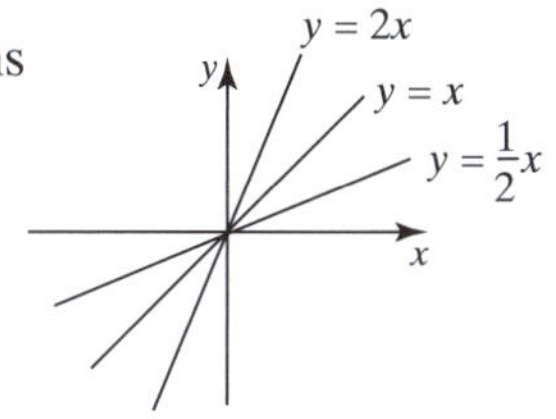

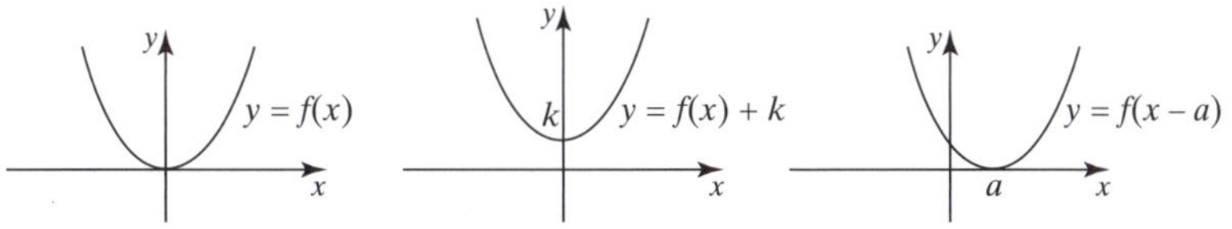

Checklist

Can you:

1 *Understand what is meant by a function?*	☐
2 *Use the vertical line test to determine whether a graph represents a function?*	☐
3 *Use the notation f(x)?*	☐
4 *Find the value of a given function for different values of x?*	☐
5 *Find the permissible x- and y-values for a function?*	☐
6 *Determine the inverse function of a function?*	☐
7 *Sketch the graph of y = f(x) + k and the graph of y = f(x – a) given the graph of y = f(x)?*	☐
8 *Recognise that the graphs of functions and their inverses are reflections in the line y = x?*	☐

FUNCTIONS
Functions and Other Graphs

SKILLS CHECK

1 Complete:
A function is a rule or relationship where for each input there is ____________________ output.

2 Use the vertical line test to determine whether the relationship shown in the graph is a function:

a
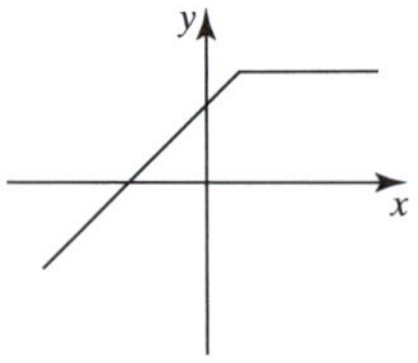

b
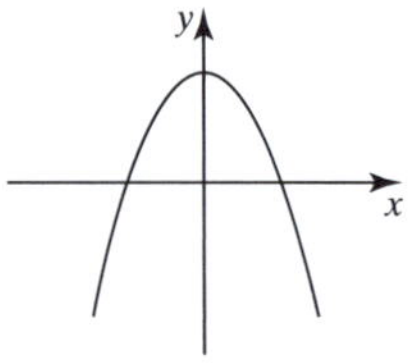

c
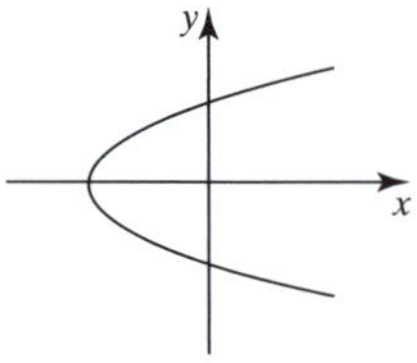

d
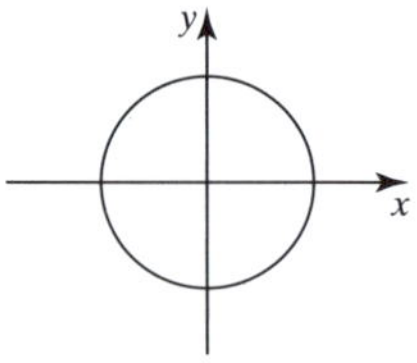

3 If $f(x) = 2x + 3$, find:

a $f(4)$ **b** $f(0)$ **c** $f(-1)$

4 Find $f(3)$ if:

a $f(x) = 2x^2$ **b** $f(x) = 5x - x^2$ **c** $f(x) = \frac{x^2 - 1}{x^2 + 1}$

5 Find $f(-2)$ if:

a $f(x) = x$ **b** $f(x) = x^2 - 6x + 5$ **c** $f(x) = \frac{4}{x}$

6 What are the possible x-values for each function?

a $y = 2x + 3$ **b** $y = \frac{6}{x}$ **c** $y = x^2$

d $y = \sqrt{x}$ **e** $y = 2^x$

7 What are the possible y-values for each function?

a $y = 2x + 3$ **b** $y = \frac{6}{x}$ **c** $y = x^2$

d $y = \sqrt{x}$ **e** $y = 2^x$

8 Determine the inverse function:

a $y = x + 2$ **b** $y = \sqrt{x}$ $(x \geq 0)$ **c** $y = \frac{1}{x}$ **d** $y = x^3$

PAGES 113–114

Answers **1** only one **2 a** function **b** function **c** not a function **d** not a function **3 a** 11 **b** 3 **c** 1 **4 a** 18 **b** 6 **c** 0.8 **5 a** −2 **b** 21 **c** −2 **6 a** all real numbers **b** $x \neq 0$ **c** all real numbers **d** $x \geq 0$ **e** all real numbers **7 a** all real numbers **b** $y \neq 0$ **c** $y \geq 0$ **d** $y \geq 0$ **e** $y > 0$ **8 a** $y = x - 2$ **b** $y = x^2$ $(x \geq 0)$ **c** $y = \frac{1}{x}$ **d** $y = \sqrt[3]{x}$

FUNCTIONS

Functions and Other Graphs

25 MINUTES

INTERMEDIATE TEST

Part A Multiple Choice

1 If $f(x) = 2x^2 - x$ then $f(-1) = ?$

A 3 **B** 1 **C** –1 **D** –3 (1 mark)

2 Which is not a function?

A

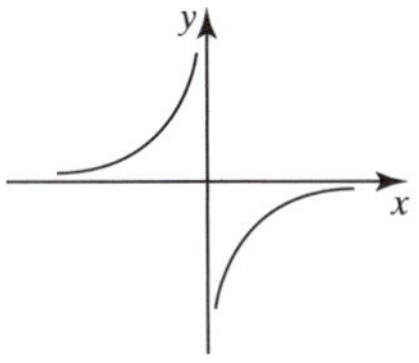

B

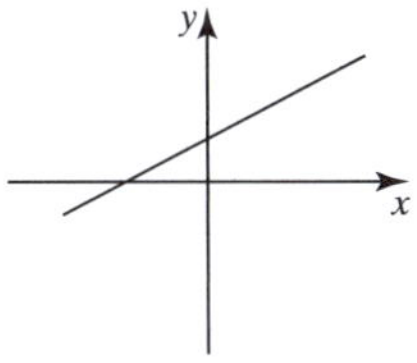

C

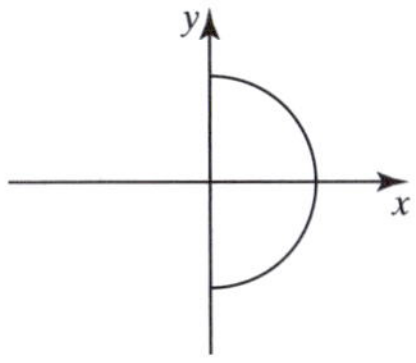

D

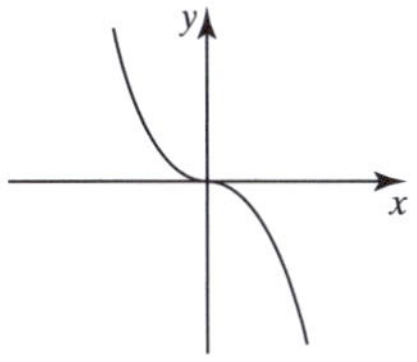

(1 mark)

3 The inverse function of $y = 2x - 4$ is:

A $y = \frac{1}{2}x + 2$ **B** $y = \frac{x}{2} - 4$ **C** $y = 2x + 4$ **D** $y = x - 2$ (1 mark)

4 For which function is the set of permissible x-values all real numbers?

A $y = \frac{3}{x}$ **B** $y = \sqrt{x}$ **C** $y = \frac{x}{7}$ **D** $y = x^{-0.5}$ (1 mark)

5 For which function is $f(0) \neq 0$?

A $f(x) = x$ **B** $f(x) = x^2 + x$ **C** $f(x) = x^5$ **D** $f(x) = 2^x$ (1 mark)

Part B Written Response

6 Find $f(3)$ if:

a $f(x) = \sqrt{25 - x^2}$ **b** $f(x) = \frac{1}{2x - 1}$ **c** $f(x) = 2x^3 - 5x^2 - 7x + 4$ (1 mark each)

7 If $f(x) = \begin{cases} 3x^2 & \text{if } x < 2 \\ x^2 + 3 & \text{if } x \geq 2 \end{cases}$ find:

a $f(-1)$ **b** $f(7) - 1$ **c** $f(2)$ (1 mark each)

8 Given $f(x) = x^2 - 5x$ find the values of a for which $f(a) = -4$. (2 marks)

9 The diagram shows the graph of $y = f(x)$. Sketch the graph of:

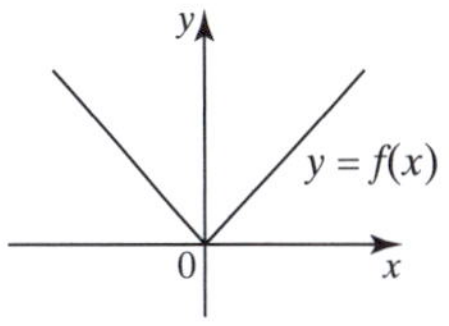

a $y = f(x) + 3$ *Hint 1* (2 marks)

b $y = f(x + 3)$ (2 marks)

10 If $f(x) = 8 - 4x$:

a find $f(-2)$ (1 mark)

b find the value of x for which $f(x) = 0$ *Hint 2* (1 mark)

c what are the permissible x and y values? (2 marks)

d find the inverse function (1 mark)

e On the same set of axes sketch the line $y = x$, the function $f(x)$ and the inverse function. (3 marks)

Hint 1: It doesn't matter what the equation of the function is. What will happen to the graph when you add three to every y-value?

Hint 2: Solve a simple equation.

Your Feedback

$\frac{\square}{25} \times 100\% = \square\%$

PAGES 114–115

PAGE 182

FUNCTIONS

Functions and Other Graphs

40 MINUTES

ADVANCED TEST

1 Determine whether or not the following are functions:

a $y = x^3 - 2$ **b** $x^2 + y^2 = 16$

c $x = y^2$ **d** $y = 4^{-x}$ (1 mark each)

2 Jill said 'All straight lines are functions.' Is she correct? Justify your answer. (1 mark)

3 If $f(x) = \begin{cases} 7 - x & \text{for } x \le -1 \\ 5 & \text{for } -1 < x < 0: \\ x^2 & \text{for } x \ge 0 \end{cases}$

a find:

i $f(-4)$ **ii** $f\left(-\dfrac{1}{2}\right) + f\left(\dfrac{1}{2}\right)$ **iii** $f(\sqrt{a})$ (1 mark each)

b sketch $y = f(x)$ (3 marks)

4 If $f(x) = ax^2 + bx + c$ find $f(-x) + f(x)$. (2 marks)

5 Given $f(x) = x^2 + x$:

a find $f(3)$ (1 mark)

b find the value(s) of c if $f(c) = 6$ (2 marks)

c what are the permissible x-values? (1 mark)

d sketch the graph of $y = f(x)$ (2 marks)

e what are the permissible y-values? (1 mark)

f what restrictions would need to be placed on this function in order for it to have an inverse function? (1 mark)

6 Consider the function $f(x) = \sqrt{9 - x^2}$.

a What are the permissible x-values? (1 mark)

b If $y = f(x)$, what are the permissible y-values? (1 mark)

c Will $f(x)$ have an inverse function? Justify your answer. (1 mark)

7 Determine the inverse function for:

a $y = \dfrac{4}{x}$ **b** $y = 2^x$ **c** $y = \sqrt{x}$ (2 marks each)

8 The diagram shows the function $f(x) = (x - 3)^2$.

a Explain why $f(x)$ does not have an inverse function. (1 mark)

b Let $g(x) = (x - 3)^2$, $x \le 3$. Sketch the inverse function $y = g^{-1}(x)$. (2 marks)

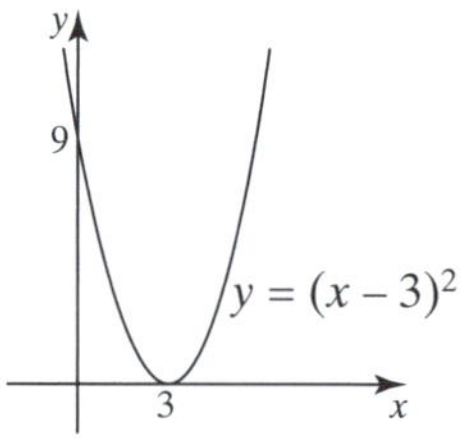

9 **a** For what values of x is $f(x) = x^3$ equal to its inverse function? (2 marks)

b Sketch the function and its inverse. (2 marks)

10 Given that the function $y = x^2 + 2x + 1$ has an inverse function for $x \ge -1$ find the equation of the inverse function. (3 marks)

Your Feedback

$\dfrac{\square}{40} \times 100\% = \square\%$

PAGES 115–116

PAGE 182

POLYNOMIAL EXPRESSIONS

Polynomials

STUDY NOTES

1 A **polynomial** is an expression of the type $a_n x^n + a_{n-1}x^{n-1} + a_{n-2}x^{n-2} + \ldots + a_1 x + a_0$.
$a_n, a_{n-1}, \ldots, a_1$ and a_0 are the **coefficients**, (the numbers by which each power of x is multiplied).

2 n is the **degree** of the polynomial. It is the highest power of the variable. The **leading term** is the term involving that highest power, ($a_n x^n$). The **leading coefficient** is the coefficient in the leading term (a_n). The **constant term** is the term involving no power of the variable (a_0). It is any term that is just a number. For example, for the polynomial $5x^3 - 8x^2 + 7x - 2$ the degree is 3, the leading term is $5x^3$, the leading coefficient is 5 and the constant term is –2. Note: the leading term may not be the first term. For example, in the polynomial $1 - 3x - 4x^2$ the leading term is $-4x^2$.

3 If the leading coefficient is 1 the polynomial is called a **monic polynomial**.

4 The **notation** $P(x)$ is used for polynomials. $P(c)$ is the value of the polynomial when $x = c$. For example, if $P(x) = x^2 - 8x + 3$ then $P(1) = 1^2 - 8 \times 1 + 3 = -4$

5 Polynomials can be **added or subtracted** by collecting like terms.

6 Any two polynomials can be **multiplied** together. Every term in one polynomial must be multiplied by every term in the other polynomial. (In this course you are only required to multiply by a linear expression, $mx + b$). For example, $(4x - 3)(x^4 + 3x^2 - 5x + 2) = 4x(x^4 + 3x^2 - 5x + 2) - 3(x^4 + 3x^2 - 5x + 2)$

7 One polynomial can be **divided** by another polynomial. (Note: you are only required to divide by a linear expression.) The answer to this division is the quotient and there may be a remainder. If when the polynomial $P(x)$ is divided by $(x - a)$, the quotient is $Q(x)$ and the remainder c, then $P(x) = (x - a)Q(x) + c$.

8 A method of **long division** is used with polynomials. For example, to divide $3x^2 - 5x + 4$ by $(x - 2)$ first ask yourself how many times does x (the leading term of the divisor) divide into $3x^2$. The answer is $3x$ so the first term in the quotient is $3x$. Now multiply each term of the divisor by $3x$, giving $3x^2 - 6x$. Subtract this answer from the first two terms of the polynomial leaving x and bring down the last term giving the expression $x + 4$. How many times does x divide into x? Once. So the second term in the quotient is 1. Multiply the divisor by 1 giving $x - 2$ and subtract from $x + 4$ leaving 6. The quotient is $3x + 1$ and the remainder is 6.

$$\begin{array}{r} 3x + 1 \\ x - 2 \overline{\smash{)}\, 3x^2 - 5x + 4} \\ \underline{3x^2 - 6x \phantom{{}+4}} \\ x + 4 \\ \underline{x - 2} \\ 6 \end{array}$$

9 The **remainder theorem** tells us that the remainder when a polynomial $P(x)$ is divided by $(x - a)$ will be $P(a)$. It is not necessary to divide to find the remainder. For example, the remainder when $P(x) = x^2 + 9x - 2$ is divided by $(x - 3)$ is 34 because $P(3) = 3^2 + 9 \times 3 - 2 = 34$

10 If $P(a) = 0$, the remainder when $P(x)$ is divided by $(x - a)$ is 0 and hence $(x - a)$ divides evenly into the polynomial. This is the **factor theorem**. $(x - a)$ is a factor of $P(x)$ if $P(a) = 0$.

11 A polynomial is **factorised** when expressed as the product of all of its factors.

Checklist

Can you:

1 *Understand what is meant by a polynomial?* ☐

2 *Understand the terms degree, leading term, leading coefficient, constant term and monic polynomial?* ☐

3 *Add, subtract, multiply and divide polynomials?* ☐

4 *Use the remainder and factor theorems?* ☐

5 *Factorise polynomials?* ☐

POLYNOMIAL EXPRESSIONS

Polynomials

SKILLS CHECK

1 The given expression is a polynomial. True or false?

a $2x^2 + 7x - 5$ **b** $1 - 7x + 4x^2 - x^3$ **c** $7 - \sqrt{x}$

d x^6 **e** $9x^8 + 7x^3 + \dfrac{2}{x}$ **f** $x + 3$

2 For the polynomial $3x^5 - 2x^4 + 7x^3 + 4x^2 - 8x + 9$ write down:

a the degree

b the leading term

c the leading coefficient

d the constant term

3 Write down the degree of each polynomial:

a $4x^7 + 5x^6 - 2x^3 + 6x$ **b** $1 - x^2$

c $x^3 + 2x^2 - 9x - 4$ **d** $7x + 2$

4 For the polynomial $P(x) = 5x^3 - 7x^2 + 8x + 1$ find:

a $P(0)$ **b** $P(1)$ **c** $P(-2)$ **d** $P(c)$

5 Find:

a $(7x^3 + x^2 + 9x - 5) + (3x^2 + 17x - 2)$

b $(x^4 - 2x^3 + 5x^2 - 9) - (7x^3 + 2x^2 - 6x + 3)$

c $(3x + 2)(x^3 - x^2 + 4x - 1)$

d $(x - 1)(x^2 - x - 1)$

6 If $P(x) = 6x^2 + 11x + 3$ and $Q(x) = 2x + 3$ find:

a $P(x) + Q(x)$ **b** $P(x) - Q(x)$ **c** $P(x) \times Q(x)$ **d** $\dfrac{P(x)}{Q(x)}$

7 Find the quotient when $x^4 - 4x^3 + 7x^2 - 10x + 8$ is divided by $x - 2$.

8 If $P(x) = 2x^3 - 9x^2 + 4$

a find:

i $P(1)$ **ii** $P(3)$ **iii** $P(-2)$

b write down the remainder when $P(x)$ is divided by:

i $x - 1$ **ii** $x - 3$ **iii** $x + 2$

9 Given $P(x) = x^3 - 2x^2 - x + 2$

a find:

i $P(1)$ **ii** $P(-1)$ **iii** $P(2)$

b write $P(x)$ as the product of its factors

PAGES 116–117

Answers **1 a** true **b** true **c** false **d** true **e** false **f** true **2 a** 5 **b** $3x^5$ **c** 3 **d** 9 **3 a** 7 **b** 2 **c** 3 **d** 1 **4 a** 1 **b** 7 **c** −83 **d** $5c^3 - 7c^2 + 8c + 1$ **5 a** $7x^3 + 4x^2 + 26x - 7$ **b** $x^4 - 9x^3 + 3x^2 + 6x - 12$ **c** $3x^4 - x^3 + 10x^2 + 5x - 2$ **d** $x^3 - 2x^2 + 1$ **6 a** $6x^2 + 13x + 6$ **b** $6x^2 + 9x$ **c** $12x^3 + 40x^2 + 39x + 9$ **d** $3x + 1$ **7** $x^3 - 2x^2 + 3x - 4$ **8 a i** −3 **ii** −23 **iii** −48 **b i** −3 **ii** −23 **iii** −48 **9 a i** 0 **ii** 0 **iii** 0 **b** $(x - 1)(x + 1)(x - 2)$

POLYNOMIAL EXPRESSIONS

Polynomials

30 MINUTES

INTERMEDIATE TEST

Part A Multiple Choice

1 The degree of the polynomial $P(x) = (x - 1)(x - 2)(x + 3)$ is:
A 1 **B** 2 **C** 3 **D** 6 (1 mark)

2 If $P(x) = 2x^3 - 7x^2 + 5x - 4$ then $P(-1) = ?$
A -18 **B** -4 **C** -66 **D** 32 (1 mark)

3 Which is not a polynomial?
A 7 **B** $9x^3$ **C** $5x^2 - 6x + x^{-1}$ **D** $1 - x^2$ (1 mark)

4 $P(x) = x^4 - 5x^3 + 2x^2 - 7x + 1$ and $Q(x) = 6x^3 - 3x^2 + 8x - 2$. What is the sum of $P(x)$ and $Q(x)$?
A $x^4 + x^3 - x^2 + x - 1$ **B** $x^4 - x^3 + x^2 - x - 1$
C $x^4 + x^3 + x^2 + x + 1$ **D** $x^4 + x^3 - x^2 - x - 1$ (1 mark)

5 Which is a factor of $P(x) = x^4 - 7x^3 + 8x^2 + 28x - 48$?
A x **B** $x - 1$ **C** $x + 1$ **D** $x - 4$ (1 mark)

6 What is the remainder when $5x^3 + x^2 - 7x - 3$ is divided by $x + 3$?
A -108 **B** -126 **C** 116 **D** 120 (1 mark)

7 The coefficient of x in the expansion $(x + 3)(x^2 - 2x + 5)$ is: *Hint 1*
A 1 **B** -1 **C** 2 **D** 8 (1 mark)

8 Which polynomial has a factor of $x - 2$?
A $x^2 - 3x - 10$ **B** $x^3 - 8x - 8$ **C** $x^4 - 7x^3 + 40$ **D** $2x^3 + 5x^2 - 7x - 6$ (1 mark)

9 Which polynomial does not have a factor of $x + 4$?
A $x^2 + 3x - 4$ **B** $x^3 + 5x^2 + 4x$ **C** $x^4 - 11x^2 - 80$ **D** $x^3 + 7x^2 + 12x - 1$ (1 mark)

Part B Written Response

10 Given the polynomial $P(x) = x^3 - 4x^2 - 61x + 280$:
a find $P(5)$ (1 mark)
b what conclusion can be drawn from your answer to part **a**? (1 mark)
c fully factorise $P(x)$ *Hint 2* (4 marks)

11 Find the remainder when $P(x) = 2x^4 - 3x^3 - 7x^2 + x + 9$ is divided by $x + 3$. *Hint 3* (2 marks)

12 Given that $6x^3 - 29x^2 + 23x - 10 = (x - 4)Q(x) + c$, find $Q(x)$ and c. *Hint 4* (4 marks)

13 Subtract $2x^3 - 3x^2 - 5x + 4$ from $3x^3 + 8x^2 - 2x - 9$. (2 marks)

14 Find the value of a if $P(x) = x^3 + 3x^2 + ax + 7$ has a remainder of 1 when divided by $(x + 2)$.
Hint 5 (3 marks)

15 **a** Find the product of $(x - 2)$ and $(x^3 - 4x^2 + 3x)$. (2 marks)
b Find all the factors of $P(x) = x^4 - 6x^3 + 11x^2 - 6x$. *Hint 6* (2 marks)

Hint 1: There is no need to expand the whole expression.
Hint 2: Divide by the factor that you know. The quotient should be easily factorised.
Hint 3: There is no need to divide.
Hint 4: Divide the original polynomial by $(x - 4)$.
Hint 5: Use the remainder theorem.
Hint 6: Refer to part **a**.

Your Feedback

$\frac{\square}{30} \times 100\% = \square\%$

PAGES 117–118

PAGE 182

POLYNOMIAL EXPRESSIONS

Polynomials

40 MINUTES

ADVANCED TEST

1 $P(x) = 6x^4 - 8x^3 + 7x^2 - 9x - 2$

- **a** What is the degree? (1 mark)
- **b** What is the leading term? (1 mark)
- **c** What is the constant term? (1 mark)
- **d** What is the leading coefficient? (1 mark)
- **e** Is $P(x)$ a monic polynomial? Explain why or why not. (1 mark)

2 $A(x) = 3x^3 - 9x^2 + 5x - 7$, $B(x) = 8x^4 - 2x^3 - 11x^2 + x - 2$, $C(x) = 4x^3 - 6x + 8$

- **a** Find $A(x) + B(x) + C(x)$. (2 marks)
- **b** Find $B(x) - C(x)$. (2 marks)

3 Expand and simplify:

- **a** $(x + 5)(x^3 - 2x^2 - x + 3)$ (2 marks)
- **b** $(2x - 3)(3x^4 - 5x^3 + 4x - 6)$ (2 marks)

4 Is $x - 4$ a factor of $x^3 - 3x^2 + 4x$? Justify your answer. (2 marks)

5 Divide $2x^4 - 5x^2 + 7x - 1$ by $(x + 2)$. (3 marks)

6 $P(x) = x^3 - 5x^2 - 8x + 12$

- **a** Given that $P(x) = (x - 6)Q(x) + c$ find $Q(x)$. (2 marks)
- **b** What is the value of c? What is the significance of this? (1 mark)

7 $(x - 1)$ is a factor of $P(x) = x^4 - 5x^3 + x^2 - 2x - k$. Find k. (2 marks)

8 When a polynomial $P(x)$ is divided by $(x + 2)$ the quotient is $(x - 7)$ and the remainder -5. What is the remainder when $P(x)$ is divided by $(x + 3)$? (2 marks)

9 $(x - 3)$ is a factor of $x^2 + bx + c$. The remainder when $x^2 + bx + c$ is divided by $(x + 3)$ is -30. Find the other factor of $x^2 + bx + c$. (4 marks)

10 $P(x) = (x^2 - x - 6)Q(x) + ax + b$. $P(3) = -5$ and $P(-2) = 15$. Find a and b. (3 marks)

11
- **a** Find the quotient $Q(x)$ and remainder $R(x)$, when $P(x) = x^3 - 7x^2 - 6x + 5$ is divided by $(x + 3)$. (3 marks)
- **b** Briefly explain why $P(x)$ will have that same remainder $R(x)$ when divided by $(x - 4)$ or $(x - 6)$. (1 mark)

12 $P(x) = x^3 + 4x^2 + x - 6$

- **a** Fully factorise $P(x)$. (3 marks)
- **b** For what values of x does $P(x) = 0$? (1 mark)

Your Feedback

$\dfrac{\square}{40} \times 100\% = \square\%$

PAGES 118–119

PAGE 182

FURTHER POLYNOMIALS

Polynomials

STUDY NOTES

1 In order to **factorise polynomials** the factor theorem can be used. The **factor theorem** says that if $x - a$ is a factor of the polynomial then $P(a) = 0$. Once one factor has been found, the polynomial can be divided by this factor and then the factor theorem can be used again to find a factor of the quotient and so on until the polynomial is **fully factorised**. For example, if $P(x) = x^3 + 5x^2 + 2x - 8$ it can easily be found that $P(1) = 0$. Dividing $P(x)$ by $(x - 1)$ leaves $x^2 + 6x + 8$ and this can be factorised to $(x + 2)(x + 4)$ so $P(x) = (x - 1)(x + 2)(x + 4)$. Note: it may require some trial and error to find the first factor. (It is not the intention of any examiner to have you spend hours trying to find a factor. If you have to use this method try the easiest numbers first; 1, 2, –1, –2).

2 **Polynomial equations** can be solved after factorising the polynomial. For example, if $P(x) = 0$ where $P(x) = x^3 + 5x^2 + 2x - 8$ then $(x - 1)(x + 2)(x + 4) = 0$ and $x - 1 = 0$ or $x + 2 = 0$ or $x + 4 = 0$. The **solutions to the polynomial equation** are $x = 1$ or $x = -2$ or $x = -4$.

3 A **zero of the polynomial** is a value for which the polynomial is zero. For example, the zeros of the polynomial $P(x) = x^3 + 5x^2 + 2x - 8$ are 1, –2 and –4.

4 A **polynomial of degree *n*** can have at most n zeros.

5 The **roots of an equation** are the values of x for which the equation equals zero. For example, the roots of the equation $x^3 + 5x^2 + 2x - 8 = 0$ are $x = 1$, $x = -2$ and $x = -4$. Note: the roots of a polynomial equation and the zeros of the polynomial will always be the same.

6 If a polynomial has a factor that is squared then the corresponding polynomial equation has a **double root**. For example, if $x^3 - 3x^2 = 0$ then $x^2(x - 3) = 0$ and $x = 0$ or $x = 3$. $x = 3$ is a single root but $x = 0$ is a double root. If a polynomial has a factor that is cubed, then the polynomial equation has a triple root. For example, if $(x - 2)^3 = 0$ then $x = 2$ is a **triple root.**

7 The **graph of a polynomial** can be sketched after factorising and finding the zeros. The zeros of the polynomial will be the ***x*-intercepts**. The y-intercept is the constant term. For example, when sketching the polynomial $y = x^3 + 5x^2 + 2x - 8$ the x-intercepts will be 1, –2 and –4. The ***y*-intercept** will be –8.

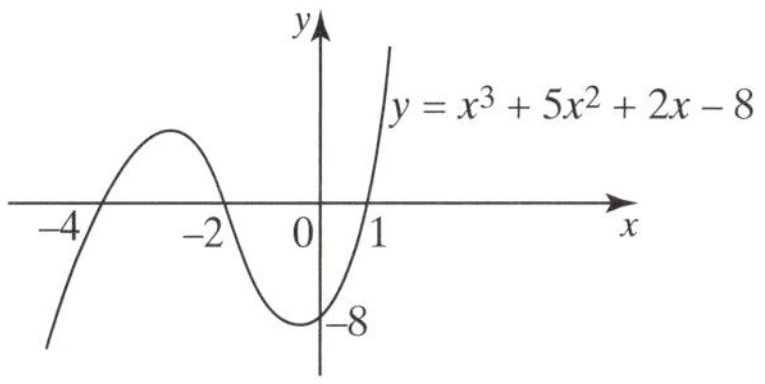

8 Information about the shape of the curve can also be gained from the **leading term**. It will determine the **behaviour of the curve** as $x \to \pm\infty$. ($x \to \infty$ means as x approaches infinity. In other words as x becomes very, very large.)

9 If the **polynomial has a double root**, the graph will **touch** but not cross the x-axis at that point. For example, a cubic polynomial with three single roots (e.g. $y = x^3 + 5x^2 + 2x - 8$) will cross the x-axis in three separate points, one with a double root (e.g. $y = x^2(x - 3)$) will cross the x-axis in one point and touch it at another and one with a triple root (e.g. $y = (x - 2)^3$) will cross the x-axis in just one point.

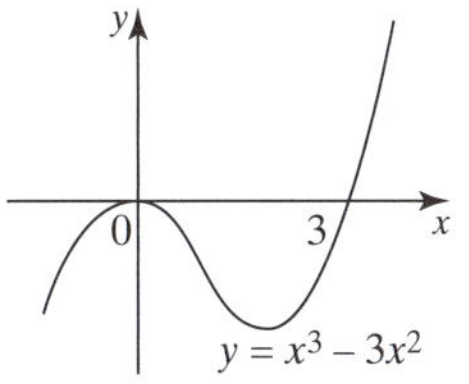

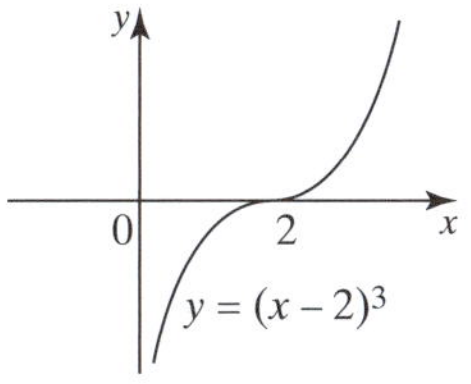

Checklist

Can you:

1. *Factorise polynomials?* ☐
2. *Solve polynomial equations?* ☐
3. *Understand zeros of a polynomial and roots of an equation?* ☐
4. *Determine the effects of single, double or triple roots?* ☐
5. *Sketch the graphs of polynomials?* ☐

FURTHER POLYNOMIALS

Polynomials

SKILLS CHECK

1 Find the remainder when $P(x) = x^3 - 4x^2 + 7x + 1$ is divided by $x - 3$.

2 The given expression is a factor of the polynomial $P(x) = x^4 - 2x^3 - x^2 + 2x$. True or false?

a $x - 1$ **b** x **c** $x + 2$ **d** $x - 2$

3 Divide $5x^2 - 3x + 7$ by $x - 2$.

4 Solve:

a $(x + 3)(x - 2) = 0$
b $x^2(x - 5)(x + 4) = 0$
c $(2x - 1)(3x + 2)(x - 4) = 0$
d $4(2x + 3)(x - 7) = 0$

5 State the maximum number of zeros that a polynomial of the given degree might have:

a degree 4 **b** degree 3 **c** degree 5 **d** degree 2

6 Factorise:

a $x^2 + 7x - 18$
b $x^3 + 5x^2 + 6x$
c $x^4 - 16$
d $x^3 - x^2$
e $x^4 - 25x^2$
f $4x^2 - 4x - 15$

7 State whether the polynomial equation has single, double or triple roots at $x = 2$:

a $(x - 2)^2 = 0$
b $(x + 8)(x - 2) = 0$
c $x^3(x - 2) = 0$
d $x(x + 3)(x - 2)(x - 4) = 0$
e $7x(x - 2)^3 = 0$

8 Consider the polynomial $y = x^2 + x - 2$.

a What is the leading term?
b What is the y-intercept?
c What are the roots of the equation $y = 0$?
d What are the x-intercepts?
e Sketch the graph of $y = x^2 + x - 2$.

9 Given $y = x^4 + 3x^3 - 2x + 5$ what can you say about the value of y for:

a very large positive values of x
b very large negative values of x.

10 Given $y = 3 - 2x + x^2 - x^3$ what can you say about the value of y for:

a very large positive values of x
b very large negative values of x.

PAGE 119

Answers **1** 13 **2 a** true **b** true **c** false **d** true **3** $5x^2 - 3x + 7 = (x - 2)(5x + 7) + 21$ **4 a** $x = -3$ or $x = 2$ **b** $x = 0, x = 5$ or $x = -4$ **c** $x = \frac{1}{2}$ or $x = -\frac{2}{3}$ or $x = 4$ **d** $x = -1\frac{1}{2}$ or $x = 2$ **5 a** 4 **b** 3 **c** 5 **d** 2 **6 a** $(x + 9)(x - 2)$ **b** $x(x + 3)(x + 2)$ **c** $(x^2 + 4)(x + 2)(x - 2)$ **d** $x^2(x - 1)$ **e** $x^2(x + 5)(x - 5)$ **f** $(2x - 5)(2x + 3)$ **7 a** double **b** single **c** single **d** single **e** triple **8 a** x^2 **b** -2 **c** $x = -2$ or $x = 1$ **d** -2 or 1 **e** see solutions **9 a** very large positive **b** very large positive **10 a** very large negative **b** very large positive

FURTHER POLYNOMIALS
Polynomials

30 MINUTES

INTERMEDIATE TEST

Part A Multiple Choice

1 The diagram shows a sketch of the graph of which polynomial?

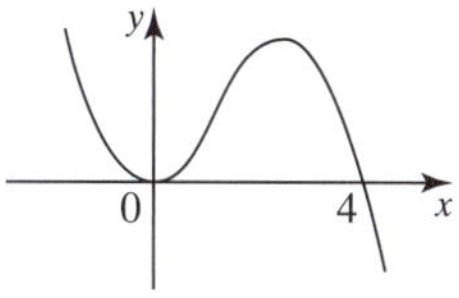

A $y = x^2(x - 4)$
B $y = x(x - 4)^2$
C $y = x^2(4 - x)$
D $y = -x(x - 4)^2$

(1 mark)

2 Which diagram shows the graph of $y = x^3(x - 1)$?

A
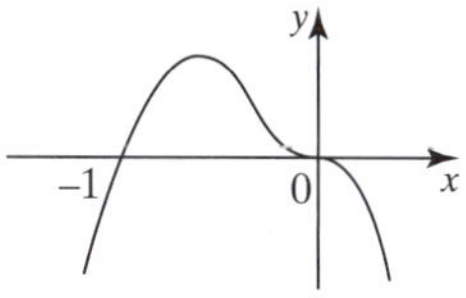

B
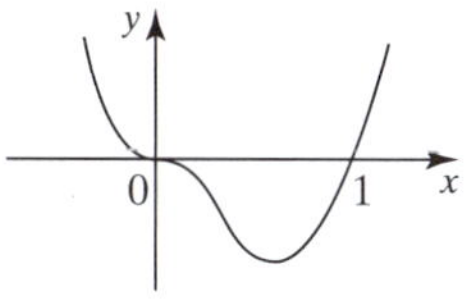

C
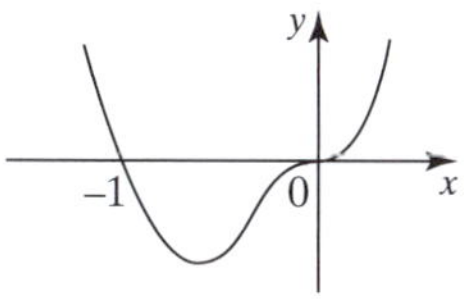

D
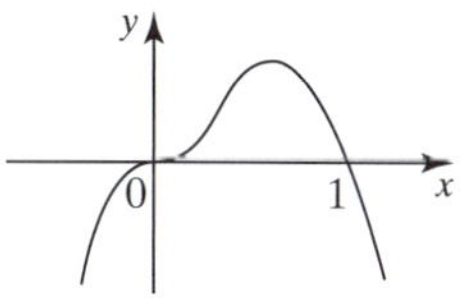

(1 mark)

3 Which polynomial has factors of both $(x - 3)$ and $(x + 2)$?

A $P(x) = x^3 - 5x^2 + x + 6$
B $P(x) = x^3 - 2x^2 - 5x + 6$
C $P(x) = x^3 + x^2 + x - 6$
D $P(x) = x^3 - 7x + 6$

(1 mark)

4 If $y = 8 - 2x + 3x^2 - 4x^3 + 7x^4 - 5x^5$ and x has a large negative value, then y will have:

A a large positive value
B a large negative value
C a small positive value
D a small negative value

(1 mark)

Part B Written Response

5 $P(x) = x^3 - 5x^2 + 2x + 8$

a Show that $x + 1$ is a factor of $P(x)$. (1 mark)
b Hence solve $x^3 - 5x^2 + 2x + 8 = 0$. *Hint 1* (3 marks)
c Draw a sketch of the graph of $y = P(x)$. (4 marks)

6 **a** Factorise $x^3 - 9x^2 + 15x + 25$. *Hint 2* (4 marks)
b Describe the features of the graph of $y = x^3 - 9x^2 + 15x + 25$. (There is no need to actually draw the graph.) *Hint 3* (4 marks)

7 $P(x) = 9x^2 - x^4$

a Solve $P(x) = 0$. *Hint 4* (2 marks)
b Draw the graph of $y = P(x)$. (4 marks)
c Sketch:
i $y = -P(x)$. *Hint 5* (1 mark)
ii $y = P(-x)$ (1 mark)
iii $y = P(x) + 3$ (1 mark)
iv $y = 2P(x)$ (1 mark)

Hint 1: Divide the polynomial by the given factor.
Hint 2: Use the factor theorem to find one factor. You may need several attempts before finding a factor.
Hint 3: What are the x- and y-intercepts? Are the roots single or double? What happens when x is large?
Hint 4: The polynomial has a common factor.
Hint 5: Refer to your graph from part **b**.

Your Feedback

$\frac{\square}{30} \times 100\% = \square\%$

PAGES 119–120
PAGE 182

FURTHER POLYNOMIALS

Polynomials

40 MINUTES

ADVANCED TEST

1 $P(x) = 3x - 4$. Sketch $y = P(x)$. (2 marks)

2 Consider $P(x) = 2x^2 + 9x - 5$
- **a** Find the zeroes of the polynomial. (2 marks)
- **b** Find the minimum value of $P(x)$. (2 marks)
- **c** Sketch $y = P(x)$ showing essential features. (2 marks)

3 It is known that $x + 1$ is a factor of $P(x) = x^3 - 6x^2 + 3x + 10$.
- **a** Fully factorise $P(x)$. (2 marks)
- **b** Sketch the graph of $y = P(x)$. (3 marks)

4 $P(x) = x^3 + 3x^2 - 9x - 27$
- **a** By grouping, or otherwise, fully factorise $P(x)$. (2 marks)
- **b** Where will $y = P(x)$ meet the x-axis? (1 mark)
- **c** Will it cut the x-axis at those points? Justify your answer. (2 marks)
- **d** Sketch $y = P(x)$. (3 marks)

5 The diagram shows $y = P(x)$ where $P(x)$ is a monic polynomial of degree 4.

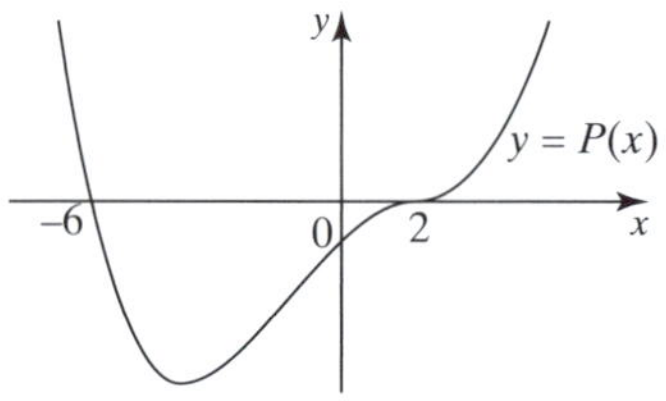

- **a** What is special about the root at $x = 2$? (1 mark)
- **b** Write $P(x)$ in factorised form. (1 mark)
- **c** What is the y-intercept? (1 mark)

6 $A(x) = ax^3 + bx^2 + cx + d$
- **a** A polynomial is odd if $P(-x) = -P(x)$ for every value of x. Show that, if $A(x)$ is odd, $d = 0$. (1 mark)
- **b** Show that, if $A(x)$ is odd, $b = 0$. (2 marks)
- **c** $B(x)$ is an odd polynomial of degree 3. $B(x)$ is divisible by $(x - 1)$. Find a possible polynomial $B(x)$. (2 marks)
- **d** Is $B(x)$ unique? Justify your answer. (1 mark)

7
- **a** Show that $x + 1$ is a factor of $P(x) = x^4 - 6x^3 + x^2 + 24x + 16$. (1 mark)
- **b** Show that $x = -1$ is a double root of $P(x) = 0$. (3 marks)
- **c** Find all the roots of $P(x) = 0$. (2 marks)
- **d** Sketch $y = P(x)$. (3 marks)
- **e** For what values of x is $P(x) < 0$? (1 mark)

Your Feedback

$\frac{\square}{40} \times 100\% = \square\%$

PAGES 121–122

PAGE 182

PARABOLAS

Linear and Non-linear Relationships

STUDY NOTES

1 The graph of $y = ax^2 + bx + c$ (where a, b and c are constants) is a **parabola**. If $a > 0$ the parabola is **concave up** ($\cup$) and if $a < 0$ the parabola is **concave down** ($\cap$).

2 The **turning point** of a parabola is called the **vertex**.

3 The parabola $\boldsymbol{y = ax^2}$ has its **vertex at the origin**. The value of a determines the width of the parabola. The larger the value of a, the thinner the parabola.

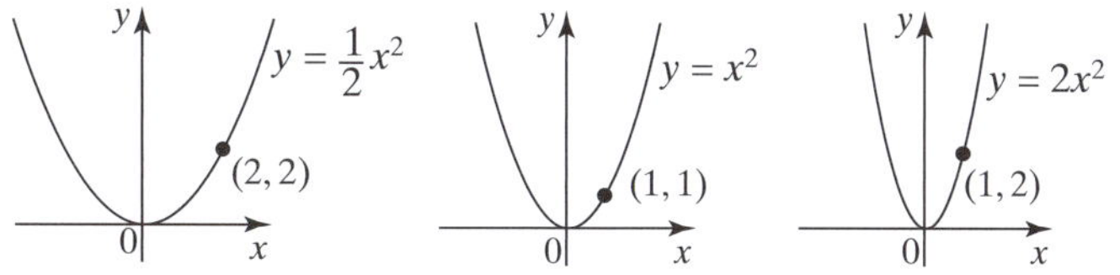

4 The **constant c**, in the equation $y = ax^2 + c$ has the effect of moving the parabola up or down the y-axis. The vertex of the parabola $y = ax^2 + c$ is at the point $(0, c)$.

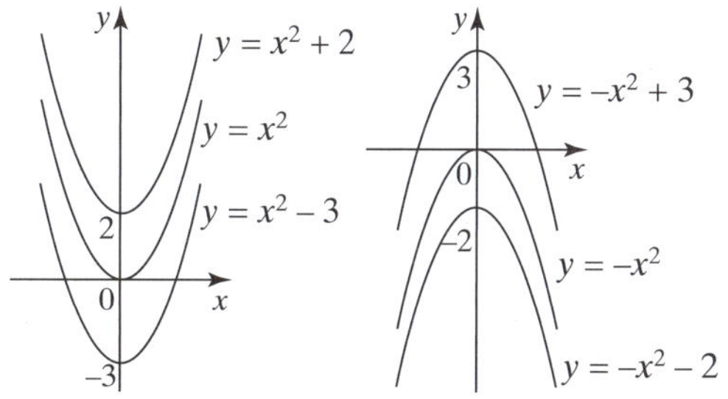

5 The ***x*-intercepts** of a parabola are the points where the curve meets the x-axis. A parabola may meet the x-axis in 0, 1 or 2 points. The x-intercepts are found when $y = 0$. The y-intercept is the point where the curve meets the y-axis. The y-intercept occurs when $x = 0$. If the equation of the parabola is of the form $y = ax^2 + bx + c$ then the y-intercept will be c.

6 The **axis of symmetry** of the parabola $y = ax^2 + bx + c$ is a vertical line (of the form $x = k$ where k is a constant). The equation of the axis of symmetry can be found because it is midway between the x-intercepts. The formula $x = -\dfrac{b}{2a}$ gives the axis of symmetry of the equation $y = ax^2 + bx + c$.
For example, if $y = 2x^2 + 8x - 7$ then $a = 2$, $b = 8$ and $c = -7$, and $-\dfrac{b}{2a} = -\dfrac{8}{2 \times 2} = -2$.
So the equation of the axis of symmetry is $x = -2$.

7 The **vertex of a parabola lies on the axis of symmetry**. The vertex can be found by first finding the equation of the axis of symmetry, and then substituting that value of x into the equation of the parabola to find the y value. If the equation of the parabola is in the form $y = a(x - k)^2 + h$ then the axis of symmetry is $x = k$ and the vertex is at (k, h). The equation $y = ax^2 + bx + c$ can be changed into the form $y = a(x - k)^2 + h$ by completing the square.

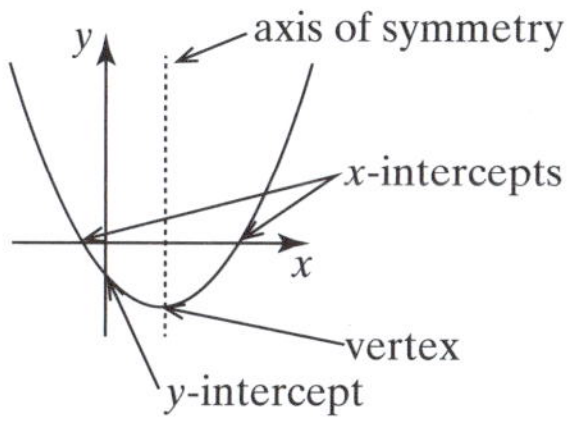

8 A sketch of the **graph of a parabola** can easily be made using the x- and y-intercepts, the axis of symmetry, the coordinates of the vertex and considering the concavity. Alternatively, a table of values can be drawn up.

Checklist

Can you:

1. *Recognise both the graph and the equation of a parabola?* ☐
2. *Draw the graphs of parabolas?* ☐
3. *Understand what is meant by the vertex of a parabola?* ☐
4. *Find the x- and y-intercepts, equation of the axis of symmetry and coordinates of the vertex of a parabola?* ☐
5. *Change the equation of a parabola from one form to another by completing the square?* ☐

PARABOLAS

Linear and Non-linear Relationships

SKILLS CHECK

1 For the parabola $y = x^2 - 8x + 12$ find:

a the x-intercepts

b the y-intercept

2 Find the y-intercept for each of these graphs:

a $y = x^2 + 3$

b $y = x^2 - 7x - 8$

c $y = 15 + 2x - x^2$

3 Determine whether each graph in question 2 meets the x-axis and if so find the x-intercept(s).

4 Sketch a graph of each of the equations given in question **2**.

5 Use the parabola shown to answer the following questions.

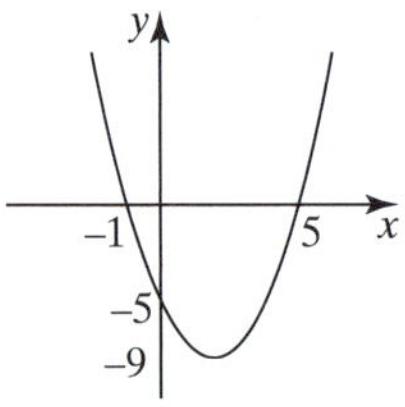

a What is the y-intercept?

b At what points does the parabola cut the x-axis?

c What is the equation of the axis of symmetry?

d What are the coordinates of the vertex?

e Is the parabola concave up or concave down?

6 Find the equation of the axis of symmetry of the parabola:

a which cuts the x-axis at $x = 1$ and $x = 5$

b with equation $y = (x + 4)^2 - 9$

7 For the parabola $y = 3x^2 - 12x + 7$ find:

a the equation of the axis of symmetry, using $x = -\dfrac{b}{2a}$

b the coordinates of the vertex

8 Determine the equation of the quadratic expression for each table of values.

a

x	−3	−2	−1	0	1	2	3
y	12	7	4	3	4	7	12

b

x	−3	−2	−1	0	1	2	3
y	27	12	3	0	3	12	27

c

x	−3	−2	−1	0	1	2	3
y	4	1	0	1	4	9	16

d

x	−3	−2	−1	0	1	2	3
y	6	2	0	0	2	6	12

9 Sketch graphs of:

a $y = (x - 3)^2 + 5$

b $y = 2x^2 + 7x - 4$

PAGES 122–123

Answers **1 a** 2 and 6 **b** 12 **2 a** 3 **b** −8 **c** 15 **3 a** does not meet the x-axis **b** −1 and 8 **c** 5 and −3 **4** see solutions **5 a** −5 **b** (−1, 0) and (5, 0) **c** $x = 2$ **d** (2, −9) **e** concave up **6 a** $x = 3$ **b** $x = -4$ **7 a** $x = 2$ **b** (2, −5) **8 a** $y = x^2 + 3$ **b** $y = 3x^2$ **c** $y = (x + 1)^2$ **d** $y = x^2 + x$ **9** see solutions

PARABOLAS

Linear and Non-linear Relationships

25 MINUTES

INTERMEDIATE TEST

Part A Multiple Choice

1 Which diagram could show the graph of $y = (x - 2)^2$?

A
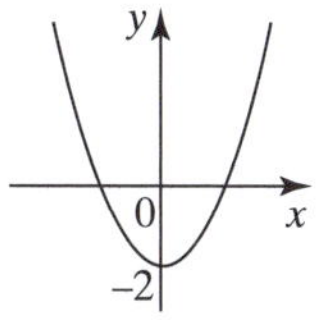

B
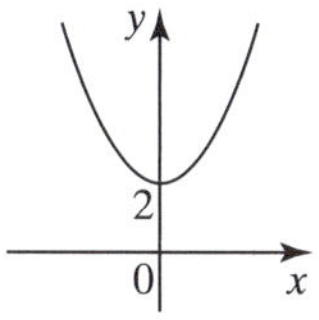

C
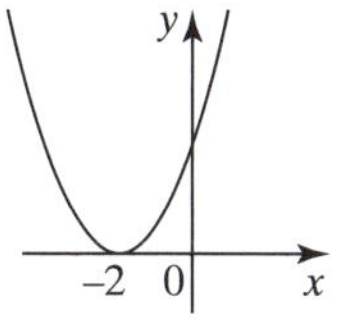

D
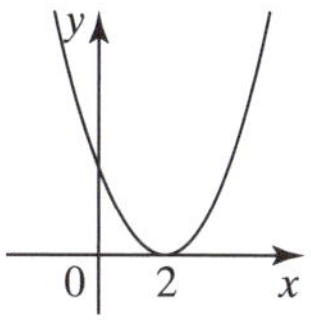

(1 mark)

2 The vertex of the parabola $y = (x + 3)^2 - 4$ is at:

A $(-3, -4)$ **B** $(-3, 4)$ **C** $(3, -4)$ **D** $(3, 4)$ (1 mark)

3 Which is not the equation of a parabola?

A $y = x - x^2$ **B** $y = 2x^2 + 7x - 5$ **C** $y = x^2 + x^3$ **D** $x^2 - y = 0$ (1 mark)

4 Which graph will pass through the point $(-2, 3)$?

A $y = 2x^2 + 3x - 5$ **B** $y = 3x^2 + 5x + 2$ **C** $y = 2x^2 + 5x + 3$ **D** $y = 3x^2 + 2x - 5$ (1 mark)

5 The equation of the axis of symmetry of $y = 3x^2 - 3x + 2$ is:

A $x = -\frac{1}{2}$ **B** $x = \frac{1}{2}$ **C** $x = -2$ **D** $x = 2$ (1 mark)

Part B Written Response

6 For the parabola $y = x^2 + 2x - 3$:

- **a** what is the y-intercept? (1 mark)
- **b** where does the parabola cut the x-axis? (2 marks)
- **c** what is the equation of the axis of symmetry? (1 mark)
- **d** what are the coordinates of the vertex? (1 mark)
- **e** sketch the graph of $y = x^2 + 2x - 3$ *Hint 1* (1 mark)

7
- **a** Find the vertex of the parabola $y = 2x^2 - 8x + 11$. *Hint 2* (2 marks)
- **b** Will the parabola $y = 2x^2 - 8x + 11$ cut the x-axis? Briefly justify your answer. *Hint 3* (2 marks)

8 What is the maximum value of $9 + 5x - 2x^2$? *Hint 4* (2 marks)

9
- **a** Rearrange the equation $4x^2 - 20x + 11$ into the form $(px + q)^2 + r$. (2 marks)
- **b** Sketch the graph of $y = 4x^2 - 20x + 11$. *Hint 5* (2 marks)

10 The diagram shows a sketch of a parabola. The equation of the parabola is of the form $y = ax^2 + c$.

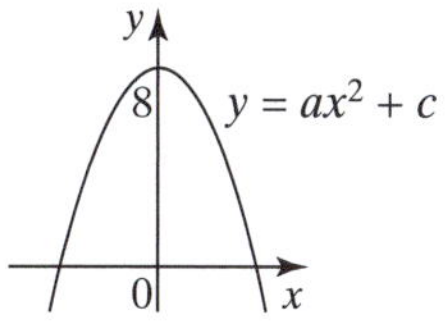

- **a** What is the value of the constant c? (1 mark)
- **b** The graph passes through the point $(1, 6)$. What is the value of a? *Hint 6* (1 mark)
- **c** Where does the graph cross the x-axis? (2 marks)

Hint 1: Use all the information gathered in the other parts of the question.
Hint 2: Use the formula to find the axis of symmetry.
Hint 3: Is the curve concave up or concave down? Where is the vertex?
Hint 4: Find the axis of symmetry.
Hint 5: Use the rearrangement from part **a**.
Hint 6: Put the value of c into the equation of the parabola and then substitute the coordinates of the point through which the parabola passes.

Your Feedback

$\frac{\square}{25} \times 100\% = \square\,\%$

PAGE 123
PAGE 182

PARABOLAS

Linear and Non-linear Relationships

40 MINUTES

ADVANCED TEST

1 **a** What is the equation of the axis of symmetry of the parabola $y = 27 + 6x - x^2$? (1 mark)

b Is the turning point of the parabola a maximum or minimum? Justify your answer. (1 mark)

c Find the coordinates of the vertex of the parabola. (1 mark)

2 **a** Sketch a parabola that has x-intercepts –2 and 5 and y-intercept –5. (2 marks)

b What is the equation of the parabola? (2 marks)

3 This table of values was drawn up when a parabola was being graphed.

x	0	1	2	3	4	5	6
y	–5	0	9	22	39	60	85

What is the equation of the parabola? (3 marks)

4 A parabola has equation $y = \frac{1}{2}x^2 + 5x - 12$. Without sketching, describe four features of the parabola. (4 marks)

5 **a** Where does the parabola $y = 4x^2 + 4x - 15$ cut the y-axis? (1 mark)

b Where does the parabola intersect the x-axis? (2 marks)

c For what values of x is $4x^2 + 4x - 15 > 0$? (2 marks)

6 The graph shows the parabola $y = x^2 - 5x - 6$ and the line $y = 18$. Find the shaded area.

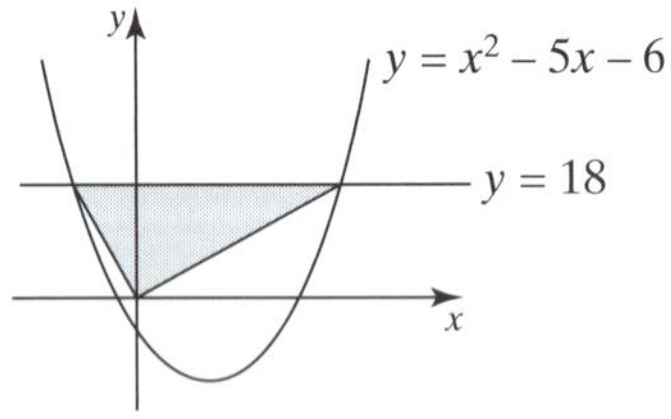

(3 marks)

7 The parabola $y = x^2 - 8x + 17$ and the line $y = x + 9$ intersect at P and Q (where P is the point with the smaller x-value).

a Find the coordinates of P and Q. (3 marks)

b The parabola $y = 2x^2 + bx + c$ also passes through P and Q. Find the values of b and c. (2 marks)

c Show that $y = 3x^2 - 26x + 33$ also passes through P and Q. (1 mark)

d If $y = 4x^2 + dx + e$ also passes through P and Q. Use a pattern to find suggested values for d and e. (1 mark)

8 **a** By completing the square, show that $ax^2 + bx + c = a\left[\left(x + \frac{b}{2a}\right)^2 - \left(\frac{b^2 - 4ac}{4a^2}\right)\right]$. (3 marks)

b Write down the formula for finding the axis of symmetry of a parabola and explain why that is the formula. (2 marks)

c Find a formula for the y-coordinate of the vertex. (2 marks)

d Use these formulas to find the vertex of $y = x^2 - 8x + 21$. (2 marks)

e Where does the parabola $y = x^2 - 8x + 21$ cut the x-axis? Justify your answer. (2 marks)

Your Feedback

$\frac{\square}{40} \times 100\% = \square\%$

PAGES 124–125

PAGE 182

FURTHER GRAPHS

Linear and Non-linear Relationships

STUDY NOTES

1 The two axes divide the number plane into four quarters or **quadrants**. For convenience these quadrants are numbered, beginning with the quadrant where both the x- and y-coordinates are positive, and moving in an anticlockwise direction.

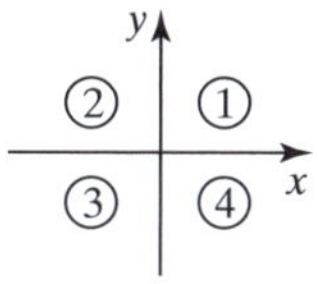

2 A **hyperbola** is a special type of curve. We refer to it as a curve even though it consists of **two** separate parts or **branches**, often in two opposite quadrants of the number plane.

3 The graph of $\boldsymbol{y = \frac{k}{x}}$ is a hyperbola. If $k > 0$ the graph is located in the first ($x > 0$ and $y > 0$) and third ($x < 0$ and $y < 0$) quadrants. If $k < 0$ the graph is located in the second ($x < 0$ and $y > 0$) and fourth ($x > 0$ and $y < 0$) quadrants. The graph does not have a value when $x = 0$ (because we cannot divide by zero).

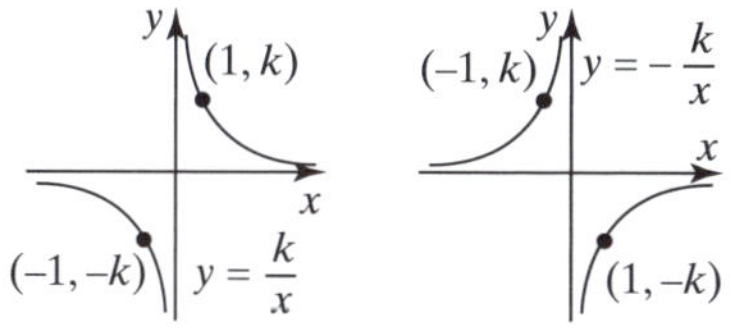

4 An **asymptote** is a line that a curve approaches but never meets. It is a line that a curve begins to imitate.

5 As **values of x get very large**, the graph of $y = \frac{k}{x}$ gets closer and closer to the x-axis, (but it will never meet, touch or cross it). When the **value of x is close to zero**, the graph of $y = \frac{k}{x}$ is close to the y-axis. If x has a very small positive value, then y will have a large positive value. If x has a very small negative value, then y will have a large negative value. The x- and y- axes are asymptotes of the curve.

6 If a **constant** is added to the equation of a hyperbola the graph will be moved across or up or down. The asymptotes will also be adjusted.

7 The graph of $y = ax^3 + d$ is a **cubic curve**. The value of a determines the width of the curve. d gives the y-intercept.

8 Any equation of the form $y = a^x$ is an **exponential curve**. $y = a^x$ has y-intercept 1 for all positive values of a.

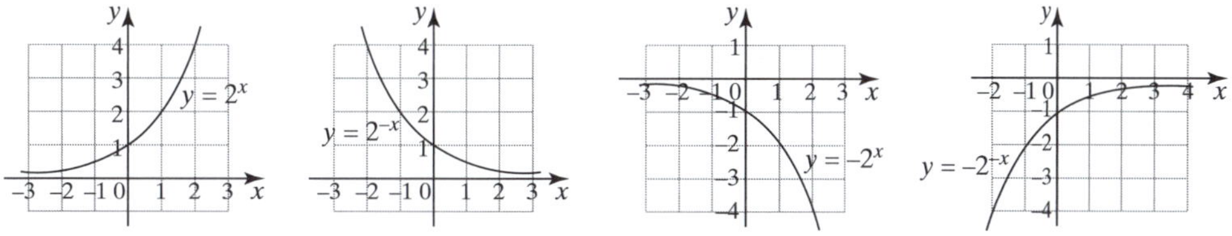

9 A **circle**, centre at the origin and radius of length r units has equation $x^2 + y^2 = r^2$. The equation $(x - h)^2 + (y - k)^2 = r^2$ is the equation of a circle, centre (h, k), radius r. Sometimes the equation of a circle is given in a different form (e.g. $x^2 + 8x + y^2 - 6y = 0$). It can be rearranged by completing the square ($(x + 4)^2 + (y - 3)^2 = 25$).

10 Any graph can be **transformed**, for example by the addition of a constant. A curve might move up or down on the y-axis, or move left or right on the x-axis or some combination of both.

Checklist

Can you:

1 *Identify the type of graph from an equation?* ☐
2 *Recognise and graph hyperbolas, cubic curves, exponential curves and circles?* ☐
3 *Find the centre and radius of a circle given its equation?* ☐
4 *Understand the ways in which graphs can be transformed by changes to their equations?* ☐

FURTHER GRAPHS

Linear and Non-linear Relationships

SKILLS CHECK

1 Name the type of graph (parabola, hyperbola, exponential curve, straight line, cubic curve or circle) with equation:

a $y = 6 - x$ **b** $y = \frac{2}{x}$ **c** $y = 3^x$ **d** $y = x^3$

e $y = \frac{x}{2}$ **f** $x^2 + y^2 = 9$ **g** $y = x^2$ **h** $y = 6 - x^2$

i $y = 2x^2 - 7x + 6$ **j** $x + y = 9$ **k** $x^2 + 4x + y^2 = 5$ **l** $y = 5^{-x}$

2 Write down the equation of a circle with centre at the origin and radius 5 units.

3 Sketch the graph of:

a $y = 2^{-x}$ **b** $y = x^3 - 1$ **c** $y = (x - 3)^2$ **d** $x^2 + y^2 = 64$

4 A circle has equation $x^2 + y^2 = 196$.

a What are the coordinates of the centre of the circle?

b What is the length of the radius of the circle?

5 Complete the table of values for $y = \frac{6}{x}$ and sketch the graph.

x	–6	–5	–4	–3	–2	–1	1	2	3	4	5	6
y												

6 What name is given to the type of graph shown?

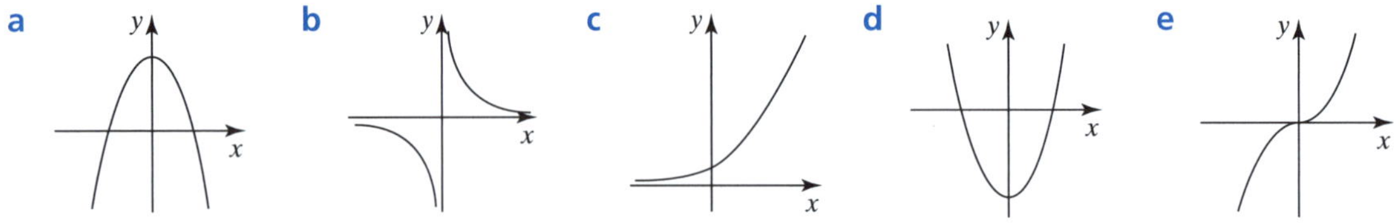

7 Choose the correct equation, from the list below, for each graph.

$y = 4x,\ y = 4x^2,\ y = 4^x,\ y = \frac{4}{x},\ y = \frac{x}{4},\ y = 4 - x,\ y = x^2 - 4,\ x^2 + y^2 = 4,\ y = \frac{-4}{x},\ y = 4 - x^2$

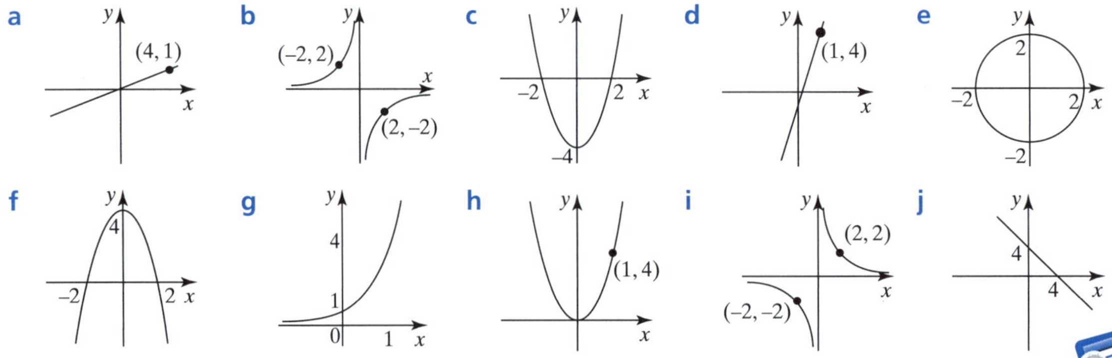

PAGES 125–126

Answers **1 a** straight line **b** hyperbola **c** exponential curve **d** cubic curve **e** straight line **f** circle **g** parabola **h** parabola **i** parabola **j** straight line **k** circle **l** exponential curve **2** $x^2 + y^2 = 25$ **3** see worked solutions **4 a** (0, 0) **b** 14 units **5** –1, –1.2, –1.5, –2, –3, –6, 6, 3, 2, 1.5, 1.2, 1 see worked solutions **6 a** parabola **b** hyperbola **c** exponential curve **d** parabola **e** cubic curve **7 a** $y = \frac{x}{4}$ **b** $y = \frac{-4}{x}$ **c** $y = x^2 - 4$ **d** $y = 4x$ **e** $x^2 + y^2 = 4$ **f** $y = 4 - x^2$ **g** $y = 4^x$ **h** $y = 4x^2$ **i** $y = \frac{4}{x}$ **j** $y = 4 - x$

FURTHER GRAPHS

Linear and Non-linear Relationships

25 MINUTES

INTERMEDIATE TEST

Part A Multiple Choice

1 Which curve does not pass through the point (3, 4)?

A $x^2 + y^2 = 25$ **B** $y = \frac{x^3}{3} - 5$ **C** $y = (x - 1)^2$ **D** $y = 2^x$ (1 mark)

2 Which diagram could show the graph of $y = -2^x$?

A

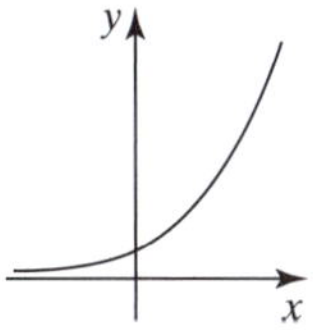

B

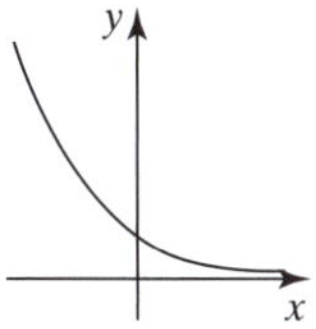

C

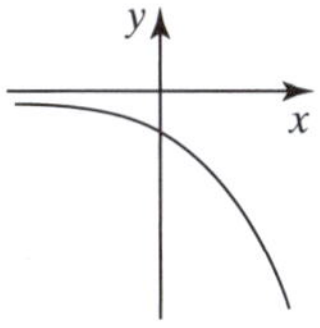

D

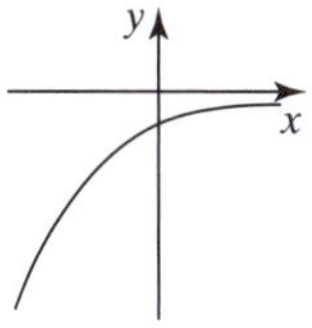

(1 mark)

3 Which point does not lie on the cubic curve $y = 2x^3 + 3$?

A (–1, –5) **B** (0, 3) **C** (1, 5) **D** (2, 19) (1 mark)

Part B Written Response

4 **a** Complete the table of values for $y = \frac{4}{x}$

x	–8	–4	–2	–1	$-\frac{1}{2}$	$\frac{1}{2}$	1	2	4	8
y										

(1 mark)

b Draw the graph of $y = \frac{4}{x}$ (2 marks)

c Briefly explain what happens to the y-values as the x-values get very large. (1 mark)

d Briefly explain what happens to the y-values as the x-values get close to zero. (2 marks)

e On separate diagrams sketch $y = \frac{4}{x} + 2$ and $y = \frac{4}{x} - 2$ *Hint 1* (2 marks)

f Briefly explain the effect of adding a constant to the equation $y = \frac{4}{x}$ (1 mark)

5 P is the point (2, –1) and Q is the point (5, 4).

a Find the length of the interval PQ in simplest surd form. (1 mark)

b Find the equation of the circle, centre P, that passes through Q. *Hint 2* (1 mark)

6 **a** On the same diagram sketch the graphs of $y = x^3$, $y = x^3 + 2$ and $y = (x + 2)^3$. (3 marks)

b Briefly explain the effect of adding constants to the equation of a cubic curve. (2 marks)

7 Find the centre and radius of the circle given by $x^2 - 6x + y^2 + 4y = 36$. *Hint 3* (2 marks)

8 Consider the equation $y = \frac{-6}{x-3}$

a What is the y-intercept? (1 mark)

b What are the equations of the asymptotes? (2 marks)

c Without any further consideration, sketch the graph of $y = \frac{-6}{x-3}$. *Hint 4* (1 mark)

Hint 1: Use the graph that is already drawn as a guide.
Hint 2: The radius of the circle will equal the length of the interval joining P to Q.
Hint 3: Complete the square.
Hint 4: It doesn't have to be an accurate graph, draw in the asymptotes, plot the y-intercept and sketch.

Your Feedback

$\frac{\square}{25} \times 100\% = \square\%$

PAGES 126–127

PAGE 182

FURTHER GRAPHS

Linear and Non-linear Relationships

40 MINUTES

ADVANCED TEST

1 **a** Describe the curve represented by the equation $(x-3)^2 + (y+1)^2 = 25$. (2 marks)

b Find the points of intersection of this curve with the line $y = -1$. (2 marks)

c Sketch the curve and the line and shade the region found above the line but beneath the curve. (2 marks)

d Find the area of the shaded part. (1 mark)

2 **a** On the same diagram sketch the curves $y = 3^x$ and $y = 3^{-x}$. (2 marks)

b Solve the equation $3^x = 3^{-x}$. (1 mark)

3 **a** Find the coordinates of the points of intersection of the cubic curve $y = x^3$ and the parabola $y = 6x - x^2$. (3 marks)

b Sketch the two curves on the same diagram. (2 marks)

4 A curve of the form $y = a^x + k$ $(a > 0)$, has x-intercept 2 and y-intercept -3. Find the values of a and k. (3 marks)

5 Consider the curve $y = \dfrac{2}{x} + 3$.

a What happens to the y-values when the x-values get very large? (1 mark)

b What happens when $x = 0$? (1 mark)

c Is there a y-value for every x-value? (1 mark)

d For what value of x does $y = 7$? (1 mark)

e What are the equations of the asymptotes? (2 marks)

f Sketch the curve. (2 marks)

g Briefly explain the effect of the constant 3 on the curve. (1 mark)

6 Consider the curve $y = \dfrac{2}{x+3}$

a What happens to the y-values when the x-values get very large? (1 mark)

b What happens when $x = 0$? (1 mark)

c Is there a y-value for every x-value? (1 mark)

d For what value of x does $y = -1$? (1 mark)

e What are the equations of the asymptotes? (2 marks)

f Sketch the curve. (2 marks)

g Briefly explain the effect of the constant 3 on the curve. (1 mark)

7 The diagram shows the curve $y = -x^4 + 18x^2 - 32$ for values of $x \geq 0$.

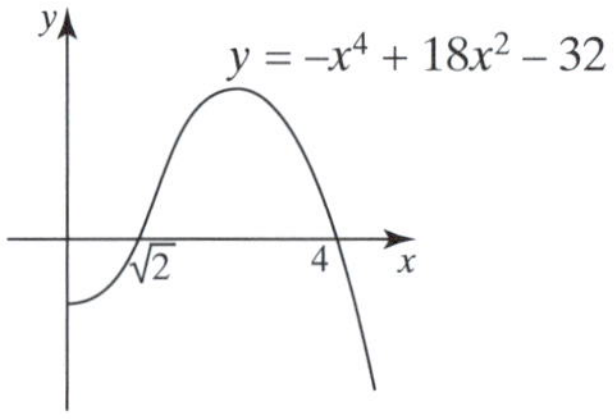

a What is the y-intercept? (1 mark)

b Briefly explain why the curve $y = -x^4 + 18x^2 - 32$ is symmetrical about the y-axis. (1 mark)

c The curve has a turning point at $(3, q)$. Find the value of q. (1 mark)

d Is q the maximum value of $-x^4 + 18x^2 - 32$? Justify your answer. (1 mark)

Your Feedback

$\dfrac{\square}{40} \times 100\% = \square\,\%$

PAGES 127–128

PAGE 182

CURVE-SKETCHING TECHNIQUES

Linear and Non-linear Relationships

STUDY NOTES

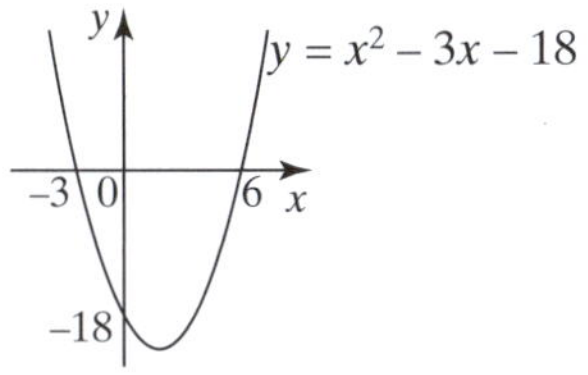

1 The **roots of the equation** give the ***x*-intercepts** of a curve. For example, if $x^2 - 3x - 18 = 0$ then $(x + 3)(x - 6) = 0$ and $x = -3$ or $x = 6$. The graph of $y = x^2 - 3x - 18$ will have x-intercepts at –3 and 6. The y-intercept will be –18.

2 The graph of $y = ax^3$ is a **cubic curve**. The value of a determines the width of the curve. The y-intercept is 0. The curve $y = ax^3 + d$ is identical to the curve $y = ax^3$ except that it has moved up (or down if d is negative) the y-axis. The y-intercept is d. The curve $y = a(x - r)(x - s)(x - t)$ is also a cubic curve. It has x-intercepts at r, s and t.

3 When sketching equations of the form $y = ax^n$ $(n > 0)$, we can consider whether n is **odd or even**. If n is even, then $(-x)^n = x^n$ for all values of x and so the curve is symmetrical about the y-axis. (Folding the graph along the y-axis the two halves of the graph will match exactly.). If n is odd $(-x)^n = -(x^n)$ for all values of x. These curves look the same if they are rotated through 180°. (They have point symmetry about the origin.)

4 The graph of $y = ax^n + k$ is just a **vertical transformation** of the curve $y = ax^n$. The y-intercept will be k. The graph will move up or down the y-axis.

5 The graph of the curve $y = a(x - r)^n$ is a **horizontal transformation** of the curve $y = ax^n$. The x-intercept is r. The curve slides along the x-axis.

6 The equation $(x - h)^2 + (y - k)^2 = r^2$ is the **equation of a circle**. The **centre** of the circle is at (h, k) and the **radius** is r units. For example, the equation $(x - 2)^2 + (y + 3)^2 = 25$ is the equation of a circle centre (2, –3) and radius 5 units.

7 By **completing the square**, an equation of the form $x^2 + gx + y^2 + jy = c$ can be changed into the form $(x - h)^2 + (y - k)^2 = r^2$. The square is completed for x and for y. For example, to complete the square for the expression $x^2 + 6x$ we need to add 9 (half of 6 is 3, $3^2 = 9$). To complete the square for the expression $y^2 - 8y$ we need to add 16 (half of 8 is 4, $4^2 = 16$). So if $x^2 + 6x + y^2 - 8y = 11$ then $x^2 + 6x + 9 + y^2 - 8y + 16 = 11 + 9 + 16$. This gives the equation $(x + 3)^2 + (y - 4)^2 = 36$ so the equation $x^2 + 6x + y^2 - 8y = 11$ is the equation of a circle with centre (–3, 4) and radius 6 units.

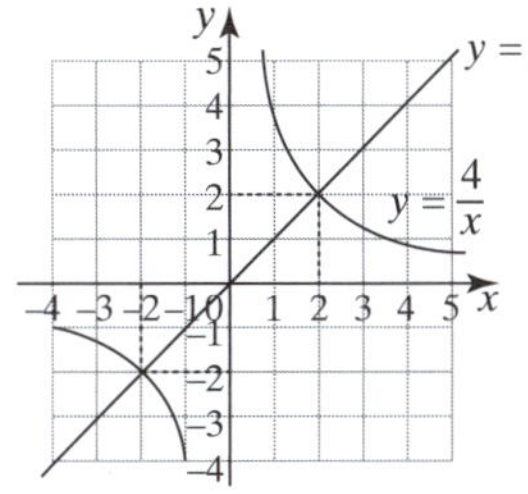

8 The **points of intersection** of a line with a **parabola**, **circle** or **hyperbola** can be found. This can be done graphically by sketching careful graphs and reading the point of intersection from the graph. Alternatively the point of intersection can be found algebraically by solving the two equations simultaneously. For example, the point of intersection of the line $y = x$ and the hyperbola $y = \frac{4}{x}$ will occur when $x = \frac{4}{x}$ and hence when $x^2 = 4$. So $x = \pm 2$ and the points of intersection are (–2, –2) and (2, 2). Remember to give the y-values as well as the x-values if asked for the points of intersection.

9 When **describing a graph** we can also consider what happens to the y-values when the x-values become large (positive or negative), and what happens when x is close to 0.

Checklist

Can you:

1 *Determine the features of a curve from its equation?* ☐
2 *Graph and compare the features of curves?* ☐
3 *Find the x- and y-intercepts of a curve?* ☐
4 *Recognise the equation of a circle and find its centre and radius?* ☐
5 *Complete the square to find the centre and radius of a circle?* ☐
6 *Determine the points of intersection of a line with a parabola, circle or hyperbola?* ☐
7 *Determine what happens to the y-values as x gets large or small or close to zero?* ☐

CURVE-SKETCHING TECHNIQUES

Linear and Non-linear Relationships

SKILLS CHECK

1 Sketch the graph of:

a $y = x^2$

b $y = x^2 + 5$

c $y = (x + 5)^2$

2 Find the y-intercept for:

a $y = x + 5$

b $y = x^2 - 2x - 3$

c $y = x^3 + 5x^2 + 6x$

3 Find the x-intercept(s) for:

a $y = x + 5$

b $y = x^2 - 2x - 3$

c $y = x^3 + 5x^2 + 6x$

4 Write down the coordinates of the centre and the length of the radius for each circle:

a $(x - 2)^2 + (y - 1)^2 = 49$

b $(x + 3)^2 + y^2 = 36$

c $(x - 1)^2 + (y + 4)^2 = 1$

5 Write the equation of each circle in the form $(x - h)^2 + (y - k)^2 = r^2$:

a centre (3, 2), radius 5 units

b centre (–1, 4), radius 2 units

6 Find the number that needs to be added to each expression to complete the square:

a $x^2 + 8x$

b $x^2 - 10x$

c $y^2 - 12y$

7 Change each equation into the form $(x - h)^2 + (y - k)^2 = r^2$:

a $x^2 - 6x + y^2 + 8y = 39$

b $x^2 + 2x + y^2 - 2y + 1 = 0$

8 Find the solutions for each pair of simultaneous equations:

a $y = \frac{6}{x}$ and $y = x + 1$

b $x^2 + y^2 = 50$ and $x = 2y + 5$

9 Consider the curve $y = 3^x$

a What happens to y when x gets very large?

b What happens to y as x gets large negatively?

c Is there a y-value for every x-value?

d What happens at $x = 0$?

PAGES 128–129

Answers **1** see worked solutions **2 a** 5 **b** –3 **c** 0 **3 a** –5 **b** –1 and 3 **c** 0, –2 and –3 **4 a** (2, 1), 7 units **b** (–3, 0), 6 units **c** (1, –4), 1 unit **5 a** $(x - 3)^2 + (y - 2)^2 = 5^2$ **b** $(x + 1)^2 + (y - 4)^2 = 2^2$ **6 a** 16 **b** 25 **c** 36 **7 a** $(x - 3)^2 + (y + 4)^2 = 8^2$ **b** $(x + 1)^2 + (y - 1)^2 = 1^2$ **8 a** $x = -3, y = -2$ and $x = 2, y = 3$ **b** $x = -5, y = -5$ and $x = 7, y = 1$ **9 a** gets very large **b** gets closer to 0 **c** yes **d** $y = 1$

CURVE-SKETCHING TECHNIQUES

Linear and Non-linear Relationships

30 MINUTES

INTERMEDIATE TEST

Part A Multiple Choice

1 The y-intercept of the curve $y = 2(x - 1)(x + 3)(x - 4)$ is:
A -12 **B** 12 **C** -24 **D** 24 (1 mark)

2 Which of these points lies outside the circle $(x - 2)^2 + y^2 = 9$?
A $(0, 0)$ **B** $(-1, -2)$ **C** $(1, -1)$ **D** $(2, 1)$ (1 mark)

3 Which is an asymptote of the hyperbola $y = \dfrac{3}{x - 6}$?
A $x = 6$ **B** $x = -6$ **C** $y = 3$ **D** $y = -3$ (1 mark)

Part B Written Response

4 Consider $y = 4x^3 - 12x^2 - 16x + 48$.
- **a** Show that $4(x - 2)(x + 2)(x - 3) = 4x^3 - 12x^2 - 16x + 48$. *Hint 1* (1 mark)
- **b** What is the y-intercept? (1 mark)
- **c** What are the x-intercepts? *Hint 2* (2 marks)
- **d** What happens to the y-values when the x-values get large? (1 mark)
- **e** Sketch the graph of $y = 4x^3 - 12x^2 - 16x + 48$. (3 marks)

5
- **a** On the same set of axes sketch the graphs of $y = \dfrac{12}{x}$ and $y = 3x - 9$. (2 marks)
- **b** Use the graph from part **a** to find the values of y for which $\dfrac{12}{x} = 3x - 9$. *Hint 3* (2 marks)

6
- **a** Change the equation of the circle $x^2 - 6x + y^2 + 8y = 0$ to the form $(x - h)^2 + (y - k)^2 = r^2$. (2 marks)
- **b** Write down the coordinates of the centre and the length of the radius of the circle. (2 marks)
- **c** Find the points of intersection of the circle with the line $x - 2y - 6 = 0$. *Hint 4* (3 marks)

7 Without sketching, briefly describe and compare the graphs of $y = x^3$, $y = 4x^3$, $y = x^3 + 4$ and $y = (x - 4)^3$. (3 marks)

8
- **a** Draw separate sketches of $y = x^2$, $y = x^3$, $y = x^4$ and $y = x^5$. (4 marks)
- **b** Which of these equations have graphs that are symmetrical about the y-axis? (1 mark)

Hint 1: You don't need to factorise. Just expand the left hand side.
Hint 2: Use the factorised expression from part **a**.
Hint 3: There is no need to solve algebraically. Just read the values from the graph.
Hint 4: Make x the subject of the equation of the line and substitute. Solve a quadratic equation to find y.

Your Feedback

$\dfrac{\quad}{30} \times 100\% = \quad\%$

PAGES 129–130

PAGE 182

CURVE-SKETCHING TECHNIQUES

Linear and Non-linear Relationships

ADVANCED TEST

1 **a** Find the points of intersection of the curve $y = \frac{12}{x}$ and the line $y = 2x + 5$. (3 marks)

b Sketch the curve and the line on the same diagram. (2 marks)

c Briefly explain why any line with negative gradient that passes through the origin will not intersect the curve $y = \frac{12}{x}$ (1 mark)

d Show algebraically why this is the case. (1 mark)

2 Sketch $y = 2(3^{-x}) - 1$. (2 marks)

3 **a** Find the centre and radius of the circle $x^2 + 6x + y^2 - 10y - 15 = 0$. (3 marks)

b Does the point (2, 10) lie inside, on, or outside the circle. Justify your answer. (2 marks)

4 The diagram, which is not to scale, shows the line $2x + y - 14 = 0$ and the parabola $y = 5 + 4x - x^2$.

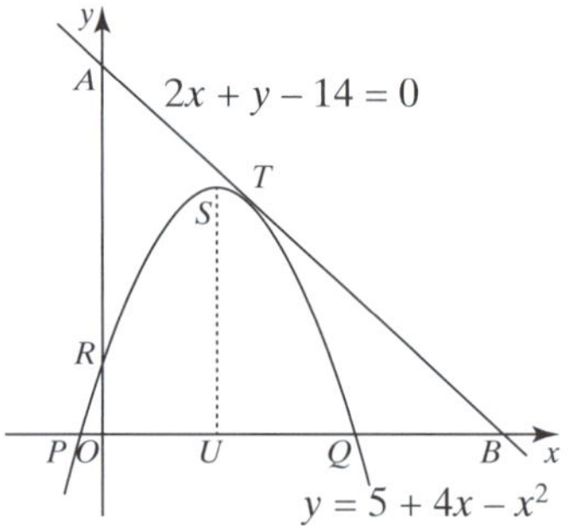

a Show algebraically that the line only meets the parabola in one point and find the coordinates of that point. (2 marks)

b Find the equation of the circle, centre O, that passes through R. (2 marks)

c Explain why that circle must also pass through Q. (1 mark)

d Show that $AR = SU$. (2 marks)

5 **a** On the same diagram sketch graphs of $y = x^3$, $y = (x + 3)^3$ and $y = (x + 3)^3 + 5$. (3 marks)

b On another diagram sketch $y = -x^3$, $y = (2 - x)^3$ and $y = (2 - x)^3 + 10$. (3 marks)

c Without sketching, describe the curve $y = (x - 4)^3 - 2$. (1 mark)

6 The diagram shows the graph of $y = \frac{x^6}{4}$

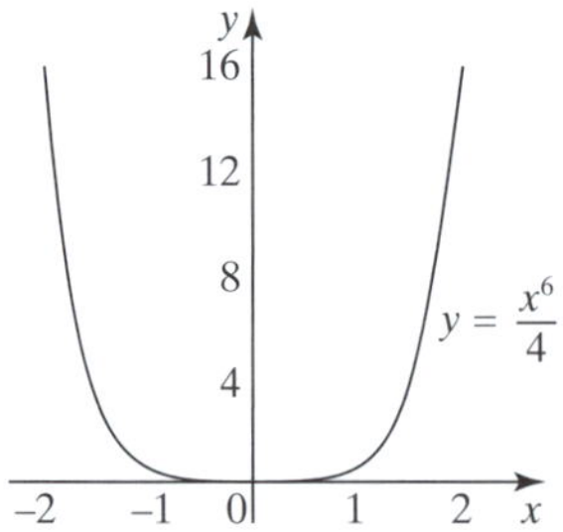

a Copy the diagram and show the graph of $y = \frac{x^6 + 8}{4}$ on the same diagram. (3 marks)

b Copy the diagram again and show the graph of $y = \frac{1}{4}(x - 2)^6$ on this same diagram. (3 marks)

c Without further sketching describe the features of the graphs $y = \frac{1}{4}x^6 - 3$ and $y = \frac{(x + 1)^6}{4}$ (2 marks)

7 The diagram shows the graph of a function of the form $y = ax^3 + bx^2 + cx + d$. Find the values of a, b, c and d. (4 marks)

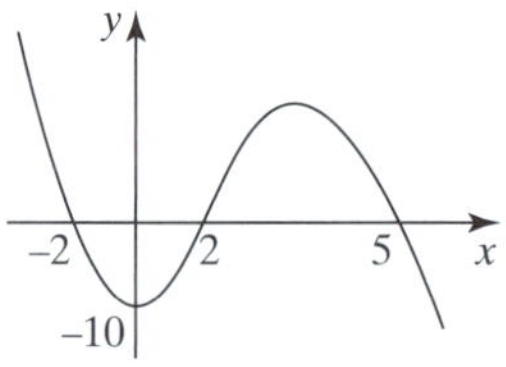

Your Feedback

$\frac{\square}{40} \times 100\% = \square\%$

PAGES 130–131

PAGE 182

THE EQUATION OF A STRAIGHT LINE

Linear and Non-linear Relationships

STUDY NOTES

1 The equation of a line is the **relationship between the x- and y-coordinates** of any point on the line. For example, if the equation of a line is $y = x + 2$, every point on that line has its y-coordinate 2 more than its x-coordinate. The line will pass through the points (1, 3), (4, 6), (10, 12), (97, 99), (–15, –13) for example, but these are only a few of the infinite number of points that lie on the line.

2 The equation $y = mx + b$ is the equation of a straight line in **gradient/intercept form**. m is the gradient or slope. It is the measure of the steepness of the line. b is the y-intercept. It tells us where the line meets the y-axis. For example, the line $y = \frac{1}{2}x - 6$ has gradient $\frac{1}{2}$ and y-intercept –6.

3 The equation $ax + by + c = 0$ gives the equation of a line in **general form**. Rearranging this equation to make y the subject will change it into gradient/intercept form.

4 The equation of a line can be found whenever we know its **gradient** and **any point on the line**. The equation of the line with gradient m and passing through the point (x_1, y_1) is given by the formula $y - y_1 = m(x - x_1)$.

5 The equation of a line can be found whenever **two points** on the line are known. If the line passes through the points (x_1, y_1) and (x_2, y_2) then the gradient could be found using the gradient formula $m = \frac{y_2 - y_1}{x_2 - x_1}$ and then use the formula $y - y_1 = m(x - x_1)$ to get the equation. Alternatively, the **two-point formula** $\frac{y - y_1}{x - x_1} = \frac{y_2 - y_1}{x_2 - x_1}$ could be used. The **two-point formula** comes from substituting the formula for m into $y - y_1 = m(x - x_1)$.

6 The ***x*-intercept** is where the line meets the x-axis. The x-axis is the line $y = 0$, so the x-intercept is found when $y = 0$. Similarly, the y-intercept is found when $x = 0$.

7 In order to **graph a line** we need to know at least two points that lie on the line. We can form a table of values and plot the resulting points, use the gradient and y-intercept, or find the x-intercept and the y-intercept and use those two points.

8 Two lines are **parallel** if their gradients are equal. If two lines have gradients m_1 and m_2 then if $m_1 = m_2$ the lines are parallel and if the lines are parallel $m_1 = m_2$.

9 Two lines are **perpendicular** if they meet at right angles. If two lines are perpendicular their gradients will be negative reciprocals. If two lines have gradients m_1 and m_2 then the lines will be perpendicular if $m_1m_2 = -1$ and if $m_1m_2 = -1$ the lines will be perpendicular.

Checklist

Can you:

1 *Understand that the equation of a line describes the relationship between the x- and y-coordinates?* ☐
2 *Understand the meaning of m and b in the equation y = mx + b?* ☐
3 *Understand the difference between an equation in gradient/intercept form and an equation in general form and change an equation in one form to an equation in the other form?* ☐
4 *Find the equation of a line with given gradient passing through a given point?* ☐
5 *Find the equation of a line through two points?* ☐
6 *Find x and y intercepts?* ☐
7 *Graph lines?* ☐
8 *Understand the conditions for two lines to be parallel or perpendicular?* ☐

THE EQUATION OF A STRAIGHT LINE

Linear and Non-linear Relationships

SKILLS CHECK

1 Give the equation of the line (in the form $y = mx + b$) with:

- **a** gradient 4 and y-intercept –2
- **b** gradient $\frac{1}{2}$ and y-intercept 5
- **c** gradient –2 and y-intercept 0

2 Use the formula $y - y_1 = m(x - x_1)$ to find the equation, (in the form $y = mx + b$), of a line with:

- **a** gradient 2 and passing through the point (3, 1)
- **b** gradient –3 and passing through the point (1, –4)
- **c** gradient $\frac{2}{3}$ and passing through the point (–3, –5)

3 A line passes through the points (3, 2) and (–1, 6). Find:

- **a** the gradient of the line
- **b** the equation of the line

4 Find the equation of the line which passes through the points (1, 3) and (2, 6).

5 Rearrange these equations of lines into gradient/intercept form:

- **a** $2x + y - 5 = 0$
- **b** $3x + 2y - 6 = 0$

6 Rearrange these equations of lines into general form:

- **a** $y = 3x - 1$
- **b** $y = -\frac{2}{3}x + 4$

7 Find the x- and y-intercepts for each equation:

- **a** $y = 2x + 8$
- **b** $x + 3y - 6 = 0$

8 The gradients of two lines are m_1 and m_2 respectively. Complete:

- **a** If $m_1 = m_2$ the two lines are ____________________ .
- **b** If $m_1m_2 = -1$ the two lines are ______________________ .

9 A line l has gradient $-\frac{1}{2}$. What is the gradient of any line:

- **a** parallel to l?
- **b** perpendicular to l?

PAGES 131–132

Answers **1 a** $y = 4x - 2$ **b** $y = \frac{1}{2}x + 5$ **c** $y = -2x$ **2 a** $y = 2x - 5$ **b** $y = -3x - 1$ **c** $y = \frac{2}{3}x - 3$ **3 a** –1 **b** $y = -x + 5$ **4** $y = 3x$ **5 a** $y = -2x + 5$ **b** $y = -\frac{3}{2}x + 3$ **6 a** $3x - y - 1 = 0$ **b** $2x + 3y - 12 = 0$ **7 a** –4, 8 **b** 6, 2 **8 a** parallel **b** perpendicular **9 a** $-\frac{1}{2}$ **b** 2

THE EQUATION OF A STRAIGHT LINE

Linear and Non-linear Relationships

25 MINUTES

INTERMEDIATE TEST

Part A Multiple Choice

1 Any line perpendicular to the line $y = 2x - 5$ has gradient:

A 2 **B** –2 **C** $\frac{1}{2}$ **D** $-\frac{1}{2}$ (1 mark)

2 The x-intercept of the line $y = -3x + 6$ is:

A 6 **B** –3 **C** 2 **D** 0 (1 mark)

3 The line which is parallel to the line $y = 2 - x$ and which passes through the origin has equation:

A $y = 2x$ **B** $y = -x$ **C** $x = 2$ **D** $x + y = 2$ (1 mark)

Part B Written Response

4 **a** Rearrange the equation $4x + 2y - 1 = 0$ into gradient/intercept form. (1 mark)

b Write down the gradient and y-intercept of the line $4x + 2y - 1 = 0$. *Hint 1* (2 marks)

c Graph the line $4x + 2y - 1 = 0$ on a number plane. *Hint 2* (2 marks)

5 The points $P(3, 7)$, $Q(-5, -5)$ and $R(1, -9)$ are the vertices of a triangle. Show that $\angle PQR = 90°$. *Hint 3* (4 marks)

6 **a** Find the y-intercept of the line $3x + 4y - 12 = 0$. (1 mark)

b Find the x-intercept of the line $3x + 4y - 12 = 0$. (1 mark)

c Graph the line $3x + 4y - 12 = 0$. *Hint 4* (2 marks)

7 The diagram shows the vertices of a triangle.

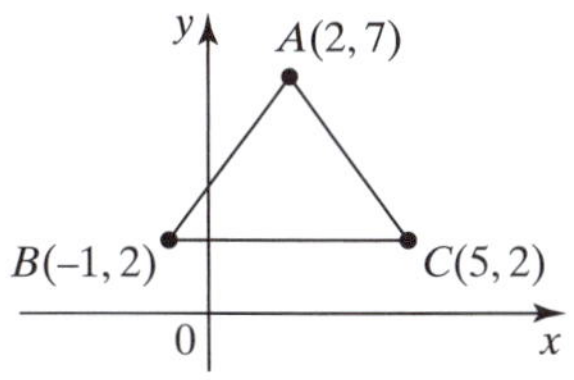

a What is the equation of the line joining B to C? *Hint 5* (1 mark)

b What is the equation of the line through A, perpendicular to BC? *Hint 6* (1 mark)

c What is the equation of the line through B perpendicular to AC? (2 marks)

d Where do the two lines described in parts **b** and **c** meet? *Hint 7* (2 marks)

e What is the equation of the line through C perpendicular to AB? (2 marks)

f Show that the line in part **e** passes through the point from part **d**. (1 mark)

Hint 1: Use the answer from part **a**.
Hint 2: Where does the line cut the y-axis? Use the gradient to find another point on the line.
Hint 3: Show that PQ is perpendicular to QR.
Hint 4: Use the intercepts found in parts **a** and **b**.
Hint 5: What do we know about the y coordinate of any point on the line.
Hint 6: What form is the equation of any vertical line?
Hint 7: Solve simultaneously.

Your Feedback

$\frac{\square}{25} \times 100\% = \square\%$

PAGES 132–133

PAGE 182

THE EQUATION OF A STRAIGHT LINE

Linear and Non-linear Relationships

40 MINUTES

ADVANCED TEST

1 P, Q and R are the points (1, 3), (4, –3) and (–2, –1) respectively. S is a point on PQ such that SR is perpendicular to PQ.

- **a** Find the coordinates of S. (4 marks)
- **b** Show that the line through P, perpendicular to QR, intersects SR at the origin. (2 marks)
- **c** Find the equation of the line that passes through both Q and O. (1 mark)
- **d** Show that line QO is perpendicular to PR. (1 mark)

2 For what values of k will the lines $2x - ky - 9 = 0$ and $x + 3y + 1 = 0$ be:

- **a** parallel? (2 marks)
- **b** perpendicular? (2 marks)

3
- **a** Find the equation of the straight line that is perpendicular to the line $2x + y - 6 = 0$ and intersects it on the x-axis. (3 marks)
- **b** Find the area of the triangle formed by these two lines and the y-axis. (2 marks)

4 Suppose a triangle ABC has sides whose equations are:

AB: $4x + 3y - 25 = 0$
AC: $3x - 4y = 0$
BC: $2x - 11y + 25 = 0$

- **a** Show that the triangle is right-angled at A. (2 marks)
- **b** Find the coordinates of B. (2 marks)
- **c** Find the coordinates of C. (2 marks)
- **d** What conclusions can be made about the triangle? Justify your answer. (2 marks)

5 The points P, Q and R have coordinates (8, 7), (–2, –1) and (–6, 5) respectively.

- **a** Find the coordinates of M, the midpoint of P and R. (1 mark)
- **b** Find, in general form, the equation of the line k that passes through P and is parallel to QR. (2 marks)
- **c** Find, in general form, the equation of the line l that passes through Q and M. (2 marks)
- **d** Find the coordinates of S, the point of intersection of k and l. (2 marks)
- **e** Show that SR is parallel to PQ. (2 marks)

6 A is the point (–3, 1) and B is the point (5, –1).

- **a** Find the coordinates of the midpoint of AB. (1 mark)
- **b** Find the gradient of the line joining A to B. (1 mark)
- **c** Find the equation of the perpendicular bisector of AB. (1 mark)
- **d** C is the point (–5, –24). Show that C lies on the perpendicular bisector of AB. (1 mark)
- **e** The point D lies on the line $2x + y - 14 = 0$ and is equidistant from A and B. Find the coordinates of D. (2 marks)

Your Feedback

$\frac{\square}{40} \times 100\% = \square\%$

PAGES 133–135

PAGE 182

EQUATIONS

Linear and Non-linear Relationships

STUDY NOTES

1 To determine whether a value of a variable is a **solution** to an equation or inequality, **substitute** the value into the equation or inequality to see whether it **satisfies** it.

2 Equations involving **fractions** can be simplified by **multiplying by the common denominator** to remove all fractions. Take care to multiply every term.

3 **Problems** can often be solved by forming an equation and solving. Let x (or some other variable) represent the unknown quantity and express the problem algebraically. Make sure that you answer the question that has been asked.

4 The **subject of a formula** is the variable before the equals sign. For example, y is the subject of the formula $y = mx + b$. If asked to **change the subject** of a formula you are required to rearrange the formula so that a different variable is before the equals sign. For example, to make x the subject of the formula $y = mx + b$, first subtract b from both sides of the equation and then divide both sides by m, giving $x = \frac{y-b}{m}$.

5 When **rearranging an equation**, you might:
- add the same term to, or subtract the same term from, both sides of the equation
- multiply, or divide, both sides of the equation by the same term
- take the square root of both sides of the equation, taking care to consider both the positive and negative square roots (if $x^2 = a$, $x = \pm\sqrt{a}$)
- take reciprocals of both sides (if $\frac{1}{x} = \frac{1}{a}$, then $x = a$).

6 In many formulae there are certain **restrictions on variables** which may not be stated but which are **implied** in the formula. For example, if $s = \sqrt{A}$, we assume that $A \geq 0$, because the formula does not make sense if A is negative. If $m = \frac{y_2 - y_1}{x_2 - x_1}$, then x_1 and x_2 cannot be equal because if they were, the denominator would be zero and we cannot divide by zero. Sometimes restrictions apply only after the subject has been changed. For example, in the formula $y - y_1 = m(x - x_1)$ if $x = x_1$ then the right hand side is zero and hence $y = y_1$. If the equation is rearranged to make m the subject then $m = \frac{y - y_1}{x - x_1}$ and $x \neq x_1$.

7 There may also be **restrictions** because of the **physical values** the variables represent. The formula $A = \pi r^2$ gives the area of a circle of radius r units. If we rearrange this formula so that the radius is the subject, then we do not want negative values of r.

8 **Exponential equations** might be able to be solved by inspection. For example, if $10^x = 1000$ then $x = 3$ because $10^3 = 1000$. If exponential equations cannot be **solved by inspection** they might be able to be **solved by equating powers**. For example, if $2^x = \frac{1}{\sqrt{2}}$ then $2^x = 2^{-\frac{1}{2}}$ so, equating powers, $x = -\frac{1}{2}$.
Exponential equations can also be solved by taking logarithms of both sides of the equation. For example, if $7^x = 5$ then $\log_{10} 7^x = \log_{10} 5$ and, using the log laws, $x \log_{10} 7 = \log_{10} 5$. So $x = \log_{10} 5 \div \log_{10} 7$ and this can be done on a calculator ($x \approx 0.827$).

Checklist

Can you:

1 *Solve harder equations including those involving fractions?* ☐
2 *Rearrange equations and change the subject of a formula?* ☐
3 *Understand the need for restrictions on variables in some equations and determine those restrictions?* ☐
4 *Solve exponential equations?* ☐

EQUATIONS

Linear and Non-linear Relationships

SKILLS CHECK

1 Solve:

a $x + 38 = 102$ **b** $x - 23 = -5$ **c** $13x = 533$

d $\dfrac{x}{18} = 21$ **e** $2x + 17 = 56$ **f** $\dfrac{x}{5} - 19 = 81$

2 Rearrange each equation so that x is the subject:

a $x + a = b$ **b** $x - y = z$ **c** $px = q$

d $\dfrac{x}{n} = k$ **e** $rx + s = t$ **f** $\dfrac{x}{m} - u = h$

3 Expand and simplify:

a $6m - (2m + 3)$ **b** $3(x + 2) + 2(x - 7)$

c $5(2a - 1) - 4(3a + 2)$ **d** $6(3p - q) - 3(5p - 2q)$

4 $x = -1$ satisfies the equation $5(x + 1) - 2(3 - 2x) = 10$. True or false?

5 If y must be a real number, what values of x cannot be used in the expression:

a $y = \sqrt{x}$? **b** $y = \dfrac{1}{x - 2}$?

6 Replace x with $2p$ in each expression:

a $3x + 2$ **b** $x^2 - x$ **c** $px - x + 2p$

7 Simplify:

a $\dfrac{x}{5} + \dfrac{3x}{5}$ **b** $\dfrac{a}{2} - \dfrac{a}{3}$ **c** $\dfrac{x + 2}{3} + \dfrac{x + 1}{4}$ **d** $\dfrac{2n - 3}{2} - \dfrac{n - 2}{3}$

8 Write as powers of 2:

a 64 **b** $\dfrac{1}{8}$ **c** $\dfrac{1}{\sqrt{2}}$ **d** $4\sqrt{2}$

9 Solve:

a $\dfrac{2x - 1}{3} = 7$ **b** $4(3x - 2) = 9 - (x + 4)$ **c** $\dfrac{x + 1}{5} = \dfrac{x - 1}{4}$

10 Evaluate:

a $27^{\frac{1}{3}}$ **b** $25^{-\frac{1}{2}}$ **c** $\left(\dfrac{1}{2}\right)^{-4}$ **d** $\log_4\left(\dfrac{1}{8}\right)$

PAGES 135–136

Answers **1 a** $x = 64$ **b** $x = 18$ **c** $x = 41$ **d** $x = 378$ **e** $x = 19.5$ **f** $x = 500$ **2 a** $x = b - a$ **b** $x = y + z$ **c** $x = \dfrac{q}{p}$ **d** $x = kn$ **e** $x = \dfrac{t - s}{r}$ **f** $x = hm + mu$ **3 a** $4m - 3$ **b** $5x - 8$ **c** $-2a - 13$ **d** $3p$ **4** false **5 a** $x < 0$ **b** $x = 2$ **6 a** $6p + 2$ **b** $4p^2 - 2p$ **c** $2p^2$ **7 a** $\dfrac{4x}{5}$ **b** $\dfrac{a}{6}$ **c** $\dfrac{7x + 11}{12}$ **d** $\dfrac{4n - 5}{6}$ **8 a** 2^6 **b** 2^{-3} **c** $2^{-0.5}$ **d** $2^{2.5}$ **9 a** $x = 11$ **b** $x = 1$ **c** $x = 9$ **10 a** 3 **b** $\dfrac{1}{5}$ **c** 16 **d** -1.5

EQUATIONS

Linear and Non-linear Relationships

25 MINUTES

INTERMEDIATE TEST

Part A Multiple Choice

1 If x is replaced by $2ap$ in the equation $x^2 = 4ay$ then $y = ?$

A p **B** ap **C** ap^2 **D** a^2p^2 (1 mark)

2 If the subject of the formula $T = \sqrt{\dfrac{4g}{l}}$ is changed to g then:

A $g = \dfrac{4T^2}{l}$ **B** $g = \dfrac{4l}{T^2}$ **C** $g = \dfrac{lT^2}{4}$ **D** $g = \dfrac{4}{lT^2}$ (1 mark)

3 If $\log_8 2 = x$ then $x = ?$

A 4 **B** $\dfrac{1}{4}$ **C** 3 **D** $\dfrac{1}{3}$ (1 mark)

4 If $1.2^n = 539$, between which two values must n lie?

A 20 and 25 **B** 25 and 30 **C** 30 and 35 **D** 35 and 40 (1 mark)

Part B Written Response

5 The formula $S = V - Dn$ is used to find the value of an item that has been decreasing in value, or depreciating. S is the present value, V is the original price, D is the amount of depreciation per year and n is the number of years.

a Find S when $V = 12\,000$, $D = 500$ and $n = 7$. (1 mark)

b Change the subject of the formula to D. (1 mark)

c Find D when $S = 1175$, $V = 1650$ and $n = 5$. *Hint 1* (1 mark)

6 Solve $\dfrac{5n-1}{3} - \dfrac{2n-3}{4} = 1$. *Hint 2* (3 marks)

7 Consider the formula $d = \sqrt{b^2 - 4ac}$

a What restriction must be placed on b^2? *Hint 3* (1 mark)

b Change the subject of the formula to c. (2 marks)

c If a is a positive number but c is a negative number, what further relationship must exist between b^2 and d^2? *Hint 4* (1 mark)

8 When 15 is subtracted from a certain number and the result multiplied by 5, the answer is seven more than three times the original number. Form an equation and solve to find the original number. *Hint 5* (2 marks)

9 Solve $9^{x-1} = \dfrac{1}{27}$ *Hint 6* (3 marks)

10 Solve $\dfrac{x-2}{2} - \dfrac{2x+1}{5} = \dfrac{3}{4}$ (3 marks)

11 Find the value of x, correct to two decimal places, if $2^x = 3.4$. *Hint 7* (3 marks)

Hint 1: Use the rearranged formula from part **b**.
Hint 2: Make sure that all parts of the equation are multiplied by the common denominator. Take care with the signs.
Hint 3: Can you take the square root of a negative number?
Hint 4: If $c < 0$, but $a > 0$ what does this mean about $b^2 - d^2$?
Hint 5: Let x be the number.
Hint 6: Both 9 and 27 are powers of 3.
Hint 7: Take logs of both sides.

Your Feedback

$\dfrac{\square}{25} \times 100\% = \square\ \%$

PAGE 136

PAGE 182

EQUATIONS

Linear and Non-linear Relationships

40 MINUTES

ADVANCED TEST

1 Solve:

a $7(x-4)=8-3(x-2)$

b $5-(6-a)=3a$

c $\frac{3t}{4}-\frac{t}{9}=2$

d $\frac{5h}{2}-\frac{h-3}{4}=h+2$

e $\frac{3m}{m+2}=\frac{5}{2}$

f $\frac{1}{2}(p-5)-\frac{1}{3}(p+8)=7$

(2 marks each)

2 Solve these inequalities:

a $\frac{2x}{5}-3\leq 1$ (2 marks)

b $6(3-2a)-5(a-4)>4$ (2 marks)

c $\frac{3(n+2)}{5}-\frac{4(n-1)}{3}<4$ (2 marks)

3 Find the exact value of x if $x=\frac{-b\pm\sqrt{b^2-4ac}}{2a}$ and $a=3$, $b=-6$ and $c=1$. (1 mark)

4 Solve, giving answers correct to three decimal places:

a $2\pi r=37$ (1 mark)

b $d^2=57$ (1 mark)

c $7x^3=18$ (1 mark)

5 **a** Solve $x^2=64$. (1 mark)

b Solve $x^3=64$. (1 mark)

c Explain why the equation in part **b** has one less solution than the equation in part **a**. (1 mark)

6 Solve simultaneously:

a $5x+2y-1=0$
$4x-3y+36=0$

b $y=x^2+11x+5$
$y=3x-2$

c $y=\frac{8}{x}$
$y=x+2$

(3 marks each)

7 Change the subject of each equation to r:

a $V=\frac{4}{3}\pi r^3$

b $A=P(1+r)^n$

c $S=\frac{a}{1-r}$

(2 marks each)

Your Feedback

$\frac{\square}{40}\times 100\% = \square\%$

PAGES 136–137

PAGE 182

FURTHER FACTORISATION

Algebraic Techniques

STUDY NOTES

1 **Factorising** an algebraic expression means rewriting it as the **product of its factors**.

2 A **common factor** is a number or term that divides evenly into each term of an algebraic expression. When finding a common factor, always look for the largest common factor. For example, $2x(4xy + 2y^2)$ is a factorisation of $8x^2y + 4xy^2$, but the part in the brackets can still be factorised. When **factorised fully** $8x^2y + 4xy^2 = 4xy(2x + y)$.

3 A **trinomial** is an algebraic expression involving the sum or difference of **three terms**. The binomial product $(x + p)(x + q) = x^2 + px + qx + pq = x^2 + (p + q)x + pq$ so, as factorising is the opposite of expanding, any expression of the form $x^2 + (p + q)x + pq$ can be factorised to give $(x + p)(x + q)$. To factorise a trinomial where the coefficient (the number in front) of x^2 is 1 look for two numbers which add to the coefficient of the middle term and which multiply to the end term. For example, to factorise $x^2 + 9x + 20$ look for two numbers that add to 9 and multiply to 20. The numbers are 4 and 5 so $x^2 + 9x + 20 = (x + 4)(x + 5)$ [or $(x + 5)(x + 4)$, the factors may be written in either order.] To factorise $x^2 - 5x - 14$ look for two numbers that add to –5 and multiply to –14. The numbers are –7 and 2 so $x^2 - 5x - 14 = (x - 7)(x + 2)$. Note: some trinomials cannot be factorised.

4 If an algebraic expression consists of two terms both of which are perfect squares and if one is subtracted from the other then we have the difference of two squares. When factorising the **difference of two squares** the result is the sum by difference, i.e. $a^2 - b^2 = (a + b)(a - b)$.
For example, $16 - 9p^2 = 4^2 - (3p)^2$ so when factorised $16 - 9p^2 = (4 + 3p)(4 - 3p)$.

5 **Perfect squares** are a special type of trinomial: $a^2 + 2ab + b^2 = (a + b)^2$ $\quad a^2 - 2ab + b^2 = (a - b)^2$
For example, $x^2 + 14x + 49 = (x + 7)^2$, $9m^2 - 30mn + 25n^2 = (3m - 5n)^2$

6 Some expressions involving four terms can be factorised by grouping in pairs. Find a common factor for the first 2 terms, and a different common factor for the last 2 terms. Then you may find that you have a new common factor for these two factorised expressions.
For example,
$$\begin{aligned} x^2 + ax + bx + ab &= x(x + a) + b(x + a) \\ &= (x + a)(x + b) \end{aligned}$$

7 Trinomials of the form $ax^2 + bx + c$ where $a \neq 1$ may also be able to be factorised. This might be done by **inspection**, using the **cross method** or by a **grouping method**. Using the grouping method, first look for two terms that add to bx and multiply to acx^2. By expressing the middle term as the sum of these two terms it can be grouped.
For example,
$$\begin{aligned} &6p^2 + 19p + 10 && \text{[What 2 numbers add to 19 and multiply to 60 } (6 \times 10)\text{?]} \\ &= 6p^2 + 15p + 4p + 10 && \text{[the answer is 15 and 4, so rewrite } 19p \text{ as } 15p + 4p\text{]} \\ &= 3p(2p + 5) + 2(2p + 5) && \text{[now it can be factorised by grouping]} \\ &= (2p + 5)(3p + 2) \end{aligned}$$

8 An expression may contain a **combination** of the different types of factorisation. Take care to always factorise fully. For example, $3x^2 - 12 = 3(x^2 - 4) = 3(x + 2)(x - 2)$. Always look for a common factor first. Then if there are two terms the expression may be a difference of two squares. If there are three terms the expression may be a perfect square or a trinomial that can be factorised, if there are four terms it may be able to be factorised by grouping.

Checklist

Can you:

1 *Understand what is meant by a common factor?* ☐
2 *Recognise perfect squares and the difference of two squares?* ☐
3 *Factorise trinomials?* ☐
4 *Factorise by grouping?* ☐
5 *Fully factorise algebraic expressions?* ☐

FURTHER FACTORISATION

Algebraic Techniques

SKILLS CHECK

1 Factorise, by taking out a common factor:

a $9x - 3y$

b $x^2 - 6x$

c $10xy - 5y$

d $12x^2 - 8x$

e $x^2y + xy^2$

f $8ab^2 - 4abc$

g $15x^7 - 3x^2$

h $16p^3q^2 - 2p^4q + 4pq^2$

2 Factorise these differences of two squares:

a $x^2 - 49$

b $a^2 - 1$

c $4p^2 - 25$

d $x^2 - y^2$

e $9m^2 - 16$

f $36 - t^2$

g $64p^2 - 81q^2$

h $a^4 - 25b^2$

3 Complete:

a $x^2 + 8x + 16 = (x + \_\_\_)^2$

b $a^2 - 10a + 25 = (a - \_\_\_)^2$

c $4n^2 + 4n + 1 = (\_\_\_ + 1)^2$

d $9x^2 - 42x + 49 = (\_\_\_ - 7)^2$

4 Express as a perfect square:

a $x^2 + 18x + 81$

b $y^2 + 12y + 36$

c $b^2 - 14b + 49$

d $a^2 - 5a + 6.25$

e $100 + 220m + 121m^2$

f $16x^2 - 40xy + 25y^2$

5 Factorise these trinomials:

a $x^2 + 8x + 15$

b $a^2 - 14a + 48$

c $p^2 + 2p - 35$

d $m^2 - 3m - 28$

e $n^2 + n - 6$

f $k^2 - 5k - 6$

6 Factorise by grouping in pairs:

a $x^2 + xy + 2x + 2y$

b $ab + 3a - 5b - 15$

c $a^2 - ab - ac + bc$

d $2mn - 2m + 3n - 3$

e $xz - yz - x + y$

f $6kn - 3k - 2n + 1$

PAGES 137–138

Answers **1 a** $3(3x - y)$ **b** $x(x - 6)$ **c** $5y(2x - 1)$ **d** $4x(3x - 2)$ **e** $xy(x + y)$ **f** $4ab(2b - c)$ **g** $3x^2(5x^5 - 1)$ **h** $2pq(8p^2q - p^3 + 2q)$
2 a $(x + 7)(x - 7)$ **b** $(a + 1)(a - 1)$ **c** $(2p + 5)(2p - 5)$ **d** $(x + y)(x - y)$ **e** $(3m + 4)(3m - 4)$ **f** $(6 + t)(6 - t)$
g $(8p + 9q)(8p - 9q)$ **h** $(a^2 + 5b)(a^2 - 5b)$ **3 a** 4 **b** 5 **c** $2n$ **d** $3x$ **4 a** $(x + 9)^2$ **b** $(y + 6)^2$ **c** $(b - 7)^2$ **d** $(a - 2.5)^2$ **e** $(10 + 11m)^2$
f $(4x - 5y)^2$ **5 a** $(x + 5)(x + 3)$ **b** $(a - 8)(a - 6)$ **c** $(p + 7)(p - 5)$ **d** $(m - 7)(m + 4)$ **e** $(n + 3)(n - 2)$ **f** $(k - 6)(k + 1)$
6 a $(x + y)(x + 2)$ **b** $(b + 3)(a - 5)$ **c** $(a - b)(a - c)$ **d** $(n - 1)(2m + 3)$ **e** $(x - y)(z - 1)$ **f** $(2n - 1)(3k - 1)$

FURTHER FACTORISATION

Algebraic Techniques

25 MINUTES

INTERMEDIATE TEST

Part A Multiple Choice

1 Factorise $a^2 - 2a - 63$:

A $(a+9)(a+7)$ **B** $(a+9)(a-7)$ **C** $(a-9)(a+7)$ **D** $(a-9)(a-7)$ (1 mark)

2 Which is the correct factorisation of $x^2 - 5x + 6$?

A $(x-3)(x-2)$ **B** $(x-6)(x+1)$ **C** $(x-5)(x-1)$ **D** none of these (1 mark)

3 $x^2 - 18x + 81 = ?$

A $(x+9)^2$ **B** $(x-9)^2$ **C** $(x+9)(x-9)$ **D** none of these (1 mark)

4 $xy - 4x - 6y + 24 = ?$

A $(x-6)(y-4)$ **B** $(x-6)(y+4)$ **C** $(x+6)(y-4)$ **D** $(x+6)(y+4)$ (1 mark)

5 $m^2 - 1 = ?$

A $(m+1)^2$ **B** $(m-1)^2$ **C** $(m+1)(m-1)$ **D** $m(m-1)$ (1 mark)

6 $6k(k+2) - (k+2) = ?$

A $5k(k+2)$ **B** $(k+2)(5k-1)$ **C** $6k(k+2)$ **D** $(6k-1)(k+2)$ (1 mark)

7 $4n^2 + 4n - 3 = ?$

A $(2n-3)(2n+1)$ **B** $(2n+3)(2n-1)$ **C** $(4n-3)(n+1)$ **D** $(4n-1)(n+3)$ (1 mark)

8 When simplified $\dfrac{x^2-64}{8x-64} = ?$

A $\dfrac{x}{8}$ **B** $\dfrac{(x+1)(x-1)}{8x-1}$ **C** $\dfrac{x+8}{8}$ **D** x (1 mark)

9 In order to factorise $5m^2 + 8m - 4$ by the grouping method it should be rewritten as:

A $5m^2 + 2m + 6m - 4$ **B** $5m^2 + 9m - m - 4$

C $5m^2 + 10m - 2m - 4$ **D** $5m^2 + 4m + 4m - 4$ (1 mark)

10 $\dfrac{g^2+7g+6}{g+6} = ?$

A $7g$ **B** $8g$ **C** $g+7$ **D** $g+1$ (1 mark)

Part B Written Response

11 Factorise $2x^2 - 4x - 30$. *Hint 1* (2 marks)

12 Factorise $6m^2 + 17m + 12$. (2 marks)

13 Factorise fully $3x^4 - 3$. *Hint 2* (2 marks)

14 Factorise $a^2 + ab - a - b$. *Hint 3* (2 marks)

15 Factorise $9p^2 - 30pq + 25q^2$. (2 marks)

16 Simplify $\dfrac{2n+6}{6} \times \dfrac{3n}{n^2+3n}$ *Hint 4* (2 marks)

17 Express $\dfrac{5}{x^2-4} + \dfrac{3}{x^2+2x}$ as a single fraction, leaving the denominator in factorised form. *Hint 5* (3 marks)

Hint 1: First take out a common factor.

Hint 2: There is a common factor. Then you will have the difference of two squares, twice.

Hint 3: Be careful with the negative values. Remember that *a* divides into *a* once.

Hint 4: Factorise, then cancel.

Hint 5: Factorise both denominators first in order to find the lowest common denominator

Your Feedback

$\dfrac{\square}{25} \times 100\% = \square\%$

PAGES 138–139

PAGE 182

FURTHER FACTORISATION

Algebraic Techniques

40 MINUTES

ADVANCED TEST

1 Fully factorise $6a^2b^3c^5 - 4ab^5c^3 + 12a^4b^2c^6$. (1 mark)

2 Factorise:

a $ab + ac - 9b - 9c$

b $1 - 16x^2$

c $x^2 - 20x + 100$

d $x^2 + 11xy - 60y^2$

e $5a^2 - 13a + 6$

f $4m^2 - 12m - 27$ (1 mark each)

3 Factorise fully:

a $x^3 - 9x$

b $2x^2 + 22x + 48$

c $m^4 - 1$

d $\frac{4}{3}\pi r^3 + \frac{1}{3}\pi r^2 h$

e $u^4 - 13u^2 + 36$

f $12x^2 - 60x + 75$ (2 marks each)

4 Simplify:

a $\dfrac{x + 7}{x^2 + 8x + 7}$

b $\dfrac{x^2 - 81}{x^2 - x - 72}$

c $\dfrac{x^2 + 2x - 15}{x^2 + 8x + 15}$

d $\dfrac{2a^2 + 3a - 2}{2a^2 + 5a - 3}$

e $\dfrac{8n^3 - 24n^2 - 144n}{2n^2 + 6n}$

f $\dfrac{8p^2 - 6pq - 28p + 21q}{16p^2 - 24pq + 9q^2}$ (2 marks each)

5 Simplify:

a $\dfrac{4x^2 - 9}{2x^2 - 5x - 12} \times \dfrac{3x^2 - 13x + 4}{6x^2 - 11x + 3}$ (3 marks)

b $\dfrac{x^2 + xy + xz}{xy - xz - yz + z^2} \div \dfrac{x^3 + x^2y + x^2z + xyz}{xy + y^2 - xz - yz}$ (3 marks)

6 Show that $\dfrac{x + 8}{x^2 + x - 6} - \dfrac{x + 4}{x^2 + 5x + 6} = \dfrac{8}{x^2 - 4}$ (3 marks)

Your Feedback

$\dfrac{\square}{40} \times 100\% = \square\%$

PAGES 139–140

PAGE 182

QUADRATIC EQUATIONS

Algebraic Techniques

STUDY NOTES

1 A **quadratic equation** is an equation involving the **square** of the unknown quantity. (The square being the highest power of the unknown.) Some examples of quadratic equations are $x^2 = 25$, $a^2 + 8a = 20$, $4p^2 - 7p + 13 = 0$ and $(3m - 2)^2 = 7m - 1$.

2 A quadratic equation may have **2, 1 or no solutions**. It will have no solutions if a perfect square is equal to a negative number, ($x^2 = -4$ for example). It will have one solution if a perfect square is equal to zero. For example, if $(x - 3)^2 = 0$ then $x = 3$ is the only solution. A quadratic equation will have two solutions if a perfect square is equal to a positive number. For example if $n^2 = 9$ then $n = 3$ and $n = -3$ are both solutions.

3 **Simple quadratic equations** are solved by taking the **square root** of both sides of the equation, remembering to consider both the **positive and negative** square roots. If $2x^2 = 8$ then $x^2 = 4$ and $x = \pm 2$. (The symbol $\pm$ means 'plus or minus'.)

4 Quadratic equations may be solved by **factorisation**. This method relies on the fact that if two numbers multiply to zero, one (or both) of the numbers must be zero. Using this method put the expression equal to zero, factorise (if possible), let each of the factorised parts equal zero and solve these simple equations. **Not every quadratic expression can be factorised**. This doesn't necessarily mean that the equation has no solution.

5 The factorisation method only works if the equation is **equal to zero**. If $(x - 2)(x - 4) = 8$ say, it doesn't mean that one part must be 8. You must first expand, put the expression equal to zero and then factorise this new expression.

$$\begin{aligned}(x-2)(x-4) &= 8\\ x^2 - 6x + 8 &= 8\\ x^2 - 6x &= 0\\ x(x-6) &= 0\\ x = 0 \text{ or } x &= 6\end{aligned}$$

6 Quadratic equations may be solved by a method of **completing the square**. Using this method an amount is added to both sides of the equation, so that the side involving the unknown becomes a perfect square. The solution is then found by taking the square root of both sides, remembering to consider both the positive and negative square roots.

$$\begin{aligned}x^2 - 8x + 1 &= 0\\ x^2 - 8x + 16 &= 15\\ (x-4)^2 &= 15\\ x - 4 &= \pm\sqrt{15}\\ x &= 4 \pm \sqrt{15}\end{aligned}$$

So there are two solutions, $x = 4 + \sqrt{15}$ (≈ 7.87) and $x = 4 - \sqrt{15}$ (≈ 0.13)

7 Using the method of completing the squares with the quadratic equation $ax^2 + bx + c = 0$ we get the **quadratic formula**: $x = \dfrac{-b \pm \sqrt{b^2 - 4ac}}{2a}$.
This formula can be used to solve any quadratic equation for which a solution exists.

8 The **solutions** to many quadratic equations are **not rational**. When the solutions are not rational they should be left in surd form if asked for **exact** answers. Such surds might be able to be simplified (taking care to only cancel a common factor). A calculator may be used to get a **decimal approximation** to each solution.

9 If you are asked to solve a quadratic equation using a **particular method**, then you must use that method. If no method is indicated, first check to see if the expression can easily be factorised because factorisation is usually the quickest and easiest. If the expression cannot be factorised you will need to complete the square or use the quadratic formula.

Checklist

Can you:

1 *Understand what is meant by a quadratic equation?* ☐
2 *Understand that a quadratic equation may have 0, 1 or 2 solutions?* ☐
3 *Solve quadratic equations by factorising, completing the square, or using the quadratic formula?* ☐

QUADRATIC EQUATIONS

Algebraic Techniques

SKILLS CHECK

1 Solve:

a $x^2 = 16$ **b** $e^2 = 9$ **c** $4k^2 = 1$

2 Solve:

a $x(x + 4) = 0$ **b** $(x - 7)(x + 1) = 0$ **c** $(2a + 5)(3a - 2) = 0$

3 Factorise:

a $x^2 + 8x + 15$ **b** $t^2 + 3t - 10$ **c** $2m^2 - 7m + 6$

4 Given the formula $x = \dfrac{-b \pm \sqrt{b^2 - 4ac}}{2a}$, find all possible values of x if $a = 8$, $b = 10$ and $c = -3$.

5 Expand:

a $(x + 5)^2$ **b** $(x - 3)^2$ **c** $(a + b)^2$

6 Write each as a square:

a $x^2 + 14x + 49$ **b** $n^2 - 12n + 36$

7 What number must be added to each expression to make a perfect square?

a $x^2 + 8x$ **b** $a^2 - 6a$ **c** $4p^2 + 20p$

8 The given value is a solution of the quadratic equation. True or false?

a $x^2 + 7x + 12 = 0$ $(x = 3)$

b $y^2 - 4y - 21 = 0$ $(y = -3)$

9 Determine whether the given equation has no solution, one solution or two solutions:

a $(x + 2)^2 = 9$ **b** $(x + 2)^2 = -4$ **c** $(x + 2)^2 = 0$

10 Solve:

a $n^2 - 3n = 0$ **b** $x^2 + 6x + 8 = 0$ **c** $a^2 - 4a - 45 = 0$

PAGE 140

Answers **1 a** $x = \pm 4$ **b** $e = \pm 3$ **c** $k = \pm \frac{1}{2}$ **2 a** $x = 0$ or $x = -4$ **b** $x = 7$ or $x = -1$ **c** $a = -2\frac{1}{2}$ or $a = \frac{2}{3}$ **3 a** $(x + 5)(x + 3)$ **b** $(t + 5)(t - 2)$ **c** $(2m - 3)(m - 2)$ **4** $x = \frac{1}{4}$ or $x = -1\frac{1}{2}$ **5 a** $x^2 + 10x + 25$ **b** $x^2 - 6x + 9$ **c** $a^2 + 2ab + b^2$ **6 a** $(x + 7)^2$ **b** $(n - 6)^2$ **7 a** 16 **b** 9 **c** 25 **8 a** false **b** true **9 a** 2 **b** 0 **c** 1 **10 a** $n = 0$ or $n = 3$ **b** $x = -2$ or $x = -4$ **c** $a = 9$ or $a = -5$

QUADRATIC EQUATIONS

Algebraic Techniques

40 MINUTES

INTERMEDIATE TEST

Part A Multiple Choice

1 The quadratic equation $4x^2 - 4x + 1 = 0$ has how many solutions?

A 0

B 1

C 2

D There is not enough information to tell how many solutions the equation has. (1 mark)

2 Which is a solution to the equation $p^2 + 8p - 9 = 0$?

A $p = -1$ B $p = 8$ C $p = -8$ D $p = -9$ (1 mark)

3 A quadratic equation has solutions $x = 3$ and $x = -2$. The equation is:

A $x^2 + 5x - 6 = 0$ B $x^2 - 5x - 6 = 0$ C $x^2 + x - 6 = 0$ D $x^2 - x - 6 = 0$ (1 mark)

4 The solution to the equation $(x - 5)(x - 2) = 28$ is:

A $x = 2$ or $x = 5$ B $x = 7$ or $x = 4$ C $x = -2$ or $x = 9$ D $x = -5$ or $x = 3$ (1 mark)

Part B Written Response

5 Solve $4x^2 = 1$. *Hint 1* (2 marks)

6 Solve by factorising:

a $x^2 - 18x + 81 = 0$ (2 marks)

b $x(x + 5) = 14$ *Hint 2* (3 marks)

c $2m^2 + 9m + 4 = 0$ (3 marks)

7 Solve by completing the square:

a $x^2 + 6x = 16$ (3 marks)

b $a^2 + 12a = 7$ *Hint 3* (3 marks)

c $9n^2 - 6n - 5 = 0$ (3 marks)

8 Use the quadratic formula to solve the quadratic equation:

a $2x^2 - 6x + 3 = 0$ giving the answer is simplest surd form *Hint 4* (4 marks)

b $3x^2 + 7x - 1 = 0$ giving each answer correct to two decimal places (4 marks)

9 Solve $(x + 3)(x + 5) = 35$. *Hint 5* (3 marks)

10 A rectangle is 12 metres longer than it is wide. The area of the rectangle is 45 m^2.

a Let the width of the rectangle be x m. Explain why the width of the rectangle can be found by solving the quadratic equation $x^2 + 12x - 45 = 0$. (3 marks)

b Find the dimensions of the rectangle. *Hint 6* (3 marks)

Hint 1: There are two answers.
Hint 2: Put the expression equal to zero.
Hint 3: The answers will involve surds.
Hint 4: Take extra care when simplifying the answer.
Hint 5: Expand first.
Hint 6: Make sure your answers are physically possible.

Your Feedback

$\frac{\square}{40} \times 100\% = \square\%$

PAGES 140–141

PAGE 182

QUADRATIC EQUATIONS

Algebraic Techniques

40 MINUTES

ADVANCED TEST

1 Solve by factorising:

a $3x^2 + 17x + 10 = 0$ (2 marks)

b $5a^2 + 50a = 120$ (2 marks)

2 Solve by completing the square:

a $x^2 + 8x = 20$ (2 marks)

b $9x^2 = 12x + 1$ (2 marks)

3 Use the quadratic formula to solve, giving answers in simplest surd form:

a $x^2 + 6x + 7 = 0$ (3 marks)

b $4x^2 - 12x + 3 = 0$ (3 marks)

4 Solve, giving each answer correct to three decimal places:

a $3x^2 + 5x - 7 = 0$ (3 marks)

b $x^2 + x = 1$ (3 marks)

5 Solve:

a $(x - 4)(x - 10) = -8$ (2 marks)

b $2^{2x} - 12(2^x) + 32 = 0$ (3 marks)

c $(n^2 + 5n)^2 - 8(n^2 + 5n) - 84 = 0$ (3 marks)

6 **a** Solve simultaneously $x^2 - 4x + y^2 + 6y = 12$ and $y = x - 4$. (3 marks)

b Miranda tried to solve the equations $x^2 - 4x + y^2 + 6y = 12$ and $y = 8 - x$ simultaneously but was unable to find a solution. Use a graph to show why there is no simultaneous solution. (3 marks)

7 **a** By completing the square on the equation $ax^2 + bx + c = 0$, derive the quadratic formula. (3 marks)

b If $b^2 - 4ac < 0$ what does this mean about the solutions to $ax^2 + bx + c = 0$? (1 mark)

c If $b^2 - 4ac = 0$ what does this mean about the solutions to $ax^2 + bx + c = 0$? (1 mark)

d If $b^2 - 4ac$ is a perfect square what does this mean about the solutions to $ax^2 + bx + c = 0$? (1 mark)

Your Feedback

$\frac{\square}{40} \times 100\% = \square\%$

PAGES 141–142

PAGE 182

VOLUME OF PYRAMIDS, CONES AND SPHERES

Surface Area and Volume

STUDY NOTES

1 **Volume is the measurement** of the amount of space inside an object.

2 **Volume is measured** in cubic units (m^3, cm^3, mm^3, km^3 and so on).

3 A **prism** is a three-dimensional figure with constant cross-section. (If you cut a prism parallel to the base, the slice will be exactly the same as the base.) The type of prism is determined by the base. A **rectangular prism** has a rectangular base (or cross-section), a **triangular prism** has a triangular base and so on.

4 The **volume of any prism** can be found by multiplying the area of the base (cross-section) by the perpendicular height i.e. $V = Ah$

5 The **height** of a prism is the dimension at right-angles to the cross-section. It is not necessarily the distance above the ground or the vertical measurement on a diagram.

6 The volume of a **cylinder** is given by $V = \pi r^2 h$. The base of a cylinder is a circle so the area of the base is πr^2. To find the volume multiply the area of the base by the height of the cylinder.

7 A **pyramid** is a solid all of whose faces, apart from the base, are triangular and meet at a point. The type of pyramid is determined by the base. If the base is a rectangle, we say the pyramid is a **rectangular pyramid**. If the base is a square it is a **square pyramid** and so on.

8 The **volume of a pyramid** is one third of the volume of the prism of the same height with the same base. $V = \frac{1}{3}Ah$. h must be the perpendicular height, at right angles to the base. It might be necessary to use **Pythagoras' theorem** to determine the perpendicular height.

9 The volume of a **cone** is one third of the volume of a cylinder with the same height and radius. $V = \frac{1}{3}Ah$ or $V = \frac{1}{3}\pi r^2 h$

10 The volume of a **sphere** is given by $V = \frac{4}{3}\pi r^3$, where r is the radius.

11 The volume of **composite solids** can be found by adding the volumes of each comprising solid.

12 There is a direct relationship between volume and **capacity**. A space with volume 1 m^3 will hold 1000 L. 1 cm^3 holds 1 mL.

13 If **two similar figures have matching intervals in the ratio 1:*k***, then the matching volumes are in the ratio $1:k^3$. For example, the two rectangular prisms are similar.
The matching intervals are in the ratio 1:2
(2:4 = 1:2, 5:10 = 1:2, 3:6 = 1:2).
The matching volumes are in the ratio $1:2^3 = 1:8$
($V_1 = 2 \times 5 \times 3 = 30$, $V_2 = 4 \times 10 \times 6 = 240$,
ratio of volumes = 30:240 = 1:8).

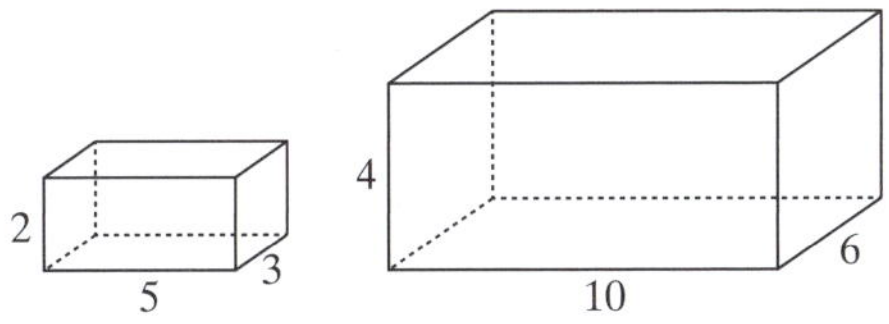

Checklist

Can you:

1 *Understand what is meant by volume?* ☐
2 *Know and understand the units associated with volume?* ☐
3 *Recognise and name prisms and pyramids?* ☐
4 *Know which dimension is the height?* ☐
5 *Find the volume of a prism, pyramid, cylinder, cone or sphere?* ☐
6 *Find volumes of composite solids?* ☐
7 *Find the capacity of a solid?* ☐

VOLUME OF PYRAMIDS, CONES AND SPHERES

Surface Area and Volume

SKILLS CHECK

1 Find the volume of each prism, given the area A of the shaded face and the perpendicular height h:

a

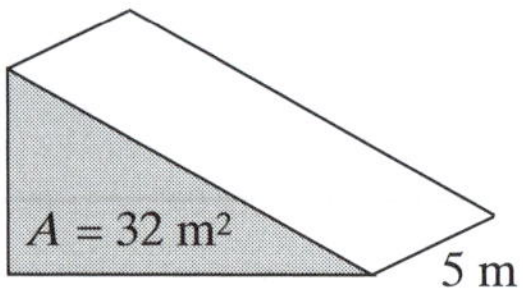

b

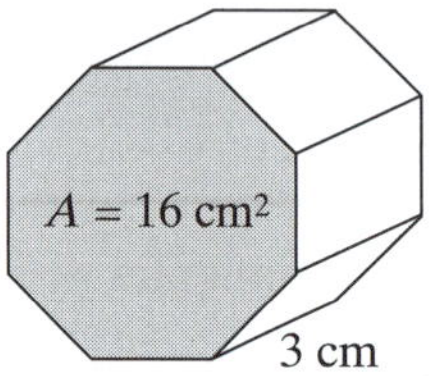

c

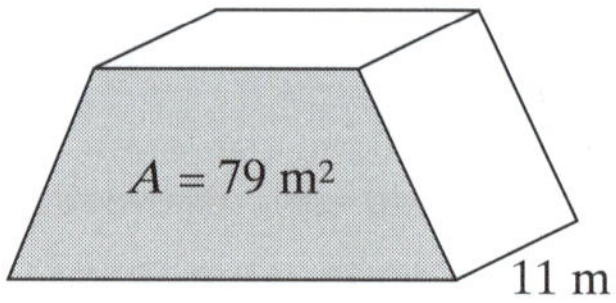

2 Find the volume of each prism:

a

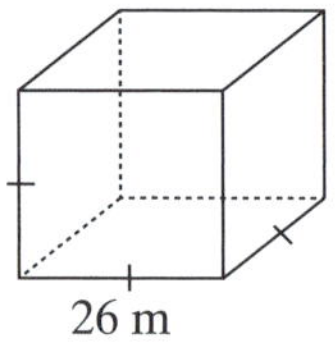

b

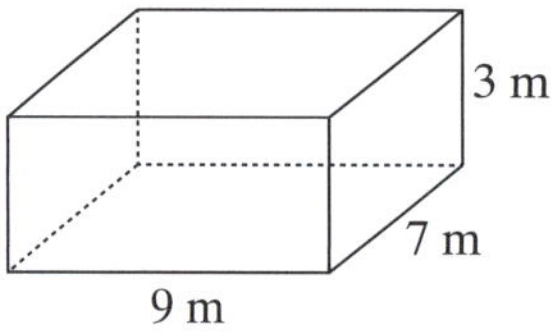

c

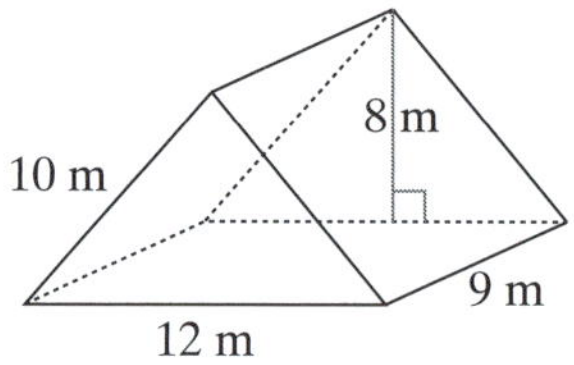

3 Find x:

a

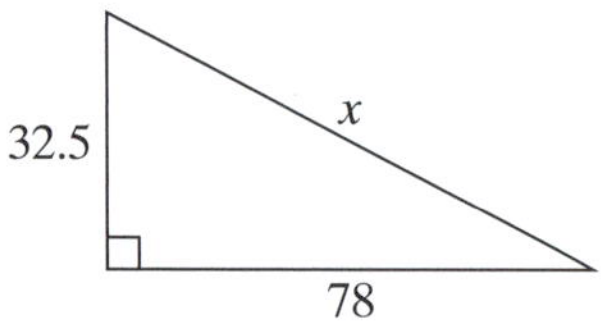

b

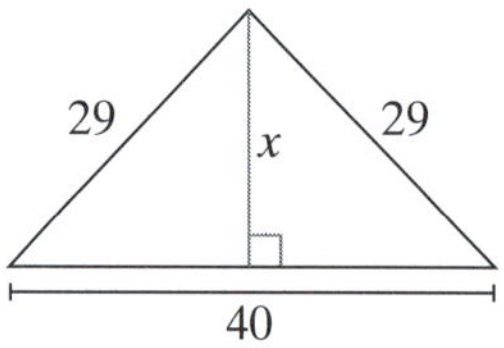

4 Find the volume of each pyramid given the area of the base and the perpendicular height:

a rectangular pyramid with base area 18 cm^2 and perpendicular height 10 cm

b triangular pyramid with base area 6 m^2 and perpendicular height 2 m

c square pyramid with base area 78 mm^2 and perpendicular height 39 mm

5 Find the volume of a:

a cylinder with radius 9 m and height 2 m

b cone with radius 5 m and height 6 m

6 Find the volume of each cone, giving the answer correct to one decimal place:

a

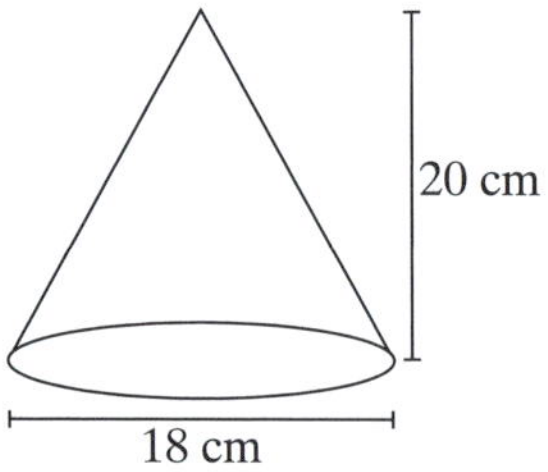

b

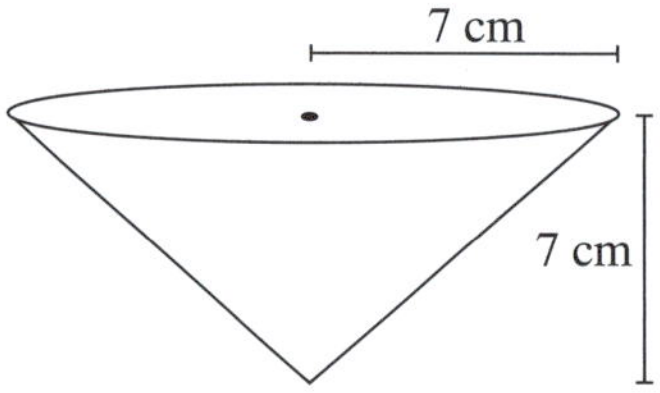

7 Find the volume of a sphere of:

a radius 9 m **b** diameter 35 cm

8 How many litres will fill a space of 8 m^3?

PAGES 142–143

Answers **1 a** 160 m^3 **b** 48 cm^3 **c** 869 m^3 **2 a** 17 576 m^3 **b** 189 m^3 **c** 432 m^3 **3 a** 84.5 **b** 21 **4 a** 60 cm^3 **b** 4 m^3 **c** 1014 mm^3
5 a 508.9 m^3 (1 d.p.) **b** 157.1 m^3 (1 d.p.) **6 a** 1696.5 cm^2 **b** 359.2 cm^3 **7 a** 3053.6 m^3 (1 d.p.) **b** 22 449.3 cm^3 (1 d.p.) **8** 8000 L

VOLUME OF PYRAMIDS, CONES AND SPHERES

Surface Area and Volume

25 MINUTES

INTERMEDIATE TEST

Part A Multiple Choice

1 If A is the area of the shaded face and h the dimension shown, in which diagram is the volume found by multiplying A by h?

A
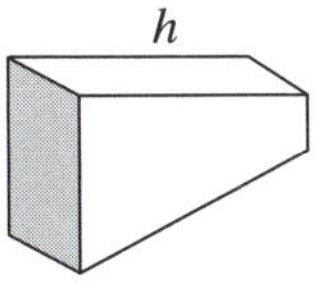

B
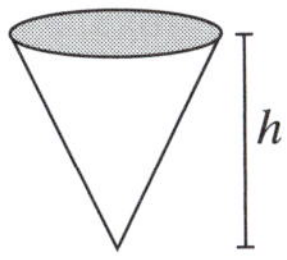

C
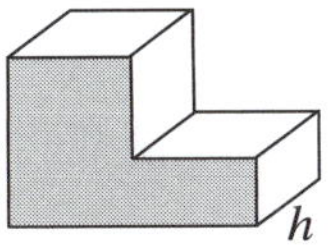

D
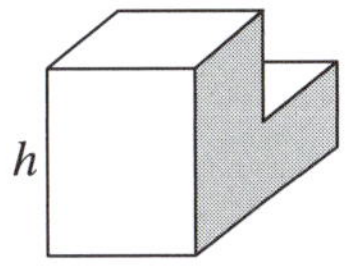

(1 mark)

2 The bases of a cone and a cylinder have the same diameter. The height of the cylinder is three times the height of the cone. The cylinder's volume is how many times that of the cone?

A 1 **B** 3 **C** 9 **D** 27 (1 mark)

3 The volume of a sphere of radius 2.7 m is closest to:

A 20 m^3 **B** 30 m^3 **C** 80 m^3 **D** 90 m^3 (1 mark)

Part B Written Response

4 **a** Find the perpendicular height of the shaded face. *Hint 1* (2 marks)

b Find the volume of the prism. (2 marks)

c What would be the volume of a pyramid that has the shaded face as its base and which has the same perpendicular height as the prism? (1 mark)

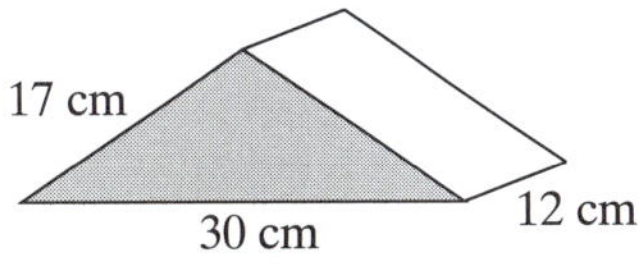

5 The volume of a hexagonal pyramid is 90 cm^3. If the height of the pyramid is 15 cm, find the area of the base. *Hint 2* (2 marks)

6 The diagram shows a right rectangular pyramid. $BC = 1.6$ m. $DC = 4$ m. The height AG of the triangle ABC is 2.5 m.

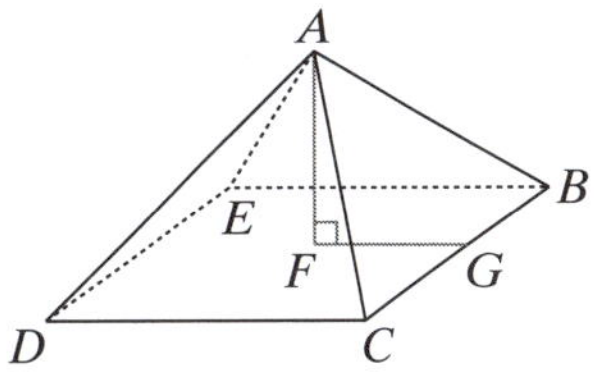

a Find the perpendicular height, AF, of the pyramid. *Hint 3* (1 mark)

b Find the volume of the pyramid. (1 mark)

7 A hemispherical bowl has radius 5 cm. Find its capacity to the nearest millilitre. (3 marks)

8 A grain storage bin is in the shape of a rectangular prism on top of a pyramid. If the total height of the bin is 2.6 m find its volume. *Hint 4* (4 marks)

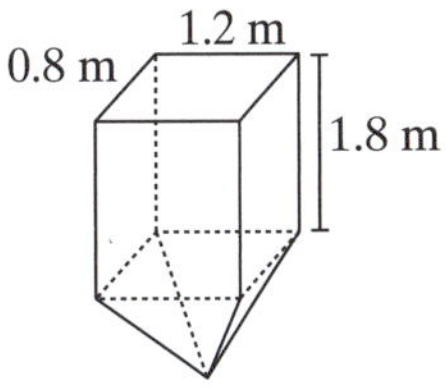

9 The volume of a sphere is 523.6 cm^3. What is the radius of the sphere? (2 marks)

10 Two similar pyramids have volumes in the ratio 27:64.

a What is the ratio of their heights? *Hint 5* (1 mark)

b What is the ratio of the areas of their bases? (1 mark)

c If the volume of the larger pyramid is 144 cm^3 and its height is 9 cm, find the area of the base of the smaller pyramid. *Hint 6* (2 marks)

Hint 1: Use Pythagoras' theorem. The bottom measurement of the right-angled triangle will not be 30 cm.
Hint 2: It doesn't matter what type of pyramid it is.
Hint 3: *FG* must be half *DC*.
Hint 4: Volume of rectangular prism plus volume of rectangular pyramid.
Hint 5: It doesn't matter that the figures are pyramids.
Hint 6: First find the area of the base of the larger pyramid. Then use the ratio from part **b**.

Your Feedback

$\frac{\square}{25} \times 100\% = \square\%$

PAGES 143–144

PAGE 182

VOLUME OF PYRAMIDS, CONES AND SPHERES

Surface Area and Volume

30 MINUTES

ADVANCED TEST

1 **a** Find the volume of a pentagonal pyramid of height 15 cm if the area of the pentagon is 18 cm^2. (1 mark)

b A hexagonal pyramid of the same height has the same volume. What is the area of the hexagon? (1 mark)

2 Find the volume of this pillar which is made up of a square-based prism and pyramid, as shown in the diagram.

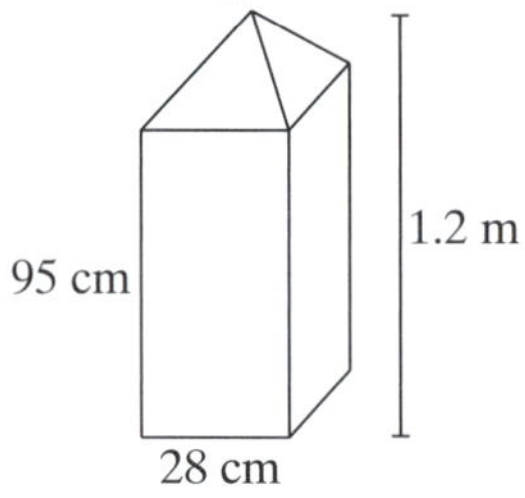

(2 marks)

3 A bollard comprises a cylinder and hemispherical cap. Find its volume.

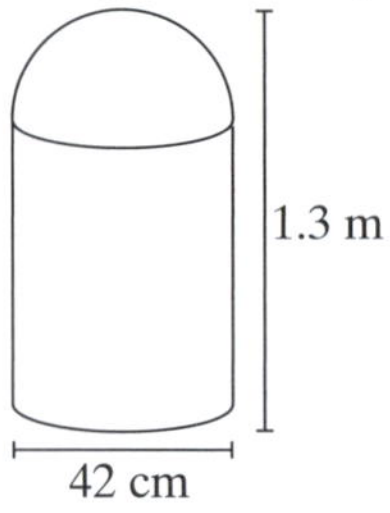

(3 marks)

4 **a** Find the volume of a cone of radius 48 cm and height 81 cm. (1 mark)

b What is the radius of a sphere with the same volume? (2 marks)

5 This tank is made up of a cylinder and cone.

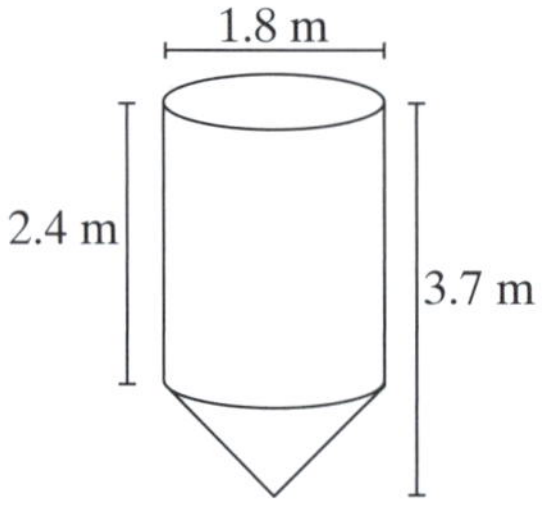

What is its capacity? Give the answer to the nearest 100 litres. (3 marks)

6 Find the volume, to the nearest cubic centimetre, of a cone with:

a diameter 150 mm and slant height 317 mm (3 marks)

b slant height 49.3 cm and perpendicular height 46.8 cm (3 marks)

7 A ball of diameter 9.6 cm fits tightly in a box (cube).

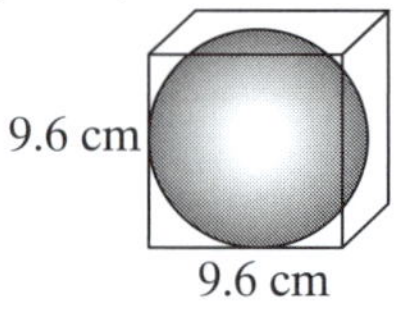

a Find the unused volume in the box. (Assume the width of the box is the same as the diameter of the ball.) (2 marks)

b How much less unused space would there be if the ball was placed tightly in a cylindrical container instead?

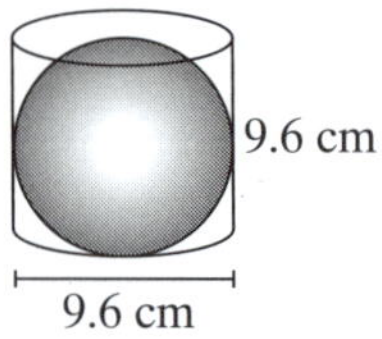

(2 marks)

8 The diagram shows the net of a right pyramid. Find the volume of the pyramid.

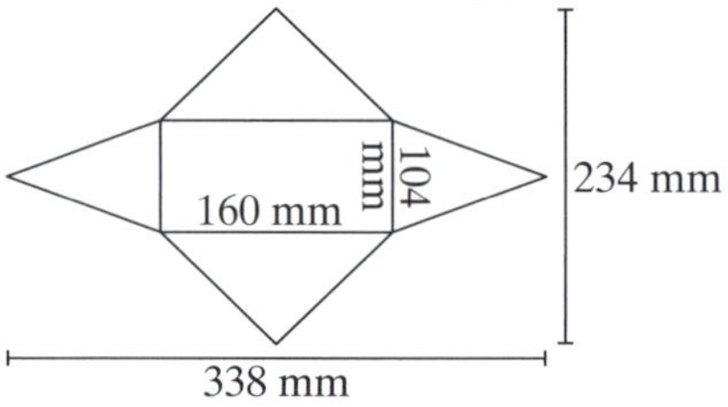

(3 marks)

9 Find the volume of this truncated cone.

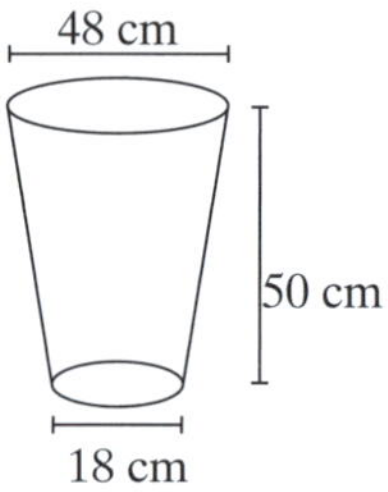

(4 marks)

Your Feedback

$\frac{\square}{30} \times 100\% = \square\ \%$

PAGES 145–146

PAGE 182

SURFACE AREA OF PYRAMIDS, CONES AND SPHERES

Surface Area and Volume

STUDY NOTES

1 The height of a right pyramid is its **perpendicular height**. It is the height from the centre of the base to the highest point, at right angles to the base.

2 The **total surface area of a pyramid** includes the area of the base and the area of all the triangular faces that make up the sides.

3 The **area of each triangular face** is found using the formula $A = \frac{1}{2}bh$ where h must be the perpendicular height of the triangle. This height is not the same as the perpendicular height of the pyramid, nor the same as the distance along an edge.

4 The perpendicular height of a triangular face of a pyramid is the hypotenuse of a right-angled triangle whose other two sides are formed by the perpendicular height of the pyramid and the distance from the base of this perpendicular to the centre of the base of the triangle. **Pythagoras' theorem** can be used to find one of these sides when the other two are known. In a right pyramid, the distance from the base of the perpendicular height to the face will be half the length of the parallel side.

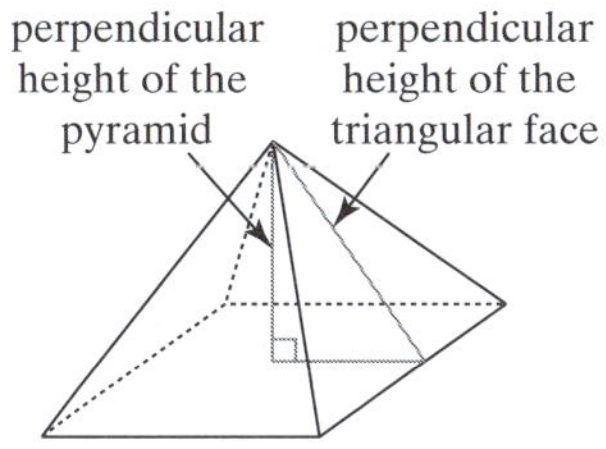

5 Unless a pyramid has a base whose sides are all equal, the **heights of the triangular faces** will not all be the same. For example in a rectangular pyramid with perpendicular height 15 units and with base 40 units long and 16 units wide, the perpendicular height of two triangles will be 25 units and of the other two triangles, 17 units.

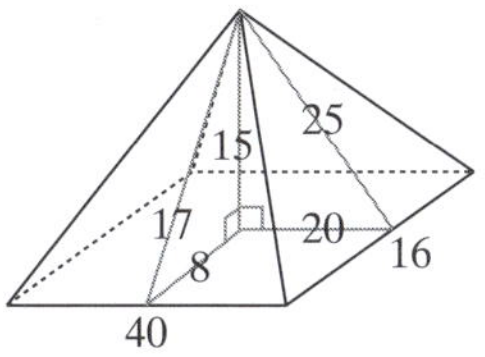

6 The **curved surface area of a right cone** is given by the formula $A = \pi rl$, where r is the radius and l the slant height. If the radius and perpendicular height are known, the slant height can be found by Pythagoras' theorem. i.e. $l^2 = h^2 + r^2$

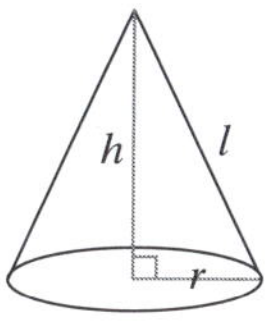

7 The base of a cone is a circle so the **total surface area of a cone** is found by adding the area of the circle to the curved surface area. i.e. $A = \pi r^2 + \pi rl$ or $A = \pi r(r + l)$

8 The **surface area of a sphere** of radius r is given by the formula $A = 4\pi r^2$.

9 A **hemisphere** is half a sphere, so its curved surface area is $2\pi r^2$. If the hemisphere is closed, the surface area also includes the area of a circle, so its total surface area is $3\pi r^2$.

10 The **surface area of composite solids** can be found by adding the areas of the composite parts, taking care only to add those areas that form part of the outer surface of the solid.

11 If the **surface area of a solid is known**, by substituting into the appropriate formula and solving the resulting equation, it may be possible to find **dimensions** of a solid.

Checklist

Can you:

1 *Identify the perpendicular height and slant height of pyramids and cones?* ☐
2 *Use Pythagoras' theorem to find slant height, perpendicular height, length of the base of a pyramid or radius of a cone?* ☐
3 *Find the surface area of pyramids?* ☐
4 *Find the curved surface area and total surface area of a cone?* ☐
5 *Find the surface area of a sphere?* ☐
6 *Find the surface area of composite solids?* ☐
7 *Find dimensions of a solid by solving equations?* ☐

SURFACE AREA OF PYRAMIDS, CONES AND SPHERES

Surface Area and Volume

SKILLS CHECK

1 For the square pyramid in the diagram, which interval gives the:

a perpendicular height of the pyramid?

b perpendicular height of a triangular face?

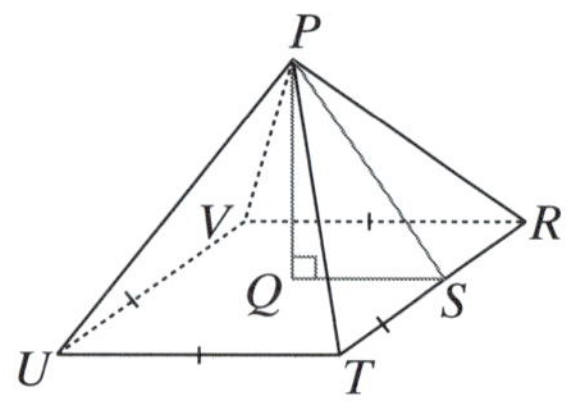

2 From the cone in the diagram write down the length of the:

a diameter

b perpendicular height

c slant height

d radius

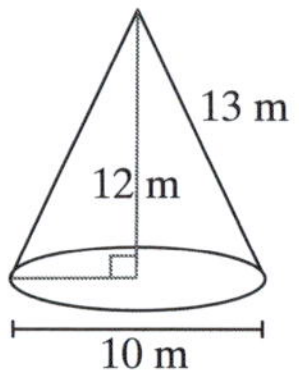

3 Find the value of x:

a

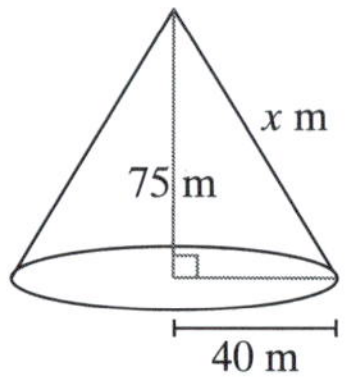

b

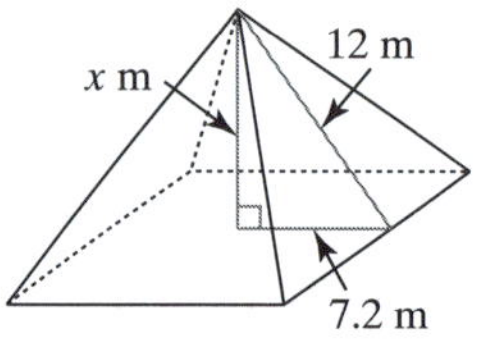

c

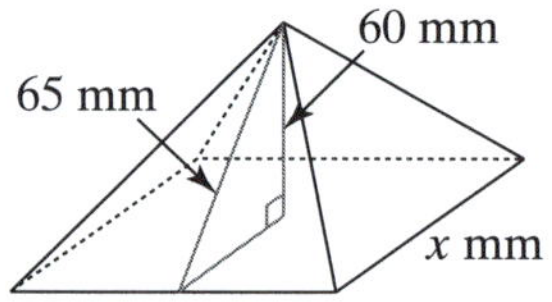

d

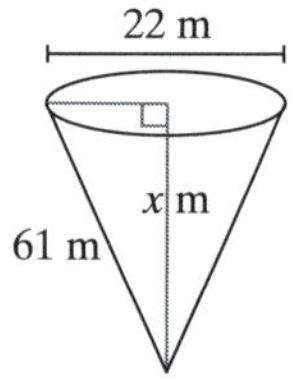

4 Find the area, correct to two decimal places, of a circle with:

a radius 1.8 cm

b diameter 3.3 m

5 Use the formula $A = \pi rl$ to find the curved surface area of a cone, correct to one decimal place, with:

a radius 48 cm, slant height 25 cm

b slant height 35 cm, diameter 15 cm

6 Use the formula $A = 4\pi r^2$ to find the surface area (to the nearest cm^2), of a sphere with:

a radius 12.6 cm

b diamater 784 mm

7 The diagram shows the net of a square pyramid.
Find its surface area.

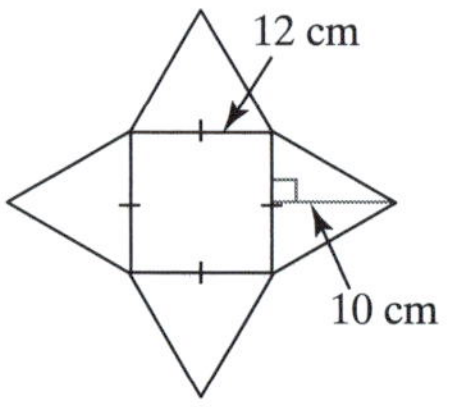

8 Find the value of r, correct to one decimal place, given:

a $A = \pi rl$ and $A = 122$ and $l = 8$

b $A = 4\pi r^2$ and $A = 1195 \quad (r > 0)$

PAGES 146–147

Answers **1 a** PQ **b** PS **2 a** 10 m **b** 12 m **c** 13 m **d** 5 m **3 a** 85 **b** 9.6 **c** 50 **d** 60 **4 a** 10.18 cm^2 **b** 8.55 m^2 **5 a** 3769.9 cm^2 **b** 824.7 cm^2 **6 a** 1995 cm^2 **b** 19310 cm^2 **7** 384 cm^2 **8 a** 4.9 **b** 9.8

SURFACE AREA OF PYRAMIDS, CONES AND SPHERES

INTERMEDIATE TEST

Part A Multiple Choice

1 The slant height of a cone of radius 44 mm is 125 mm. What is the perpendicular height of the cone?

A 90 mm **B** 109 mm **C** 117 mm **D** 133 mm (1 mark)

2 The radius of the Earth is approximately 6400 km. Which of the following is closest to the total area of the Earth's surface, in square kilometres? *Hint 1*

A 5×10^7 **B** 5×10^8 **C** 5×10^9 **D** 5×10^{10} (1 mark)

3 The surface areas of two similar figures are in the ratio 25:1. The ratio of the volumes of the two figures is:

A 5:1 **B** 25:1 **C** 125:1 **D** 625:1 (1 mark)

Part B Written Response

4 **a** Find the area (correct to one decimal place) of the circular base of the cone. (1 mark)

b Find the slant height, l cm, of the cone. (1 mark)

c Find the total surface area of the cone. *Hint 2* (2 marks)

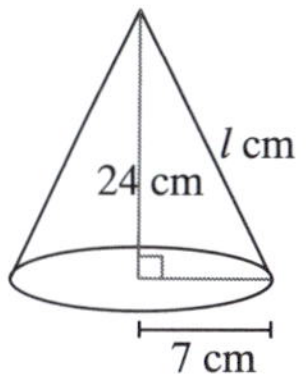

5 The diagram shows a right rectangular pyramid. The perpendicular height of the pyramid, AH, is 24 cm. $AF = 30$ cm and $AG = 26$ cm.

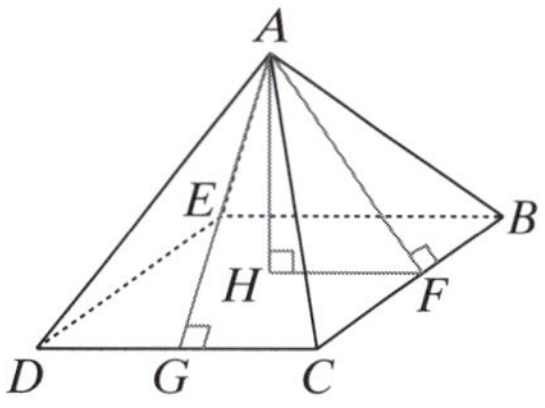

a Find the width, BC, of the rectangular base. *Hint 3* (2 marks)

b Find the area of the rectangular base. *Hint 4* (3 marks)

c Find the total surface area of the pyramid. (3 marks)

6 A sphere has a surface area of 2463 cm^2. Find its radius. (2 marks)

7 Twenty-five decorative bollards have been set in the ground beside a pathway. Each bollard is made up of a cylindrical post topped by a hemispherical dome. The diameter of each bollard is 50 cm and each reaches 1.5 m above ground level.

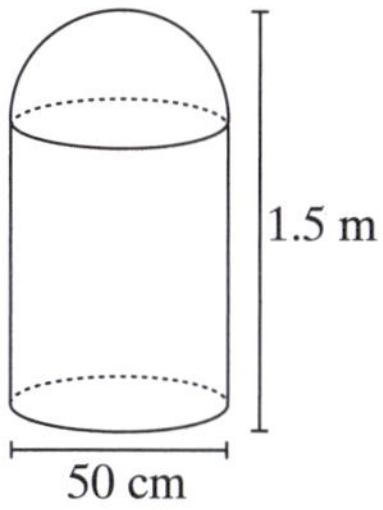

a Find the exposed surface area of each bollard in square metres, correct to two decimal places. *Hint 5* (4 marks)

b All the bollards are to be given two coats of paint. If 1 litre of paint covers 15 m^2, find the number of 4 litre cans of paint that will be required. (3 marks)

8 A pyramid has a rectangular base which is 72 m long and 28 m wide. The perpendicular height of the pyramid is 48 m. Find the total surface area of the pyramid. *Hint 6* (6 marks)

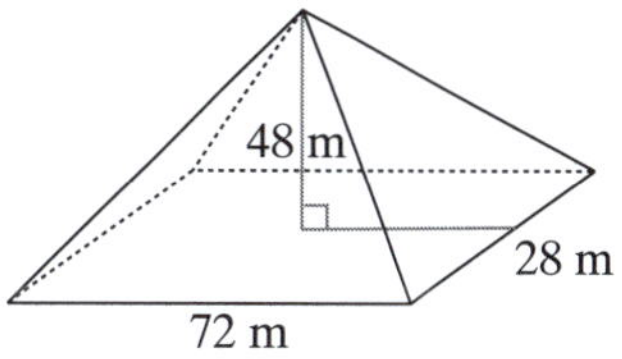

Hint 1: Assume that the Earth is a sphere.
Hint 2: The curved surface area plus the area of the base.
Hint 3: Use Pythagoras' theorem in ΔAHG to find HG.
Hint 4: Use ΔAHF to find HF and then find DC.
Hint 5: Make sure you express the radius in metres. You will need to subtract the radius from the total height to find the height of the cylinder. Because the post is set in the ground you will not have to add the area of the base of the cylinder.
Hint 6: You will need to use Pythagoras' theorem to find the slant height of both sizes of triangular face.

Your Feedback

$\frac{\square}{30} \times 100\% = \square\%$

PAGES 147–148

PAGE 182

SURFACE AREA OF PYRAMIDS, CONES AND SPHERES

Surface Area and Volume

35 MINUTES

ADVANCED TEST

1 **a** Find the surface area of a sphere of radius 12 cm. (1 mark)

b Find the surface area of the sphere with double the radius of the sphere in part **a**. (1 mark)

c What is the percentage increase in surface area? (1 mark)

d Will the percentage increase always be the same when the radius of a sphere doubles? Justify your answer. (1 mark)

e Find the curved surface area of a cone with radius 8 cm and slant height 20 cm. (1 mark)

f If both the radius and slant height of a cone double what would be the percentage increase in curved surface area? (1 mark)

g Will this percentage increase be the same for the total surface area of a cone when both the radius and slant height double? (1 mark)

h What will the percentage increase in curved surface area be if the radius and slant height of a cone triples? (1 mark)

2 Find the surface area of a sphere with volume 11.5 cm^3. (3 marks)

3 This rectangular pyramid has perpendicular height $AX = 2$ m. $BC = 2$ m and $DC = 8$ m.

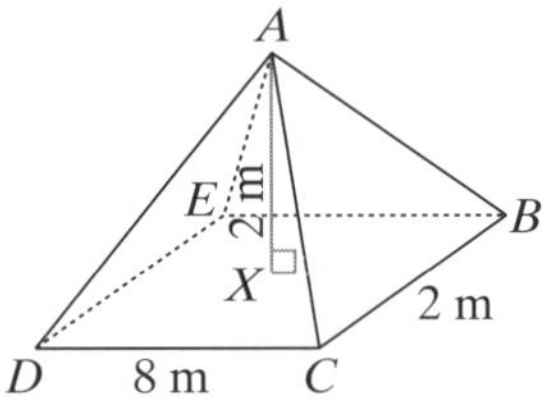

a Find the exact area of ΔABC. (2 marks)

b Show that the area of ΔADC is exactly twice that of ΔABC. (2 marks)

c Find the exact surface area of the pyramid. (1 mark)

4 Find the total surface area of a right pyramid with rectangular base 84 cm by 18 cm and perpendicular height 40 cm. (5 marks)

5 This solid was made by gluing two identical cones together. Find the surface area. (2 marks)

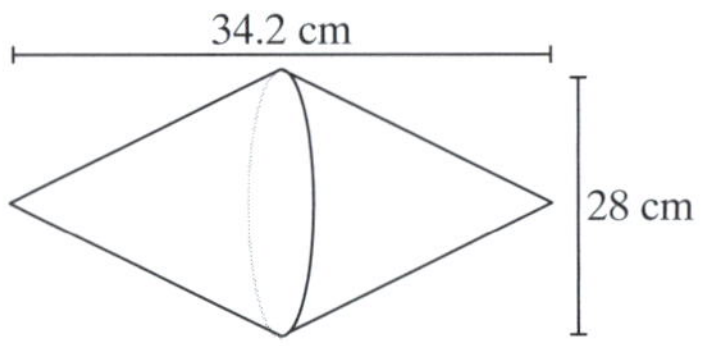

6 A rectangular pyramid has a base that is 2.5 times longer than its breadth. $AB = 10$ cm and $AC = 6.8$ cm. Let the width be $4x$ cm, length $10x$ cm and perpendicular height h cm.

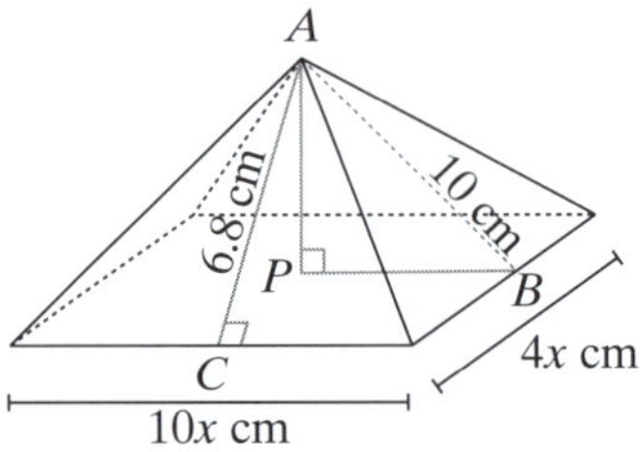

a By considering ΔAPB, show that $h^2 = 100 - 25x^2$. (1 mark)

b Find the value of x. (2 marks)

c Find the perpendicular height. (1 mark)

d Find the surface area. (2 marks)

7 The exact surface area of this cone with hemispherical cap is 115π cm^2. The slant height of the cone is 13 cm.

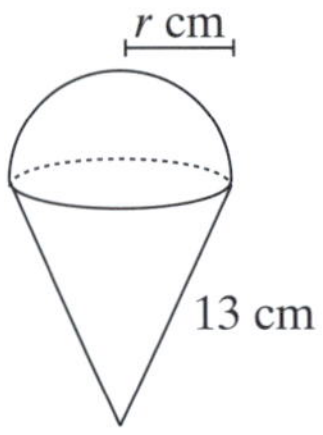

a Show that solving the equation $2r^2 + 13r - 115 = 0$ will give the radius. (2 marks)

b Find the radius. (2 marks)

c Find the total height of the capped cone. (2 marks)

Your Feedback

$\frac{\square}{35} \times 100\% = \square\%$

PAGES 148–149

PAGE 182

GEOMETRY REVIEW

Geometry

STUDY NOTES

1 **Angles** in a **right angle** add to 90°. Angles in a **straight line** add to 180°. Angles at a point add to 360°. **Vertically opposite** angles are equal.

2 If a **pair of lines is cut by a transversal**, three types of angles are formed. If the lines are parallel, the **corresponding angles** are equal, the **alternate angles** are equal and the **co-interior angles** add to 180°. If the corresponding angles are equal, or if the alternate angles are equal or if the co-interior angles are supplementary the lines are parallel.

3 The **angle sum of a triangle is 180°**. The angle sum of a quadrilateral is 360°. The exterior angle of a triangle is equal to the sum of the interior opposite angles.

4 A **scalene triangle** has no sides equal in length. An **isosceles triangle** has two sides equal. An **equilateral triangle** has three equal sides. Each angle of an equilateral triangle measures 60°.

5 If two sides of a triangle are equal the **angles opposite** those sides are equal. If two angles of a triangle are equal the **sides opposite** those angles are equal.

6 **Congruent figures** are exactly the same shape and the same size. The symbol ≡ means 'is congruent to'. If two figures are congruent the **corresponding sides are equal** and the **corresponding angles are equal**. Two triangles are congruent if:

- three sides of one triangle are equal to three sides of the other triangle (SSS)
- two sides and the included angle of one triangle are equal to two sides and the included angle of the other triangle (SAS)
- two angles and a side of one triangle are equal to two angles and the corresponding side of the other triangle (AAS)
- the hypotenuse and one side of one right-angled triangle are respectively equal to the hypotenuse and one side of another right-angled triangle (RHS).

7 A quadrilateral is a four-sided figure. The **special quadrilaterals** include squares, rectangles, rhombuses, parallelograms, trapeziums and kites. (Some quadrilaterals fit into more than one category; a rhombus for example, is a parallelogram.) These properties should be known:

- The opposite angles of a parallelogram are equal and the opposite sides are equal.
- The diagonals of a parallelogram bisect each other.
- The diagonals of a rhombus bisect each other at right angles, and they bisect the vertex angles through which they pass.
- The diagonals of a rectangle are equal.

8 **Similar figures** are exactly the same shape but not necessarily the same size. The symbol ||| means 'is similar to'. In any pair of similar figures the **corresponding angles are equal** and the **corresponding sides are in the same ratio**. Two triangles are similar if:

- three sides of one triangle are proportional to three sides of the other triangle
- two sides of one triangle are proportional to two sides of the other triangle and the included angles are equal
- two angles of one triangle are equal to two angles of the other triangle
- the hypotenuse and one side of a right-angled triangle are proportional to the hypotenuse and one side of the other right-angled triangle.

9 You need to understand and apply all these definitions and properties and use them in **proofs** and in solving problems. Draw a **large clear diagram** and mark in all the given information. Be logical. **Give reasons** for every step.

Checklist

Can you:

1 *Understand and apply the properties of angles?* ☐
2 *Understand and apply the properties of triangles, and prove triangles congruent or similar?* ☐
3 *Understand and apply the definitions and properties of the special quadrilaterals?* ☐
4 *Solve problems and complete proofs?* ☐

GEOMETRY REVIEW

Geometry

SKILLS CHECK

1 Find the value of each pronumeral:

a 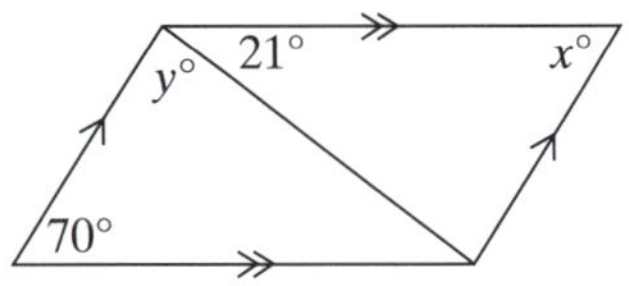

b 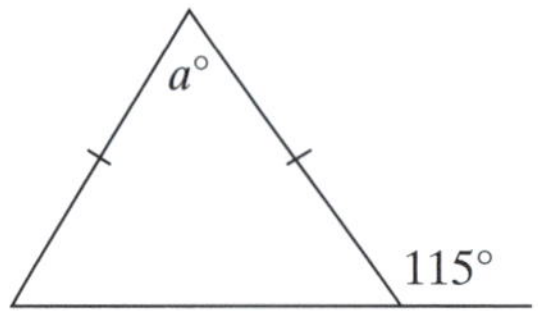

c 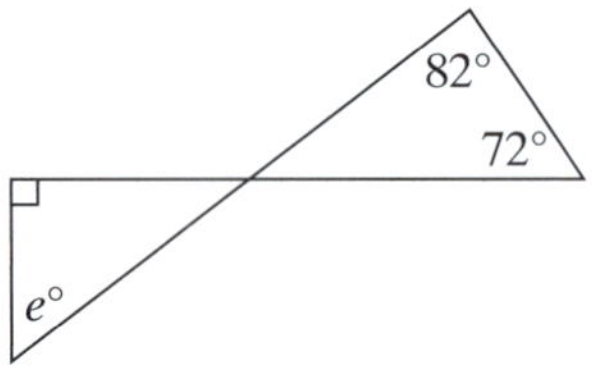

d 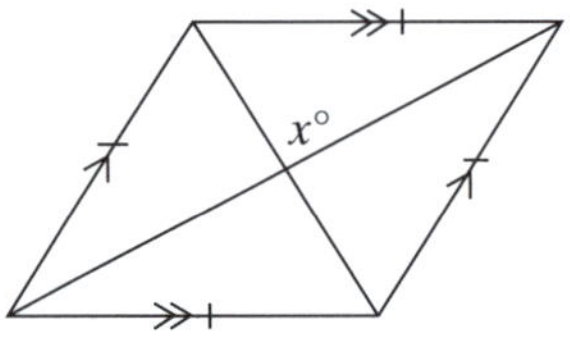

e 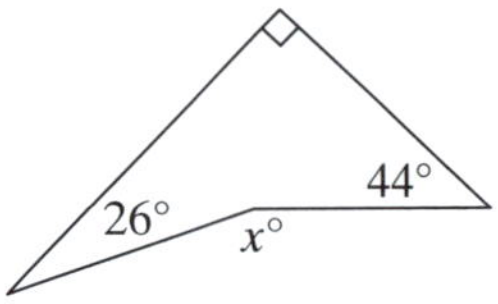

f 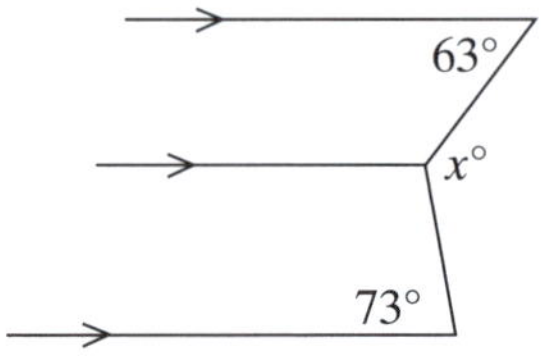

2 Which test (if any) could be used to prove the pair of triangles congruent?

a (AE and BD intersect at C)

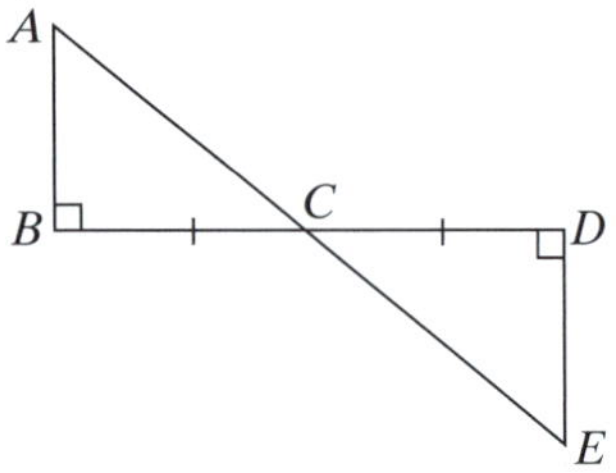

b 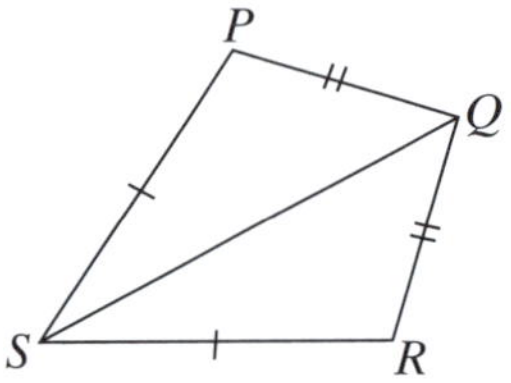

c 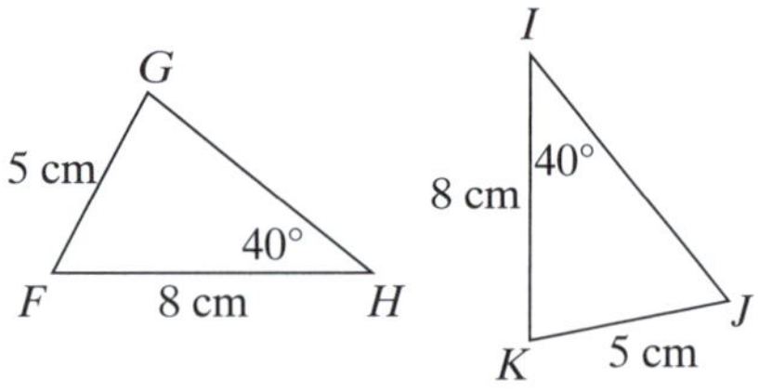

3 The pair of triangles is similar. Find the value of x:

a 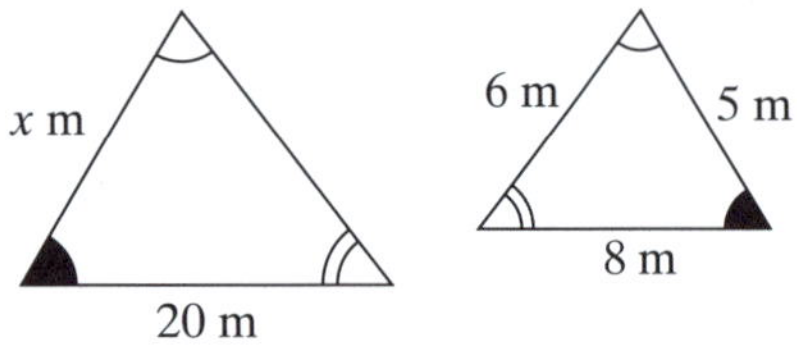

b 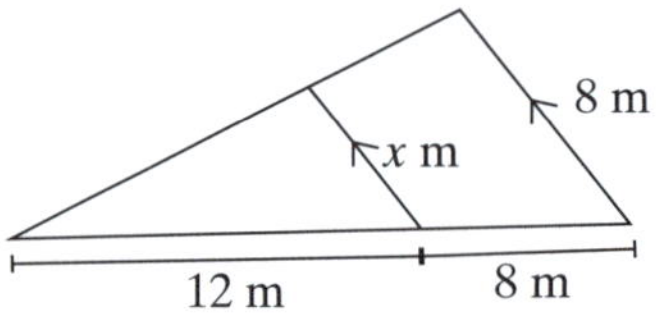

4 Complete:

a The angles of a quadrilateral add up to ______________.

b The opposite angles of a parallelogram are ______________.

c An ______________ triangle has two equal sides.

d The diagonals of a rectangle are ______________ in length.

e The diagonals of a rhombus ______________ the vertex angles through which they pass.

f If two angles of a triangle are equal, the sides opposite those angles are ______________.

g If all sides of a quadrilateral are equal the quadrilateral is a ______________.

h If two angles of one triangle are equal to two angles of another triangle the two triangles are ______________.

i If three sides of one triangle are equal to three sides of another triangle the two triangles are ______________.

PAGES 149–150

Answers **1 a** $x = 70, y = 89$ **b** $a = 50$ **c** $e = 64$ **d** $x = 90$ **e** $x = 160$ **f** $x = 136$ **2 a** AAS **b** SSS **c** not necessarily congruent **3 a** $x = 12.5$ **b** $x = 4.8$ **4 a** 360° **b** equal **c** isosceles **d** equal **e** bisect **f** equal **g** rhombus **h** similar **i** congruent

GEOMETRY REVIEW

Geometry

Part A Multiple Choice

1 Which of the following triangles are congruent?

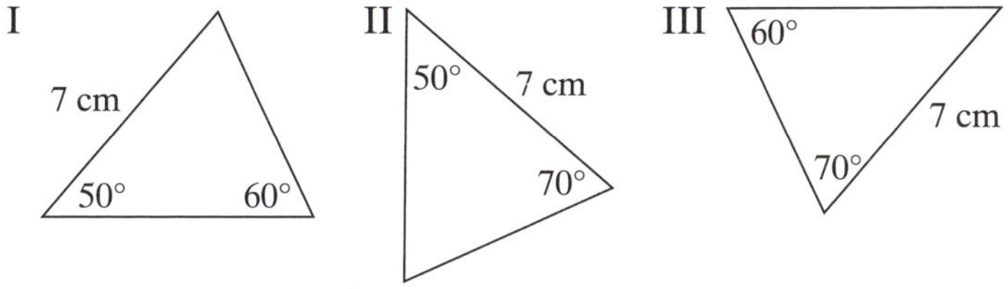

A Only I and II **B** Only II and III **C** Only I and III **D** I, II and III (1 mark)

2 $x = ?$ (Diagram not to scale and all measurements are in metres.)

A 4 **B** 9.6
C 16 **D** 21.6 (1 mark)

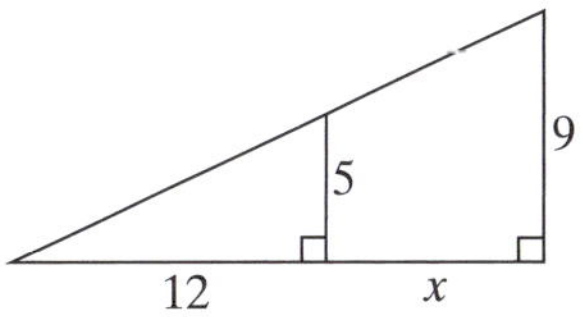

3 $PQ = SQ = SR$. $\angle QPS = 42°$.
$x = ?$

A 21 **B** 42
C 48 **D** 69 (1 mark)

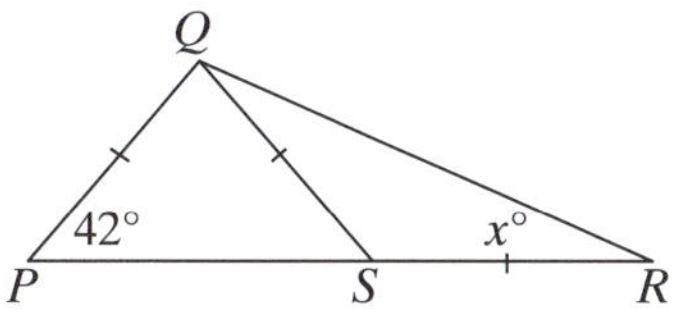

Part B Written Response

4 $PQRS$ is a square. QS is parallel to PT.
RT is a straight line.
What is the size of $\angle QPT$?
(Give reasons.) *Hint 1* (3 marks)

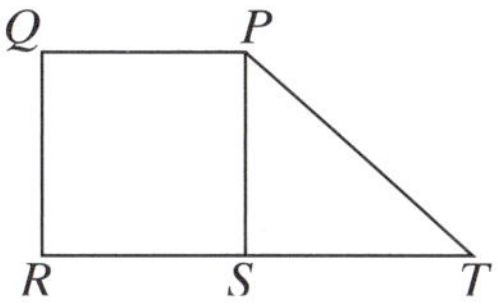

5 **a** Prove that JL bisects $\angle KJM$. *Hint 2* (3 marks)
b What property have you proven? (1 mark)

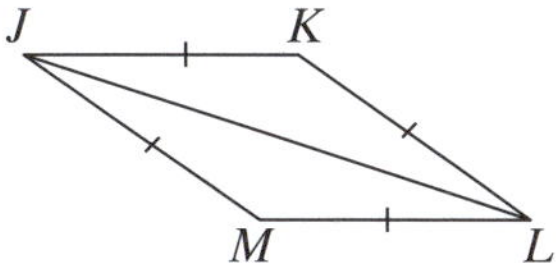

6 **a** Prove that ΔOPS and ΔOQR are similar. (3 marks)
b Is SP parallel to RQ? Justify your answer. (2 marks)

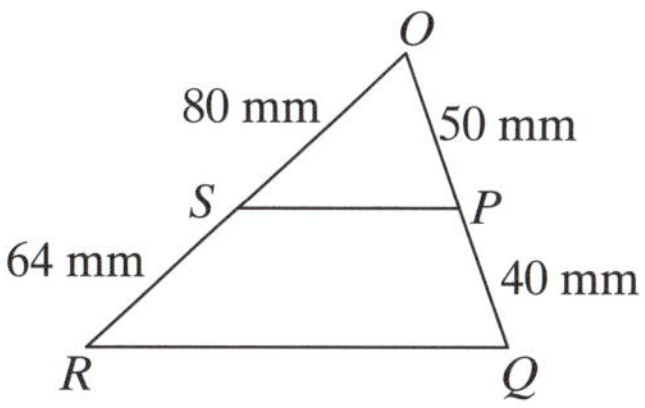

7 O is the midpoint of AC. $BO = DO = AO$.
$\angle DAO = x°$ and $\angle OAB = y°$.
a Find $\angle DOC$ in terms of x. (2 marks)
b Show that $\angle DCB$ and $\angle DAB$ are supplementary. *Hint 3* (3 marks)

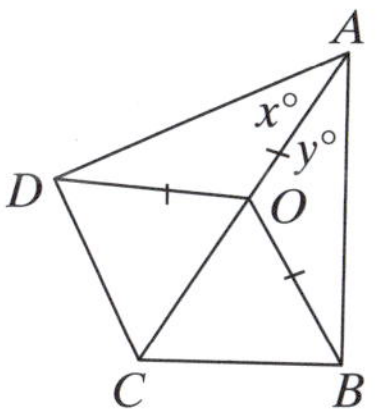

Hint 1: Mark all the given information on your diagram.
Hint 2: Prove that triangles JKL and JML are congruent.
Hint 3: Supplementary angles add to 180°. Show that $\angle DCB = 180° - (x + y)°$.

Your Feedback

$\frac{\square}{20} \times 100\% = \square\%$

PAGES 150–151
PAGE 182

GEOMETRY REVIEW

Geometry

40 MINUTES

ADVANCED TEST

1 Show that T is the midpoint of PR.

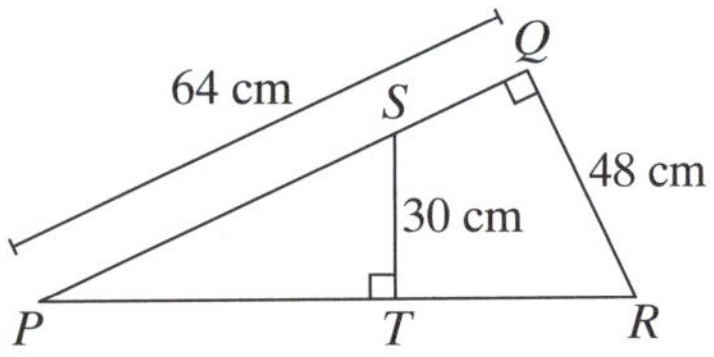

(3 marks)

2 The diagram shows a rectangle inside a regular pentagon.

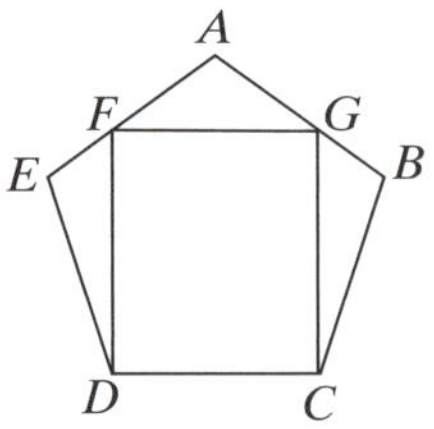

a Find the size of $\angle FDE$. (2 marks)

b Show that $\angle AFG = 2\angle FDE$. (2 marks)

3 $ABCD$ is a rectangle. $BP = 35$ cm. $PC = 28$ cm. $DC = 84$ cm.

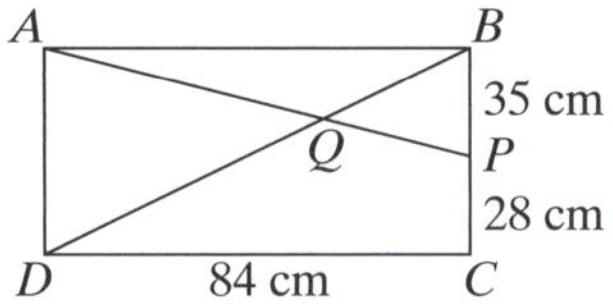

a Show that ΔAQD is similar to ΔPQB. (2 marks)

b How much longer is BD than AP? (2 marks)

c Find the length of BQ. (2 marks)

4 XM is the perpendicular bisector of BC. $AB = XC$.

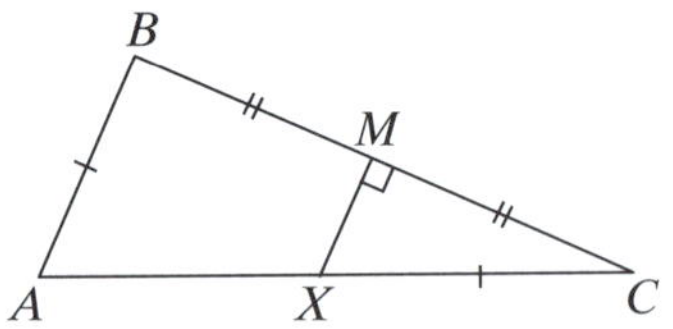

a Show that $\Delta BMX \equiv \Delta CMX$. (2 marks)

b Show that $\angle BAC = 2\angle BCA$. (3 marks)

5 $ABDE$ is a rhombus. ED is produced to C such that $BD = BC$.

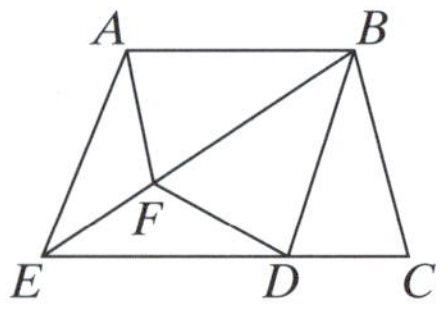

a Show that $\Delta AEF \equiv \Delta DEF$. (2 marks)

b If $\angle AEF = 35°$, find $\angle ABC$. (3 marks)

6 $PT = RS = 20$ cm. $PR = 50$ cm. $PT \perp TS$. $PR \perp RS$. PR and TS intersect at Q.

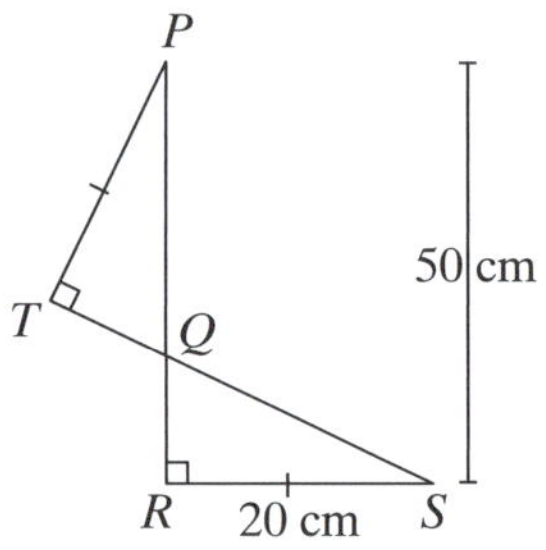

Find the length of QR. (4 marks)

7 $ABCD$ is a rhombus of side 3 cm. E is a point on diagonal BD such that $DE = CE$.

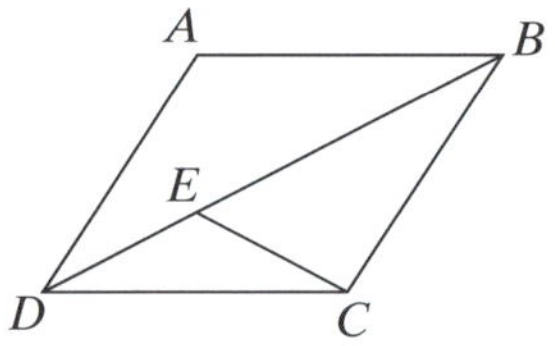

a Show that ΔEDC is similar to ΔABD. (2 marks)

b $EC = \sqrt{3}$ cm. Show that $BD = 3\sqrt{3}$ cm. (2 marks)

c Show that ΔBEC is right-angled. (2 marks)

d Show that $\angle BAD = 120°$. (3 marks)

8 BE is parallel to CD.

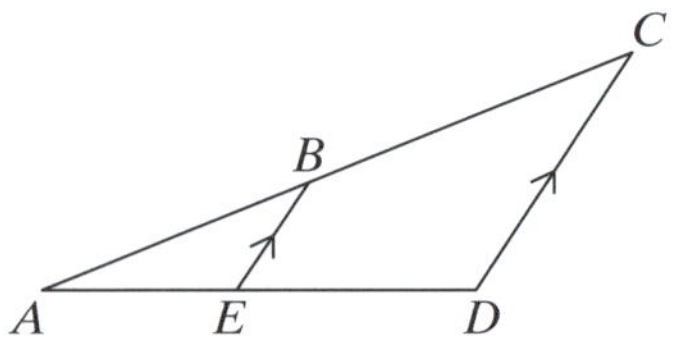

a Show that ΔABE is similar to ΔACD. (1 mark)

b Show that $BE = CD - \dfrac{BC \times CD}{AC}$ (3 marks)

Your Feedback

$\dfrac{\square}{40} \times 100\% = \square\%$

PAGES 151–153
PAGE 182

PARTS OF CIRCLES AND CHORD PROPERTIES

Circle Geometry

STUDY NOTES

1 The **parts of a circle** (centre, radius, diameter, circumference, arc, sector, segment, semi-circle, tangent and secant) should all be known.

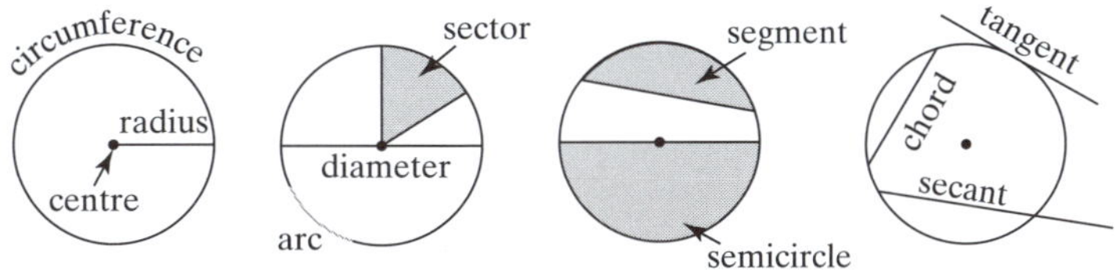

2 An **angle at the centre** is an angle whose **vertex** is at the centre of the circle. For example, $\angle AOB$ in diagram I.

3 An **angle at the circumference** is an angle whose vertex is on the circumference of a circle. For example, $\angle ACB$ in diagram II.

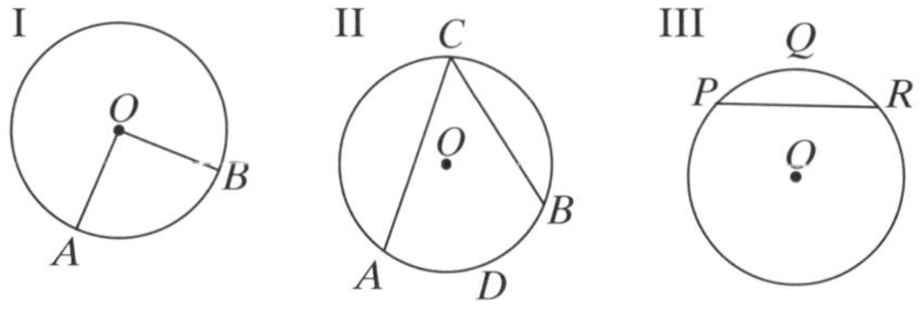

4 The word **subtends** means extending under. For example, in diagram III chord PR subtends arc PQR.

5 If an angle is **standing on an arc**, then the arms of the angle begin at the ends of the arc that subtends the angle. For example, in diagram II above, $\angle ACB$ is standing on arc ADB.

6 Any chord divides a circle into two segments. An **angle in an alternate segment** lies in the opposite segment to the one being considered.

7 At any point on a circle, there is a unique **tangent to the circle**. The tangent is **perpendicular to the radius** at the point of contact. (You are not required to prove this result. It is an assumption that may be used in proving other results.)

8 You should be able to **prove** and **apply** the following **theorems**:

- Chords of equal length in a circle subtend equal angles at the centre and are equidistant from the centre.
- The perpendicular from the centre of a circle to a chord bisects the chord.
- The line from the centre of a circle to the midpoint of a chord is perpendicular to the chord.
- The perpendicular bisector of a chord passes through the centre of the circle.
- When two circles intersect, the line joining their centres bisects their common chord at right angles.

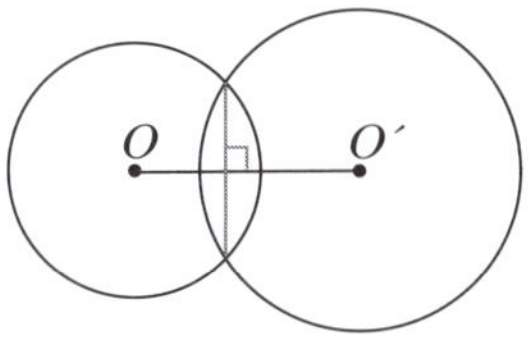

- Given any three non-collinear points, the point of intersection of the perpendicular bisectors of any two sides of the triangle formed by the three points is the centre of the circle that passes through all three points.

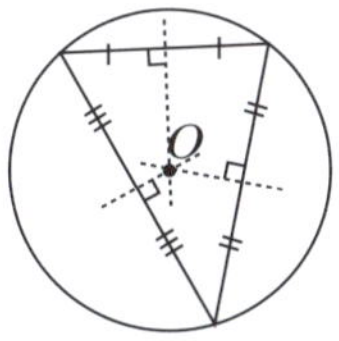

Each of the above properties can be proven by first proving that a pair of triangles is congruent.

Checklist

Can you:

1 *Name and identify all the parts of a circle?* ☐

2 *Use terminology associated with angles in circles?* ☐

3 *Identify the arc on which an angle stands and the segment in which an angle lies?* ☐

4 *Understand what is meant by a tangent to a circle and that the tangent is perpendicular to the radius at the point of contact?* ☐

5 *Prove and apply properties of chords?* ☐

PARTS OF CIRCLES AND CHORD PROPERTIES

Circle Geometry

SKILLS CHECK

1 Name the part of the circle shown in colour. (O is the centre of the circle.)

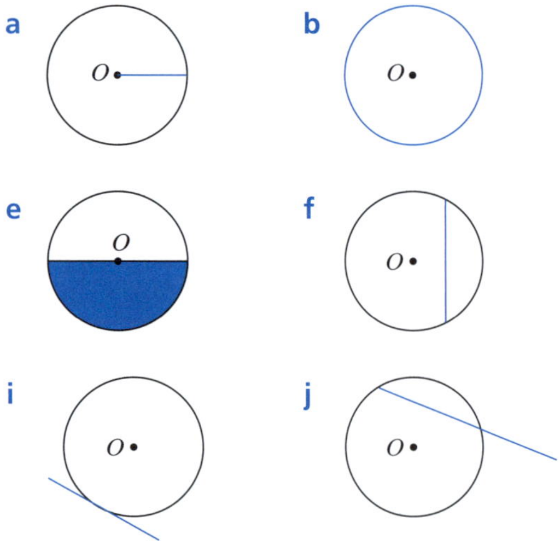

2 Find the value of x:

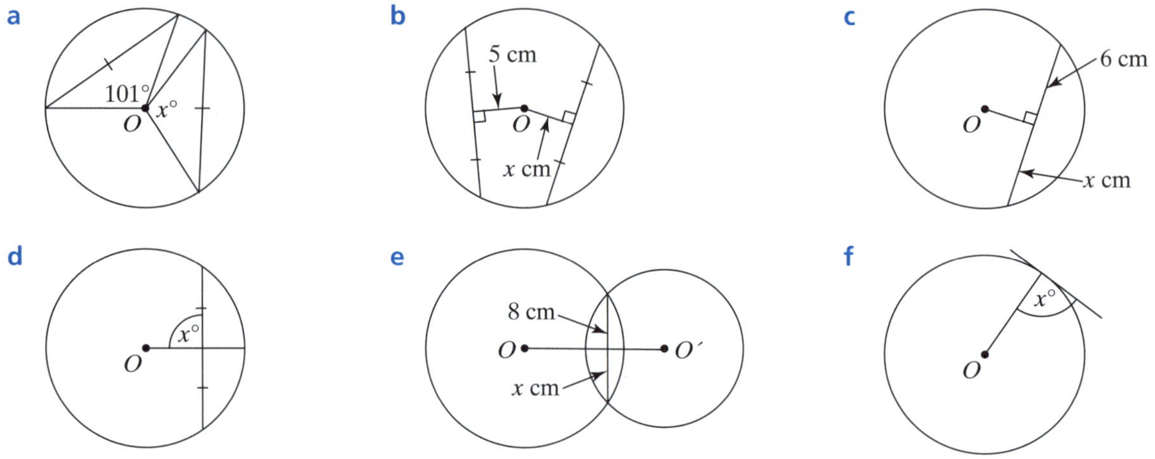

3 Complete:

a The perpendicular bisector of a chord of a circle passes through the ____________ of the circle.

b When two circles intersect, the line joining their centres bisects their common chord at ______________________ .

c When any three points form a triangle, the point of intersection of the perpendicular bisectors of any two sides of the triangle is the centre of the circle which passes through ______________ .

PAGES 153–154

Answers **1 a** radius **b** circumference **c** diameter **d** sector **e** semi-circle **f** chord **g** segment **h** arc **i** tangent **j** secant **2 a** $x = 101$ **b** $x = 5$ **c** $x = 6$ **d** $x = 90$ **e** $x = 8$ **f** $x = 90$ **3 a** centre **b** right angles **c** all three points

PARTS OF CIRCLES AND CHORD PROPERTIES

Circle Geometry

20 MINUTES

INTERMEDIATE TEST

Part A Multiple Choice

1 The part of the circle shown in blue is called a:

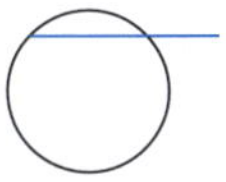

A secant **B** sectant

C sector **D** segment (1 mark)

2 PQ is a chord of a circle of radius 14.9 m. The distance of the chord from the centre of the circle is 5.1 m. What is the length of the chord?

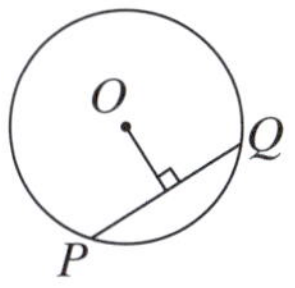

A 20 m **B** 22 m **C** 24 m **D** 28 m (1 mark)

3 P and Q are the centres of two circles as shown in the diagram. The circles intersect at A and B. Which best describes the triangle formed by points A, P and Q?

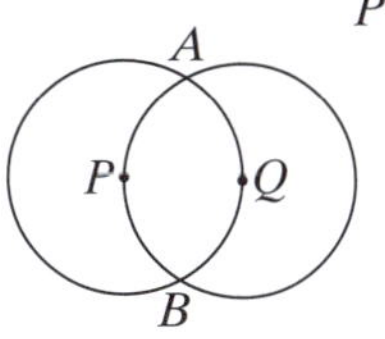

A scalene **B** isosceles

C equilateral **D** right-angled (1 mark)

Part B Written Response

4 O is the centre of the circle. Chords AD and BC are equal. Prove that triangles ADO and BCO are congruent, and hence prove that $\angle AOD = \angle BOC$. *Hint 1* (4 marks)

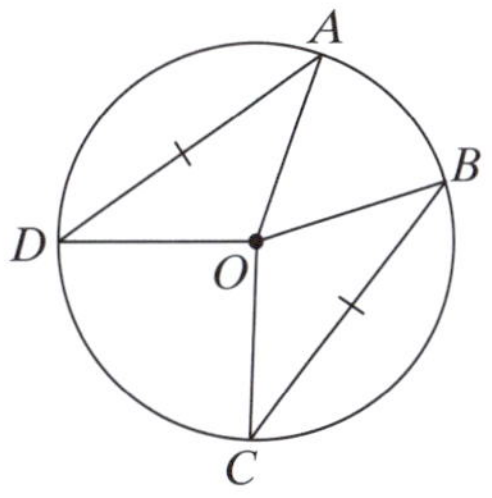

5 AB is a diameter of a circle, centre O.
C is a point on the circumference such that $\angle COB = 60°$.

a Show this information on a diagram. (1 mark)

b Explain why chord CB must have the same length as the radius. *Hint 2* (2 marks)

c Explain why the area of sector AOC is twice that of sector COB. (2 marks)

6 PQ and RS are equal chords of a circle.
PS and QR intersect at O, the centre of the circle.
If $\angle PQO = 51°$:

a find the size of $\angle ROS$ (giving reasons) (2 marks)

b show that PQ is parallel to RS *Hint 3* (3 marks)

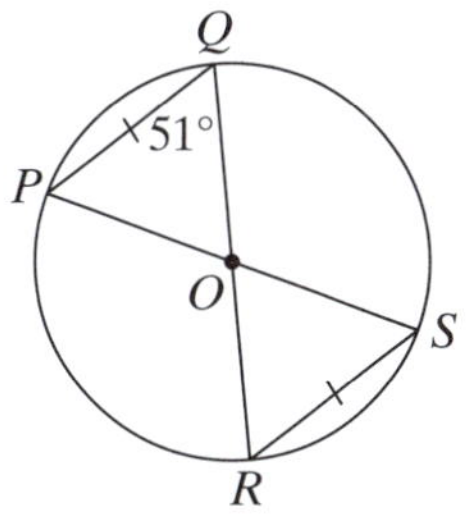

7 The perpendicular distance of a chord AB from the centre, O, of a circle is 9 cm. The length of the chord AB is 24 cm. Find the circumference of the circle. *Hint 4* (3 marks)

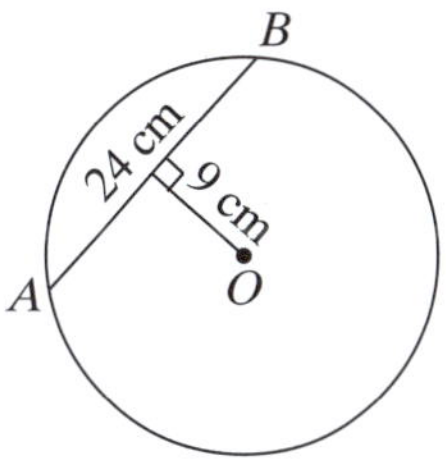

Hint 1: You cannot use the fact that $\angle AOD = \angle BOC$ in your proof.
Hint 2: What type of triangle is COB?
Hint 3: Find the size of $\angle ORS$.
Hint 4: The perpendicular from the centre to the chord bisects the chord.

Your Feedback

$\frac{\square}{20} \times 100\% = \square\%$

PAGES 154–155

PAGE 182

PARTS OF CIRCLES AND CHORD PROPERTIES

Circle Geometry

30 MINUTES

ADVANCED TEST

1 AB, BC, CD and DA are chords of equal length.

- **a** Explain why $ABCD$ must be a square. (2 marks)
- **b** If the diameter of the circle is 12 cm, find the exact length of chord AB. (2 marks)

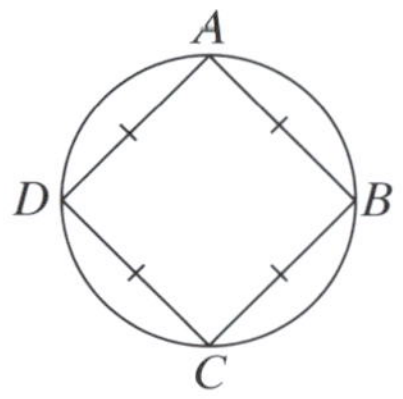

2 A and B are two points on a circle such that AB is perpendicular to chord PQ. AB and PQ intersect at M, the midpoint of PQ. X lies on AB.

- **a** Prove that ΔPMX is congruent to ΔQMX. (2 marks)
- **b** Explain why $PX = QX$. (1 mark)
- **c** Sophie concluded that X must be the centre of the circle. Is she correct? Justify your answer. (1 mark)

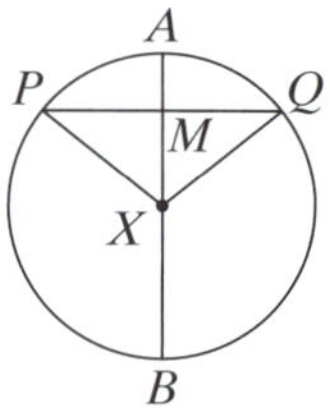

3 Two circles, with centres at P and Q, intersect at A and B.

- **a** Prove that $\Delta PAQ \equiv \Delta PBQ$. (2 marks)
- **b** If C is the point of intersection of PQ and AB, prove that $AC = CB$. (3 marks)
- **c** Prove that AB is perpendicular to PQ. (2 marks)

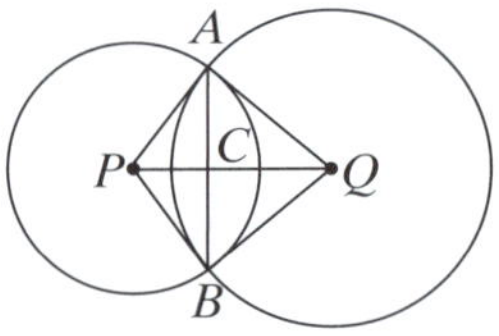

4 Chord AB is 45 cm long and 12 cm from the centre of the circle. Chord CD is 15.3 cm from the centre. How much longer is chord AB than chord CD? (3 marks)

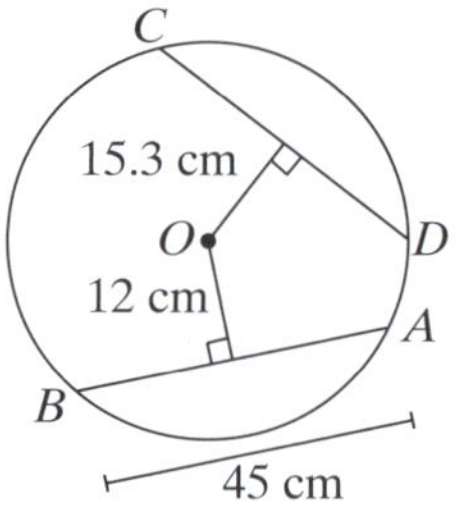

5 M is the midpoint of chord AB of a circle, centre O. Prove that $\angle AMO = 90°$. (4 marks)

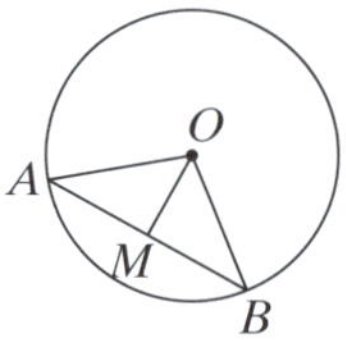

6 AB and CD are two chords of a circle, centre O. $\angle AOB = \angle COD$. Prove that $AB = CD$. (3 marks)

7 XP is the perpendicular bisector of side AB and XQ is the perpendicular bisector of side AC of ΔABC.

- **a** Show that $\Delta APX \equiv \Delta BPX$. (2 marks)
- **b** Explain why X is the centre of the circle that passes through A, B and C. (3 marks)

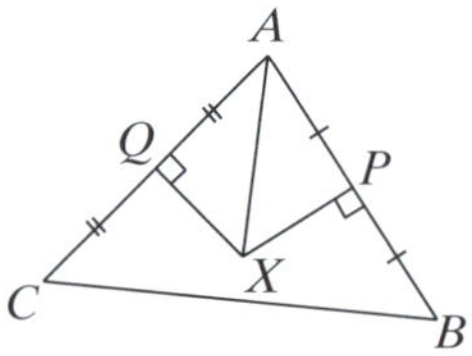

Your Feedback

$\frac{\square}{30} \times 100\% = \square\%$

PAGES 155–156
PAGE 182

ANGLE PROPERTIES, TANGENTS AND SECANTS

Circle Geometry

STUDY NOTES

1 The **angle at the centre of a circle is twice the angle at the circumference**. (This theorem can be proven by drawing the diameter from the vertex of the angle at the circumference and using the exterior angle of a triangle.)

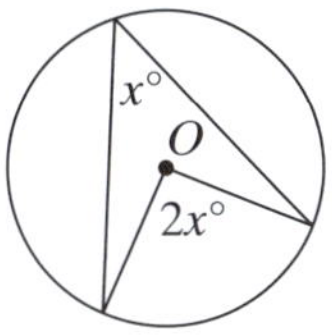

2 The **angle in a semi-circle is a right angle**. This is sometimes called Thale's theorem. (It is just a special case of the theorem in point 1, the angle at the centre is 180°.)

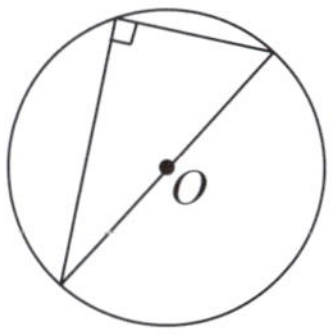

3 **Angles at the circumference, standing on the same arc, are equal**. (The angle at the centre is the same for both angles.)

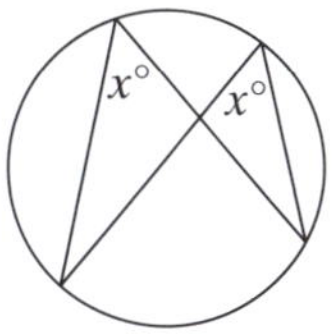

4 A **cyclic quadrilateral** is a quadrilateral whose four vertices lie on the circumference of a circle. The **opposite angles** of a cyclic quadrilateral are **supplementary**. (The angles at the centre form a revolution, so add to 360°. The angles at the circumference therefore add to 180°.) An **exterior angle** of a cyclic quadrilateral is equal to the **interior opposite angle**.

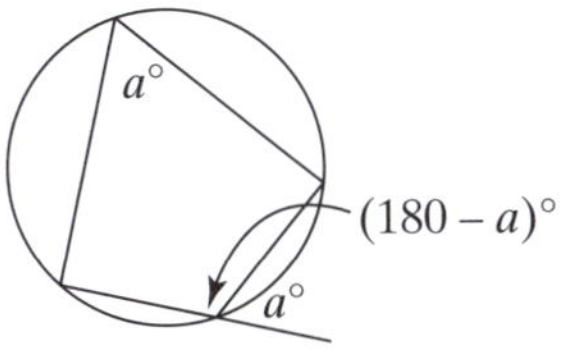

5 The **two tangents drawn to a circle from an external point are equal** in length. (This can be shown by joining the external point to the centre of the circle, and proving the two resulting triangles congruent.)

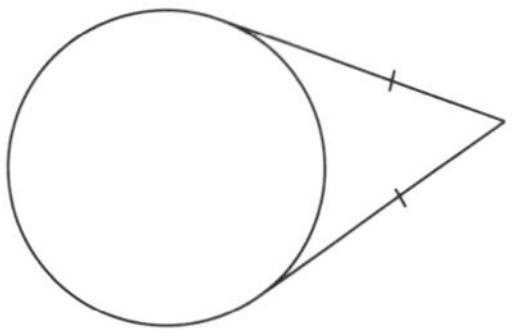

6 The **angle between a tangent and a chord** drawn to the point of contact equals the angle in the **alternate segment**.

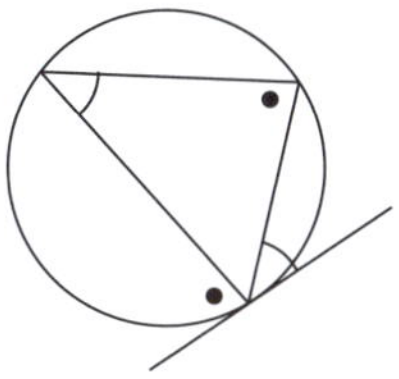

7 When **two circles touch**, their centres and the point of contact lie on the same straight line.

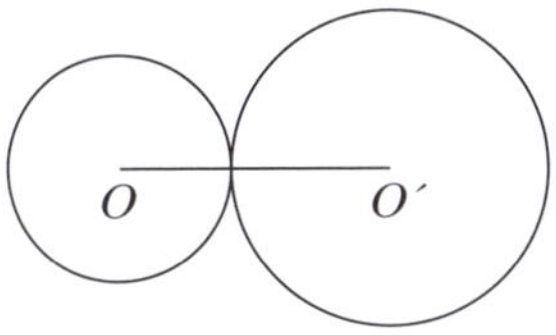

8 The **products of the intercepts of two intersecting chords of a circle are equal**. i.e. $ab = cd$

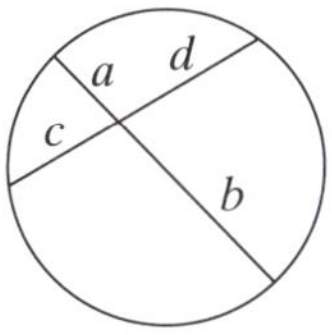

9 The **square of a tangent** to a circle from an external point equals the product of the **intercepts of any secant** from the point.
Note: the intercepts are from the external point, the distances from the external point to each side of the circle. For example, in the diagram $a^2 = b(b + c)$.

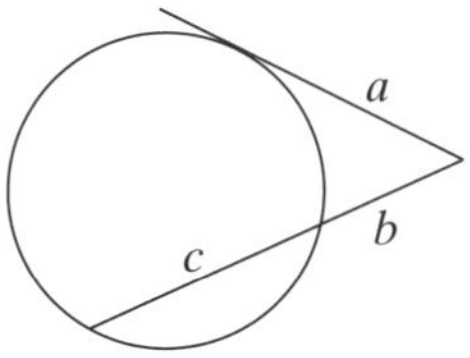

10 The **products of the intercepts of two secants** to a circle from an external point **are equal**. For example, $a(a + b) = c(c + d)$ in the diagram.

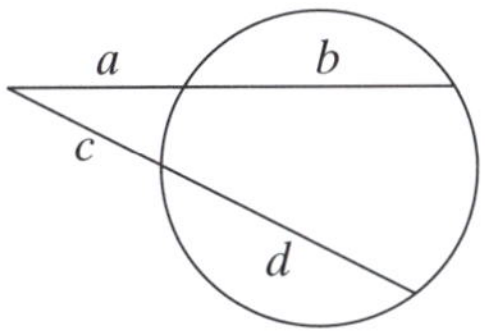

Checklist

Can you:

1 *Find angles at the circumference and angles at the centre?* ☐
2 *Understand what is meant by a cyclic quadrilateral and find angles in a cyclic quadrilateral?* ☐
3 *Find angles between tangents and chords?* ☐
4 *Use the properties of intercepts with chords, tangents and secants?* ☐
5 *Prove the above results?* ☐

ANGLE PROPERTIES, TANGENTS AND SECANTS

Circle Geometry

SKILLS CHECK

1 Find the value of x. (O is the centre of the circle).

a

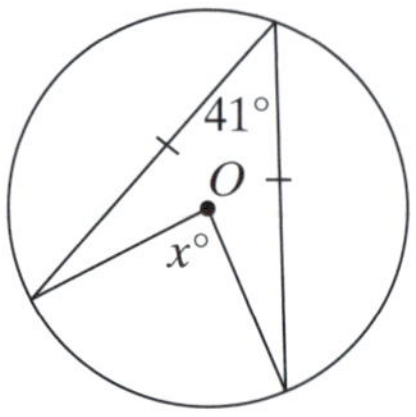

b

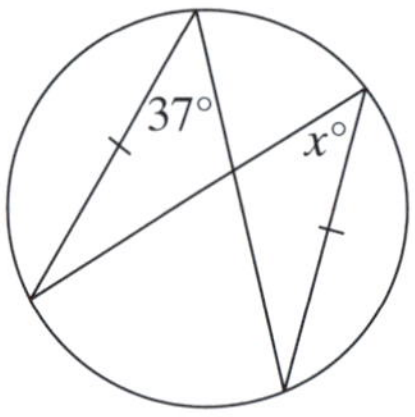

c

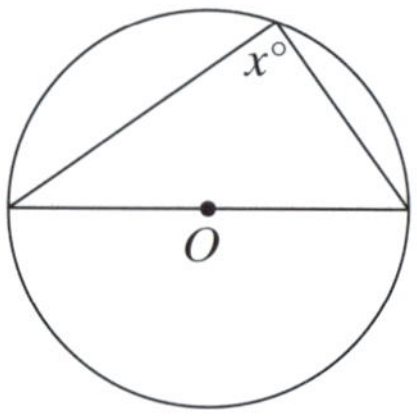

d

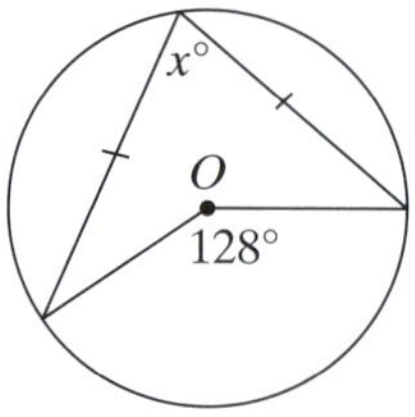

e

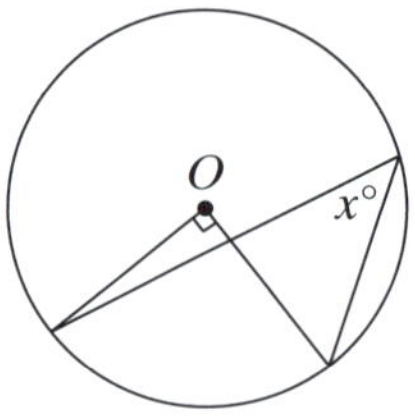

f

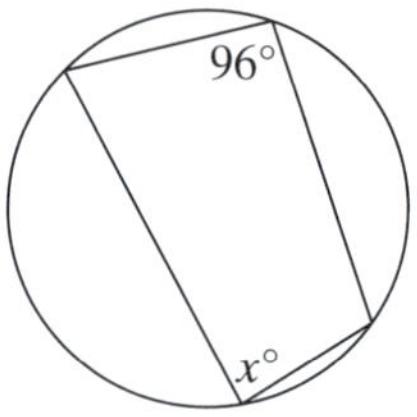

g

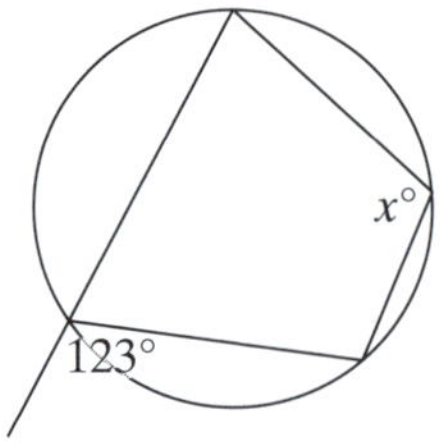

h

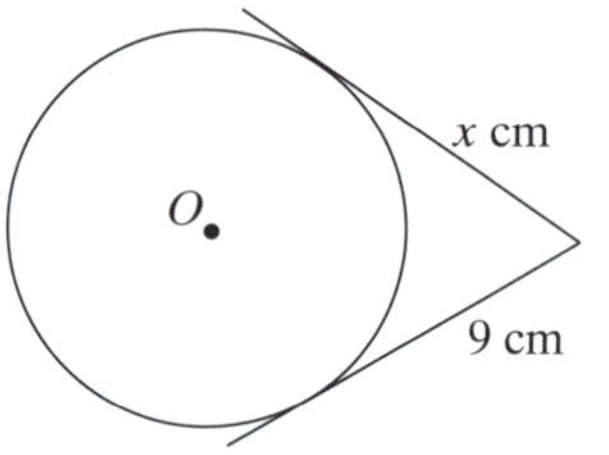

i

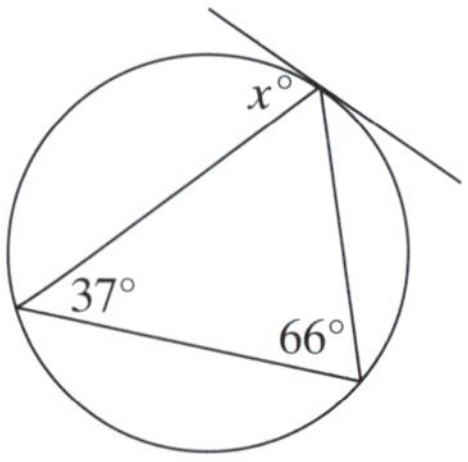

j

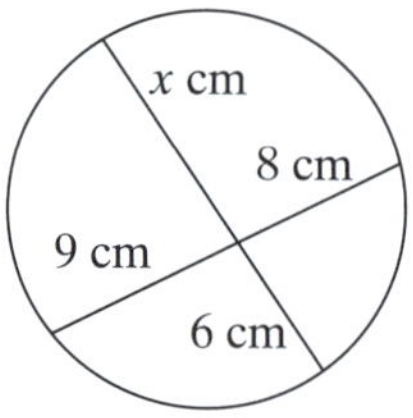

k

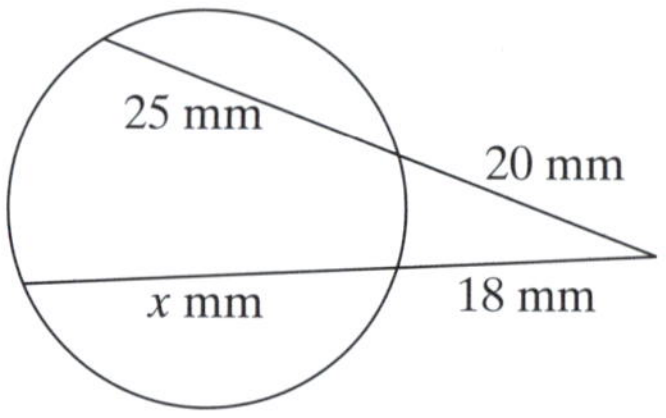

l

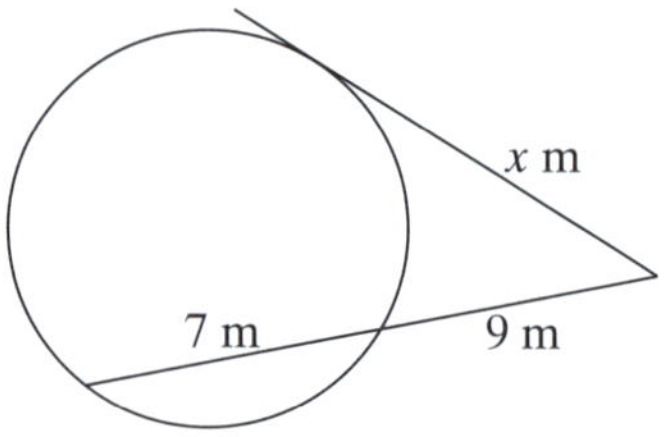

2 Two circles, with centres P and Q, touch at R.
What can you say about points P, Q and R?

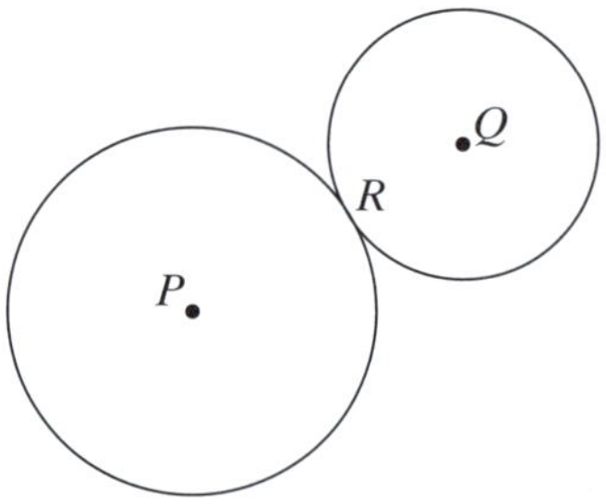

PAGES 156–157

Answers **1** **a** $x = 82$ **b** $x = 37$ **c** $x = 90$ **d** $x = 64$ **e** $x = 45$ **f** $x = 84$ **g** $x = 123$ **h** $x = 9$ **i** $x = 66$ **j** $x = 12$ **k** $x = 32$ **l** $x = 12$
2 They are collinear.

ANGLE PROPERTIES, TANGENTS AND SECANTS

Circle Geometry

25 MINUTES

INTERMEDIATE TEST

Part A Multiple Choice

1 O is the centre of the circle. What is the value of x?

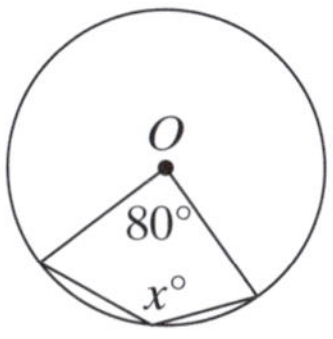

A 100
B 120
C 140
D 160

(1 mark)

2 PQ is parallel to SR. PR and QS intersect at T. $\angle QPT = 40°$. What is the size of $\angle STR$?

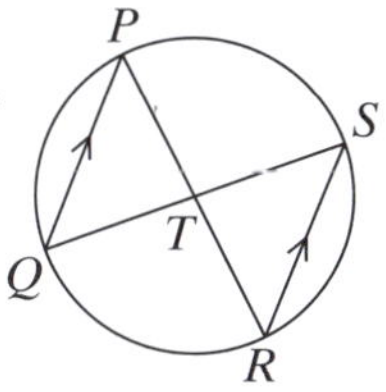

A 80°
B 90°
C 100°
D There is not enough information. (1 mark)

3 $AB = BC$. $\angle BDC = 64°$. What is the size of $\angle ABC$?

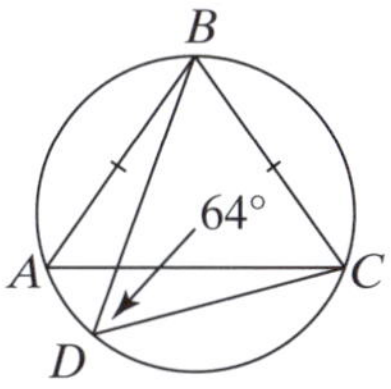

A 52°
B 58°
C 64°
D There is not enough information.

(1 mark)

Part B Written Response

4 PR is a diameter of the circle. $\angle QPR = 21°$. Find the size of $\angle PSQ$, giving reasons.

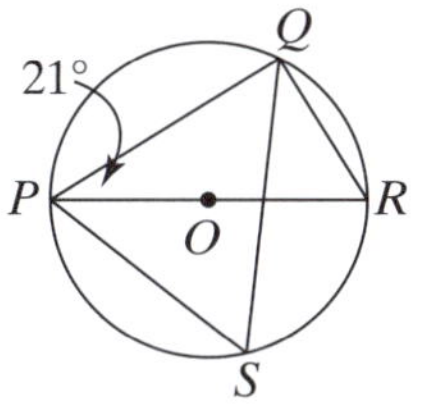

(3 marks)

5 MP is a tangent to the circle. AB is a diameter of the circle and when produced meets the tangent at P. If $MP = 18$ cm and $BP = 15$ cm find the radius of the circle. *Hint 1*

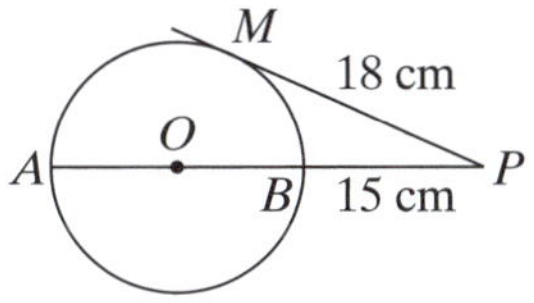

(3 marks)

6 O is the centre of the circle. AB is parallel to CO. AO and BC meet at D. $\angle BAD = 56°$. Find the size of $\angle DCO$, giving reasons.

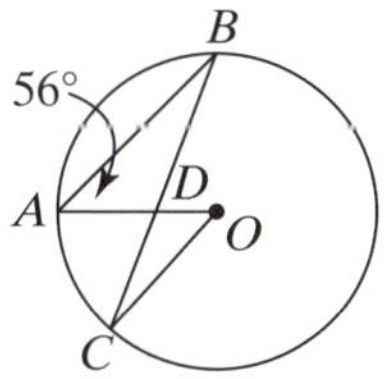

(3 marks)

7 O is the centre of the circle. P, Q, R and S are four points on the circumference of the circle. QS is a diameter of the circle.

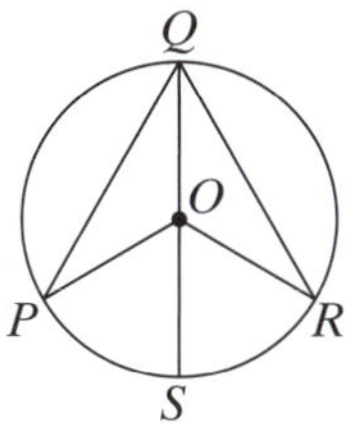

a If $\angle PQO = x°$, find the size of $\angle POS$ in terms of x, giving reasons but without using any of the circle theorems. *Hint 2* (2 marks)

b Briefly explain why it follows that $\angle POR = 2\angle PQR$. (2 marks)

c What is the size of $\angle QPS$? (1 mark)

8 PM and PN are both tangents to the circle, centre O. By proving triangles OMP and ONP congruent, show that $MP = NP$. *Hint 3*

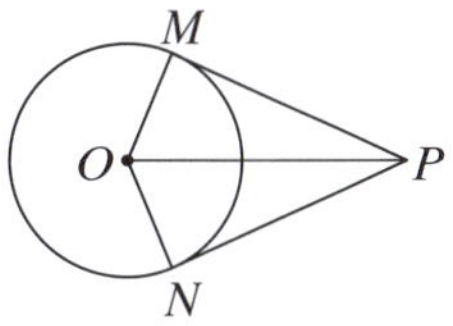

(3 marks)

9 AB and PQ are two chords of the circle that intersect at T.

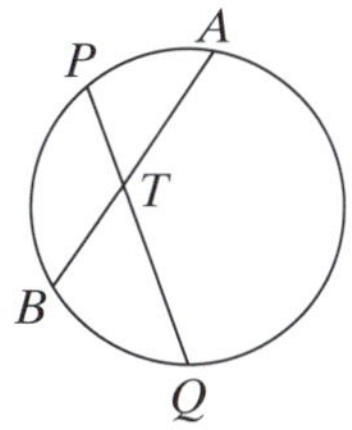

a Show that triangles AQT and PBT are similar and hence show that $AT \times TB = PT \times TQ$. (3 marks)

b If $AB = 14$ cm, $PT = 4$ cm, $QT = 12$ cm and $AT < TB$, find the length of AT. (2 marks)

Hint 1: AP is a secant.
Hint 2: PO and QO are radii of the circle.
Hint 3: The tangent is perpendicular to the radius at the point of contact.

Your Feedback

$\frac{\square}{25} \times 100\% = \square\,\%$

PAGES 157–158
PAGE 182

ANGLE PROPERTIES, TANGENTS AND SECANTS

Circle Geometry

35 MINUTES

ADVANCED TEST

1 P, Q and R are three points on a circle. The tangents at P and R meet at T. $\angle PTR = 48°$.

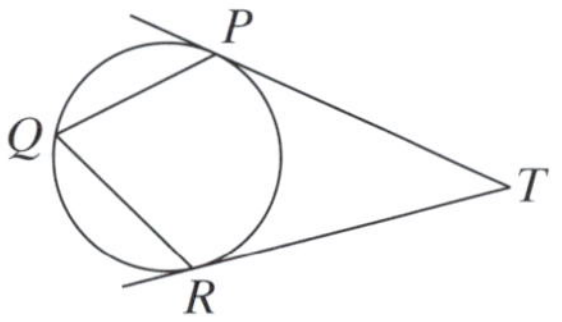

a Find the size of $\angle PQR$, giving reasons. (3 marks)

b Is it possible to draw a circle through P, Q, R and T? Justify your answer. (1 mark)

2 Chord AB is parallel to diameter CD. $\angle ABD = 114°$. Find the size of $\angle ACB$, giving reasons.

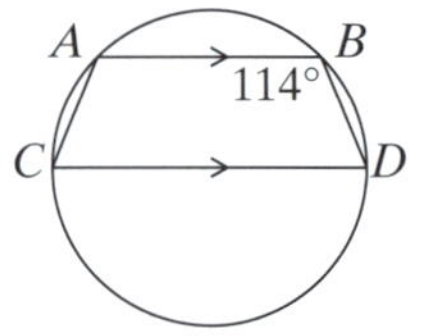

(4 marks)

3 P, Q and R are points on a circle. The tangent at P and the secant through Q and R intersect at T.

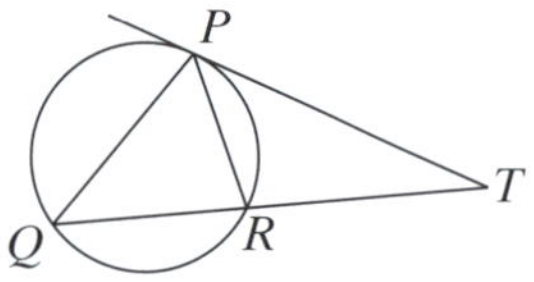

a Give a reason why $\angle TPR = \angle TQP$. (1 mark)

b Explain why triangles TPR and TQP are similar. (1 mark)

c If $PT = 12$ cm and $QR = 7$ cm, find the length of RT. (3 marks)

4 ABC is a tangent to the circle, B is the point of contact. Prove that $\angle QPB = \angle QBA$.

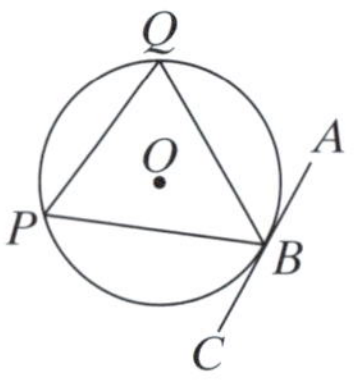

(3 marks)

5 PA and PB are diameters of two circles that intersect at P and at Q.

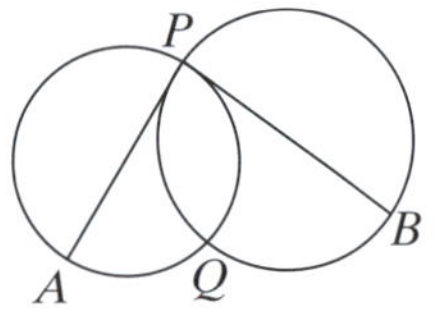

Show that Q lies on AB. (2 marks)

6 AB is a diameter of the circle. Chords AP and BQ produced meet at C.

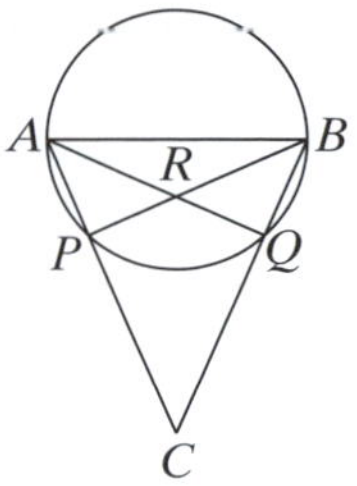

a Find the size of $\angle BPC$. (2 marks)

b Explain why $PRQC$ is a cyclic quadrilateral. (1 mark)

7 Two circles touch at Q. TP is a tangent to the smaller circle and TR is a tangent to the larger circle.

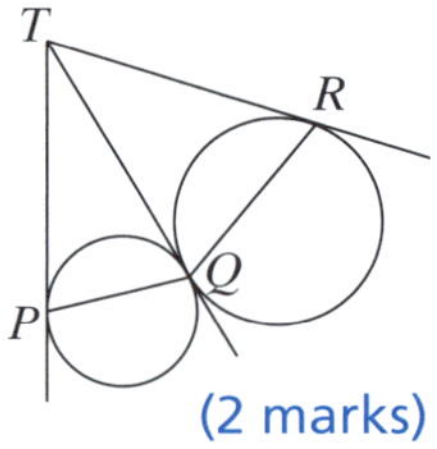

a Show that $TP = TR$. (2 marks)

b If $\angle PTR = 90°$ show that $\angle PQR = 135°$. (4 marks)

8 Two circles intersect at A and C. The tangents to the larger circle at A and at C are perpendicular and meet at a point P on the smaller circle. CB is a chord of the larger circle and intersects the smaller circle at Q.

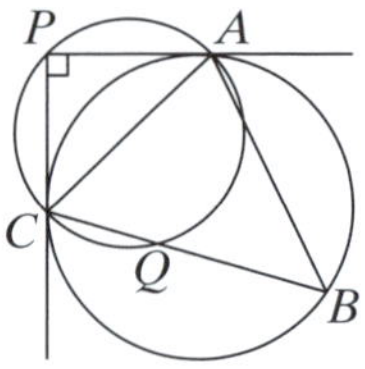

a Show that $\angle ABC = \angle PQC$. (2 marks)

b Show that AQ is perpendicular to CB. (1 mark)

9 The chord BC of a circle is produced to meet the tangent to the circle at A at T. $AT = 15$ cm. $BC = 9$ cm and $CT = 16$ cm.

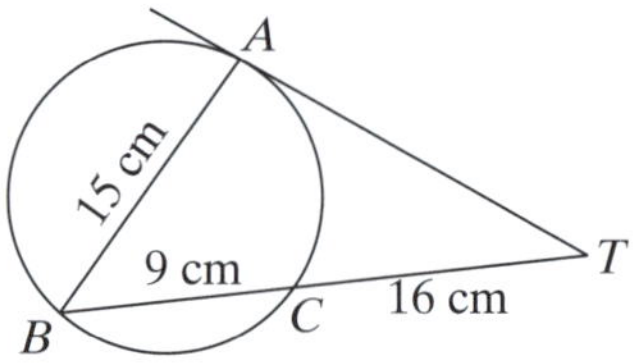

a Show that AB is a diameter of the circle. (3 marks)

b Find the length of AC. (2 marks)

Your Feedback

$\frac{\square}{35} \times 100\% = \square\%$

PAGES 158–159

PAGE 182

PROBLEMS IN TWO AND THREE DIMENSIONS

Pythagoras and Trigonometry

STUDY NOTES

1 **Pythagoras' theorem** states: In any **right-angled triangle** the square on the hypotenuse is equal to the sum of the squares on the other two sides.

2 In a right-angled triangle, the **hypotenuse** is the longest side and opposite the right-angle. The **opposite** and **adjacent sides** depend on which of the other two angles we are considering. The opposite side is opposite the angle and the adjacent side is next to the angle.

3 The **tangent**, **sine** and **cosine ratios** can be defined as ratios of the sides of a right-angled triangle. The ratio for tan is $\frac{\text{opposite}}{\text{adjacent}}$, sin is $\frac{\text{opposite}}{\text{hypotenuse}}$ and cos is $\frac{\text{adjacent}}{\text{hypotenuse}}$.

4 We use a **calculator** to find the **trigonometric ratio** of any angle, or to find the size of the angle for which we know the ratio. Make sure you know how to use your calculator with the trig ratios.

5 There are 16 points of the **compass** that should be recognised. (For example, SSE). **True bearings** are measured from north in a clockwise direction. Bearings are given using three digits, so if the angle from north is 70°, for example, then the bearing will be 070°.

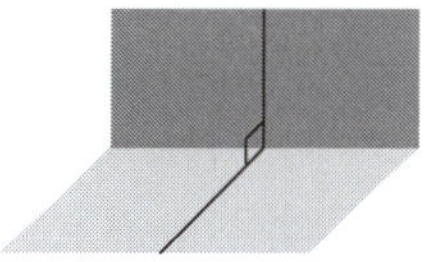

6 If **two planes** (surfaces) are **perpendicular** then any line in one of the planes will be perpendicular to any line in the other plane. So, for example, in the rectangular prism, AD, AE, AH, BC, BF and BG are all perpendicular to AB. Pythagoras' theorem can then be used in any of the right angled triangles containing these perpendicular lines to find other measurements. For example if the dimensions of the rectangular prism are known, then we can use **Pythagoras' theorem** in triangle ABF to find the length of AF and then again in triangle AFG to find the length of AG.

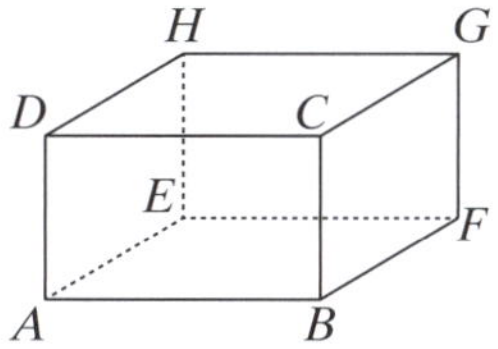

7 When **drawing diagrams** to **illustrate 3D problems** we need to do things differently than for two-dimensional diagrams. For example, if we need to show the height of a pole and a point due south of the pole we can't draw these all in a straight line.

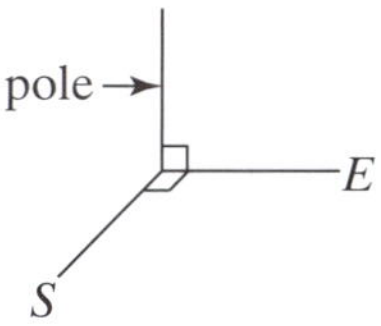

8 **Problems in three dimensions** can often be reduced to finding the lengths of sides or the size of angles in the individual triangles that can be drawn. For example, if we know the height (h m) of AB and the size of angle APB we can then find the length PB using the tan ratio in triangle ABP. Similarly, if we know the size of angle AQB we can again use the tan ratio but in triangle ABQ, to find the length of BQ. We could then use Pythagoras' theorem in triangle PBQ to find the length of PQ.

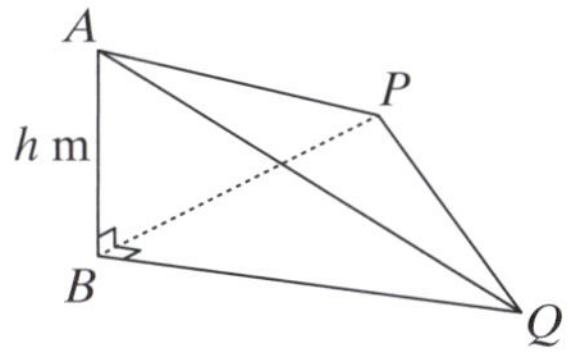

Checklist

Can you:

1 *Use Pythagoras' theorem?* ☐
2 *Identify the hypotenuse, opposite and adjacent sides in a right-angled triangle?* ☐
3 *Know the ratios for tan, sin and cos?* ☐
4 *Use your calculator with the trig ratios and find sides and angles?* ☐
5 *Identify the 16 points of a compass and understand true bearings?* ☐
6 *Draw diagrams and solve problems involving bearings and the trig ratios?* ☐
7 *Draw diagrams and solve three-dimensional problems?* ☐

PROBLEMS IN TWO AND THREE DIMENSIONS

Pythagoras and Trigonometry

SKILLS CHECK

1 Find the value of x. (Give each answer correct to one decimal place.)

a

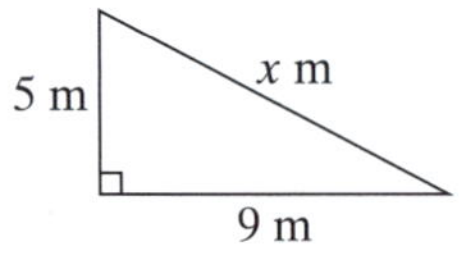

b

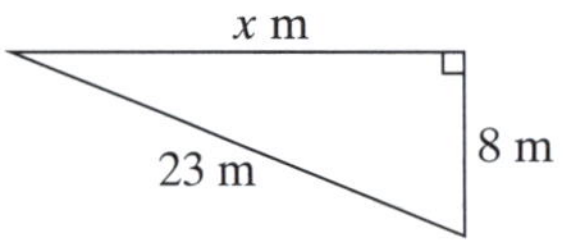

c

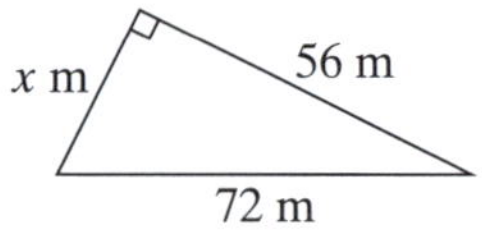

d

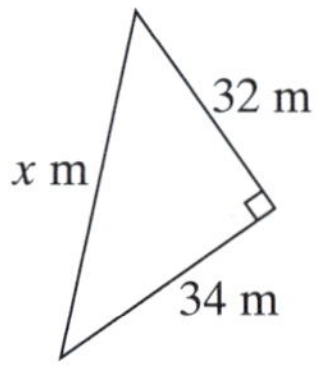

2 Using the diagram, complete:

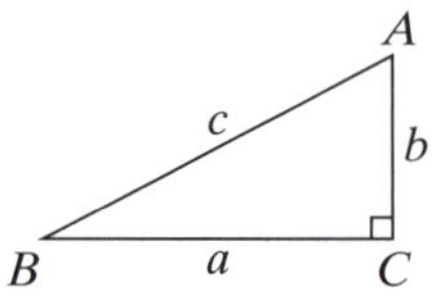

a $\tan A =$ ______ **b** $\sin B =$ ______
c $\cos B =$ ______ **d** $\cos A =$ ______
e $\sin A =$ ______ **f** $\tan B =$ ______

3 Find, correct to three decimal places:
a tan 58° **b** sin 24° **c** cos 72°35′

4 If A is an acute angle and $\sin A = 0.375$ find the size of angle A to the nearest whole degree.

5 Find the size of angle B, to the nearest minute, if:
a $\cos B = 0.2197$
b $\tan B = 3$

6 Find the length of the side marked x:

a

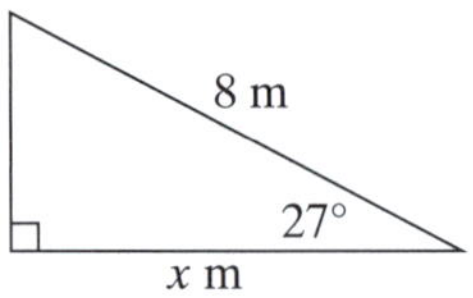

b

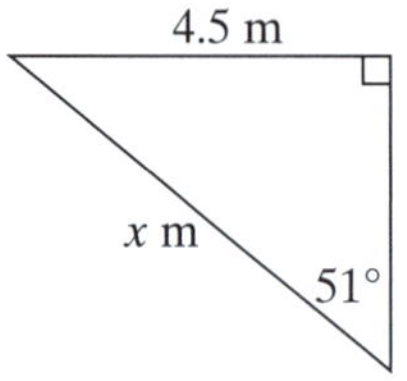

7 Find the size of angle P, to the nearest degree:

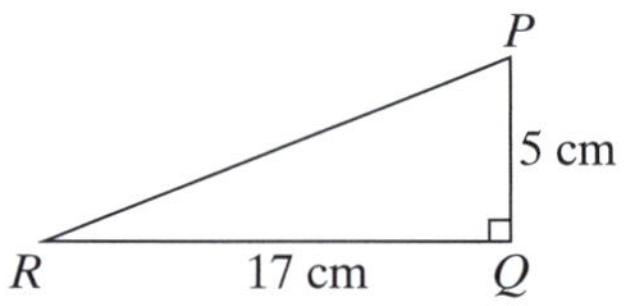

8 A pole of height 3 m casts a shadow on horizontal ground of 2.6 m. Find, to the nearest degree, the angle of elevation of the Sun.

9 In what direction are you travelling if moving on a bearing of:
a 090°? **b** 270°? **c** 180°?
d 045°? **e** 135°?

10 For each diagram, give the bearing of P from Q:

a

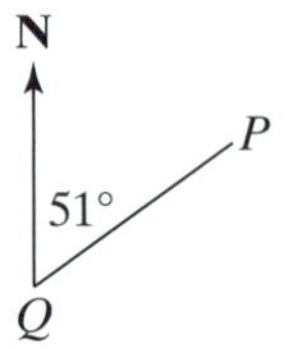

b

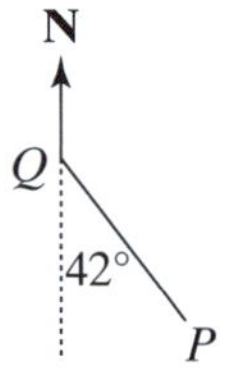

c

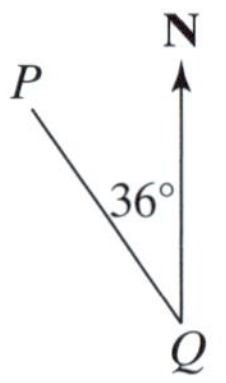

PAGES 159–160

Answers **1 a** 10.3 **b** 21.6 **c** 45.3 **d** 46.7 **2 a** $\frac{a}{b}$ **b** $\frac{b}{c}$ **c** $\frac{a}{c}$ **d** $\frac{b}{c}$ **e** $\frac{a}{c}$ **f** $\frac{b}{a}$ **3 a** 1.600 **b** 0.407 **c** 0.299 **4** 22° **5 a** 77°19′ **b** 71°34′ **6 a** 7.1 m (1 d.p.) **b** 5.79 m (2 d.p.) **7** 74° **8** 49° **9 a** east **b** west **c** south **d** NE **e** SE **10 a** 051° **b** 138° **c** 324°

PROBLEMS IN TWO AND THREE DIMENSIONS

Pythagoras and Trigonometry

25 MINUTES

INTERMEDIATE TEST

Part A Multiple Choice

1 In which diagram does $x = 4\tan 40°$?

A

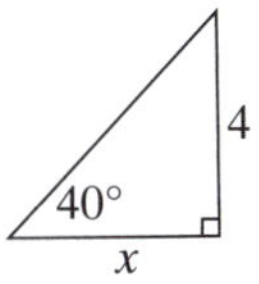

B

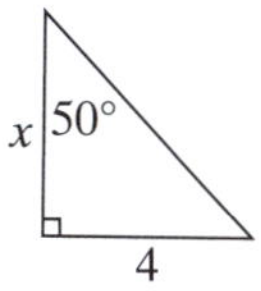

C

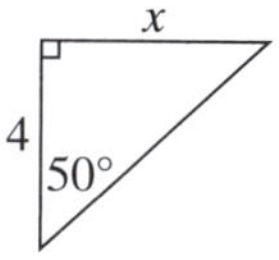

D

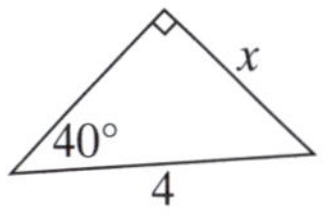

(1 mark)

2 $\cos R = ?$

A $\frac{r}{p}$ **B** $\frac{p}{q}$ **C** $\frac{q}{p}$ **D** $\frac{q}{r}$

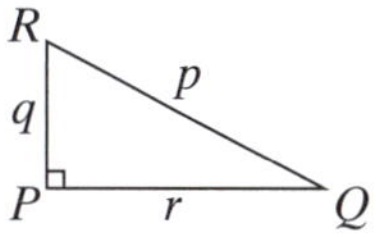

(1 mark)

3 If $NL = 3$, which expression will give LM?

A $3\sin\theta$ **B** $3\cos\theta$ **C** $\frac{3}{\sin\theta}$ **D** $\frac{3}{\cos\theta}$

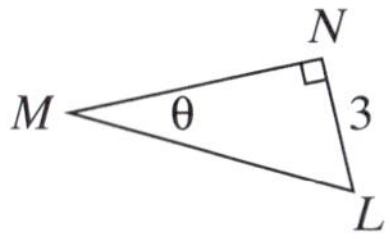

(1 mark)

Part B Written Response

4 A helicopter leaves its base at P and flies 61 km on a bearing of 110° to rescue a patient at Q. From Q it flies on a bearing of 200° to the hospital at R, which is 82 km from Q.

a Draw a diagram to show this information. (1 mark)

b What is the size of $\angle PQR$? *Hint 1* (1 mark)

c Find the size of $\angle RPQ$ to the nearest degree. (2 marks)

d What is the bearing of P from R? (1 mark)

5 A box is 28 cm long, 21 cm wide and 12 cm high. James wants to pack a cane into the box. The cane is 39 cm long and James believes he can place it diagonally as shown in the diagram. Will the cane fit in the box? Justify the answer with appropriate calculations. *Hint 2*

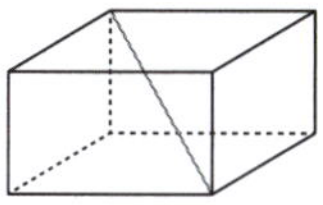

(3 marks)

6 A square-based pyramid has perpendicular height 28 cm. Each side of the base is 32 cm long. Find the angle between the base of the pyramid and any of the triangular faces. (2 marks)

7 B is 6 km due east of A. D is due south of A and C is due south of B. The bearing of C from D is 120°. The distance from B to D is 12 km.

a Show this information on a diagram. (1 mark

b Find the size of $\angle BDC$. *Hint 3* (2 marks)

c Find the size of $\angle DBC$. (1 mark

d Find the distance from B to C to the nearest 100 metres. (2 marks)

e On what bearing should a helicopter fly from B to D? (1 mark

8 A pole (AB) standing on level ground, is 37 m tall. From D, a point due south of B, the angle of elevation of A is 45°. From C, a point due east of B, the angle of elevation of A is 28°.

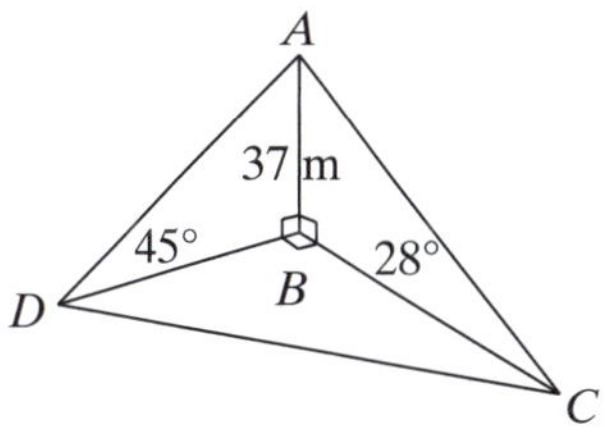

a What is the length of BD? *Hint 4* (1 mark

b Find the distance from B to C. *Hint 5* (2 marks)

c Find the distance from C to D. (2 marks)

Hint 1: There is no need to use trigonometry.
Hint 2: First find the length of the diagonal of the box base.
Hint 3: First find the size of $\angle ADB$.
Hint 4: What type of triangle is ΔABD?
Hint 5: Use ΔABC.

Your Feedback

$\frac{\square}{25} \times 100\% = \square\%$

PAGES 160–161

PAGE 182

PROBLEMS IN TWO AND THREE DIMENSIONS

Pythagoras and Trigonometry

30 MINUTES

ADVANCED TEST

1 The diagram shows a right pyramid.
$AE = 24$ cm, $AB = 25$ cm and $ED = 10$ cm.

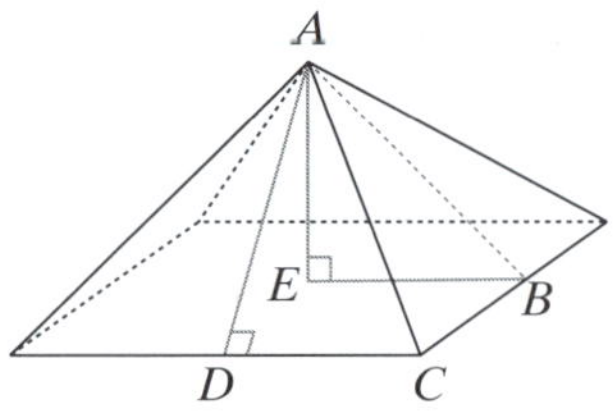

- **a** Find EB. (1 mark)
- **b** Find AD. (1 mark)
- **c** Find AC in simplest surd form. (2 marks)

2 Point P is 80 m due south of the base of a tower which is 125 m tall. Point Q is due east of P. From Q, the angle of elevation of the top of the tower is 31°.

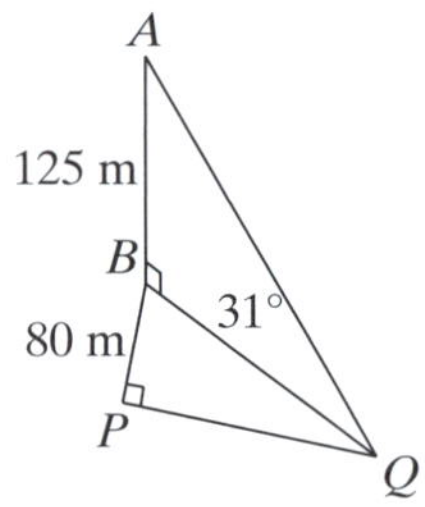

- **a** Find the length of BQ. (2 marks)
- **b** Find the length of PQ. (1 mark)
- **c** Find the bearing of B from Q. (3 marks)

3 A tree is 78 m tall. P is a point due east of the base, B, of the tree and Q is a point due south of B (all on level ground). From P the angle of elevation of the top of the tree, A, is 35° and from Q the angle of elevation of A is 22°.

- **a** Show this information on a diagram. (1 mark)
- **b** Find the distance from B to P. Give the answer correct to one decimal place. (2 marks)
- **c** Find the distance from B to Q. Give the answer correct to one decimal place. (2 marks)
- **d** Find the distance from P to Q. Give the answer correct to the nearest metre. (1 mark)

4 In the diagram, triangles ADC, CDB and ADB are all right-angled at D. $AB = 900$ m. $CD = h$ m.

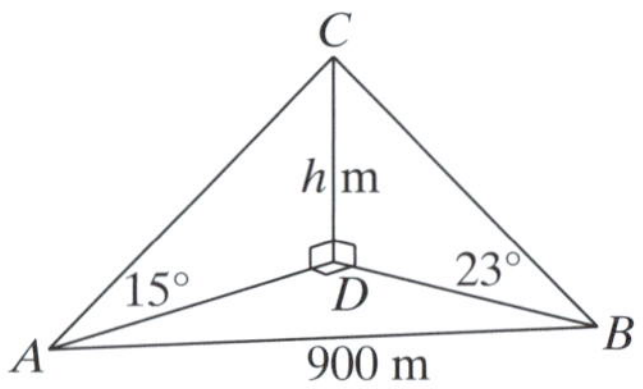

- **a** Show that $BD = \dfrac{h}{\tan 23°}$ (1 mark)
- **b** Show that $AD = \dfrac{h}{\tan 15°}$ (1 mark)
- **c** Hence find the value of h. (3 marks)

5 As shown in the diagram, two wires have been attached to a pole at different heights. The base of the lower wire is 27.5 m from the base of the pole, on level ground.

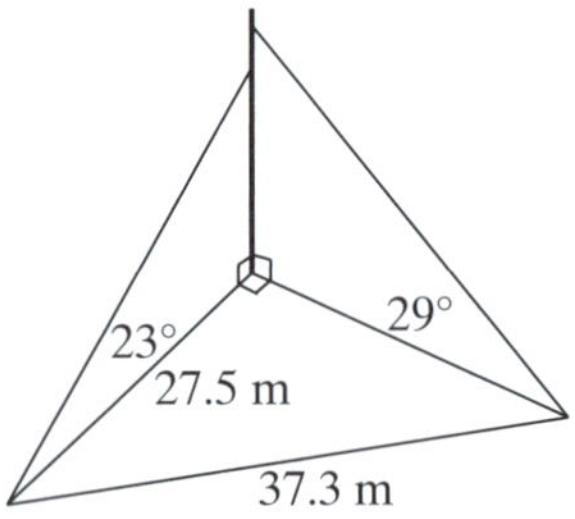

- **a** How long is the lower wire? (2 marks)
- **b** How high above the ground is the lower wire attached to the pole? (2 marks)
- **c** How far is the base of the higher wire from the base of the pole? (1 mark)
- **d** How much higher than the lower wire is the higher wire attached to the pole? (2 marks)
- **e** Which wire is longer and by how much? Give the answer to the nearest centimetre. (2 marks)

Your Feedback

$\dfrac{\square}{30} \times 100\% = \square\,\%$

PAGES 161–162

PAGE 182

COMPLEMENTARY RATIOS, EXACT VALUES & THE UNIT CIRCLE

Pythagoras and Trigonomety

STUDY NOTES

1 **Complementary angles** add to 90°. The complement of 40° is 50°, because 40° + 50° = 90°.

2 Sine and cosine are **complementary ratios** or co-ratios.

$\sin A = \cos(90° - A)$ and $\cos A = \sin(90° - A)$

For example, from the diagram, $\sin A = \cos B = \dfrac{a}{c}$ and $\sin B = \cos A = \dfrac{b}{c}$

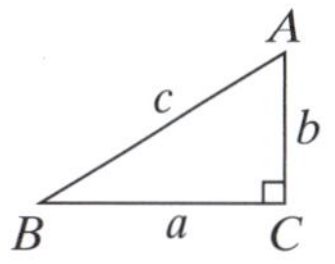

3 By drawing an **isosceles right-angled triangle** with equal sides 1 unit in length, the **exact ratios for angles of 45°** can be found. By Pythagoras' theorem the hypotenuse will be $\sqrt{2}$ units.

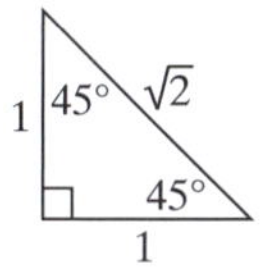

$\sin 45° = \dfrac{1}{\sqrt{2}}$ $\cos 45° = \dfrac{1}{\sqrt{2}}$ $\tan 45° = 1$

4 By drawing an **equilateral triangle**, with equal sides 2 units in length, the **exact ratios for angles of 30° and 60°** can be found. Drawing in the perpendicular from the base to the other vertex, divides the triangle into two parts. The base of each part will be 1 unit in length and the perpendicular height, by Pythagoras' theorem, will be $\sqrt{3}$ units.

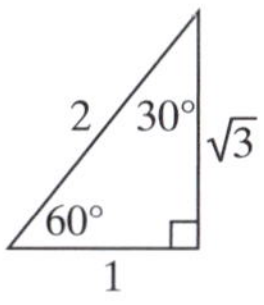

$\sin 30° = \dfrac{1}{2}$, $\cos 30° = \dfrac{\sqrt{3}}{2}$, $\tan 30° = \dfrac{1}{\sqrt{3}}$, $\sin 60° = \dfrac{\sqrt{3}}{2}$, $\cos 60° = \dfrac{1}{2}$, $\tan 60° = \sqrt{3}$

5 The **unit circle** is a circle of radius 1 unit. This circle has equation $x^2 + y^2 = 1$. The **trigonometric ratios** can be defined in terms of the rotation, in an anticlockwise direction, from the positive x-axis.

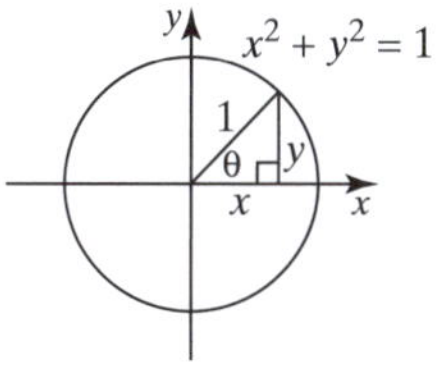

$\sin\theta = \dfrac{y}{1} = y$ $\cos\theta = \dfrac{x}{1} = x$ $\tan\theta = \dfrac{\sin\theta}{\cos\theta} = \dfrac{y}{x}$

Using these definitions the trig ratios can be used with **angles of any magnitude**.

6 The coordinate axes divide the unit circle into **4 quadrants**, numbered from 1 to 4. In the second quadrant θ is obtuse and $\sin\theta$ is positive but $\cos\theta$ and $\tan\theta$ are both negative (because y is positive but x is negative). In the third quadrant, x and y are both negative, so $\sin\theta$ and $\cos\theta$ are both negative but $\tan\theta$ is positive. In the fourth quadrant, x is positive so $\cos\theta$ is positive. Because y is negative in the fourth quadrant $\sin\theta$ and $\tan\theta$ are both negative.

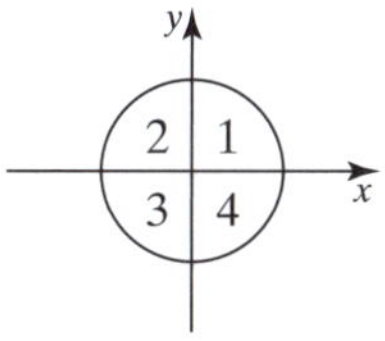

7 When **finding an angle** given a trig ratio, or **solving an equation** involving trig ratios, more than one answer may be possible. For example, if $\sin A = 0.5$ then A could be 30° or 150°. The calculator will give the acute angle, ($\sin^{-1} 0.5 = 30°$). The obtuse angle can be found by subtracting the acute value from 180°. (The obtuse angle is $(180 - 30)° = 150°$.)

8 The **graphs** of $y = \sin x°$, $y = \cos x°$ and $y = \tan x°$ are **periodic** (they repeat after a certain interval).

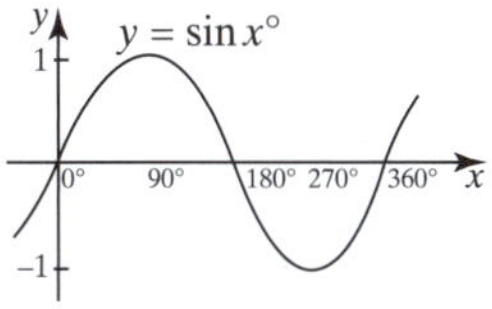

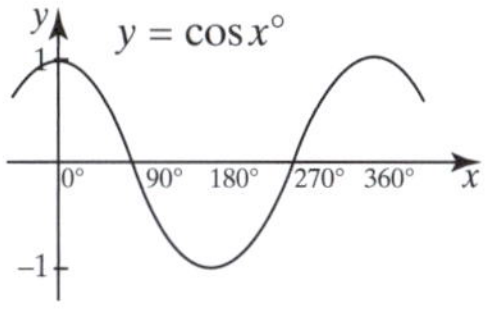

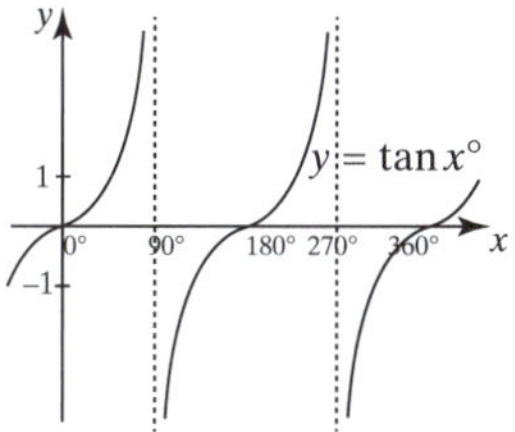

Checklist

Can you:

1 *Understand what is meant by complementary angles and complementary ratios?* ☐

2 *Understand the exact ratios for angles of 30°, 45° and 60°?* ☐

3 *Understand the definitions of the trig ratios using the unit circle and the use of quadrants?* ☐

4 *Solve simple trig equations?* ☐

5 *Draw the graphs of $y = \sin x°$, $y = \cos x°$ and $y = \tan x°$ for values of x between 0° and 360°?* ☐

COMPLEMENTARY RATIOS, EXACT VALUES & THE UNIT CIRCLE

Pythagoras and Trigonometry

SKILLS CHECK

1 Find the size of the angle marked θ:

a

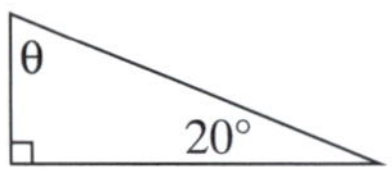

b

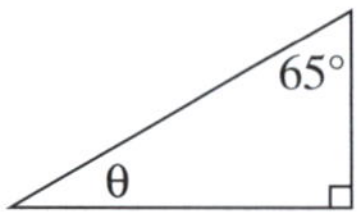

c

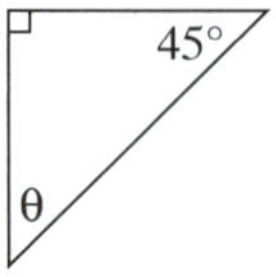

2 Find the complement of:

a 20° **b** 75° **c** 42°

3 Write down the value to three decimal places of:

a $\cos 20°$ **b** $\sin 70°$ **c** $\sin 38°$ **d** $\cos 52°$

4 Complete:

a $\sin 30° = \cos$ ____ **b** $\cos 18° = \sin$ ____ **c** $\sin 45° = \cos$ ____

5 If $\sin\theta = 0.6$ and $\cos\theta = 0.8$, find the value of $\tan\theta$.

6 Write down the exact value of:

a $\sin 30°$ **b** $\tan 45°$ **c** $\cos 60°$

d $\sin 60°$ **e** $\tan 30°$ **f** $\cos 45°$

g $\tan 60°$ **h** $\sin 45°$ **i** $\cos 30°$

7 Determine whether each statement is true or false:

a $\sin 130° = \sin 50°$ **b** $\cos 110° = \cos 70°$ **c** $\tan 175° = -\tan 5°$

8 Find, correct to four decimal places:

a $\tan 102°$ **b** $\sin 156°$ **c** $\cos 173°$

9 If A is an obtuse angle, find its size to the nearest degree, if:

a $\sin A = 0.237$ **b** $\cos A = -0.5$ **c** $\tan A = -1.2345$

10 If $0° \le \theta \le 90°$, find the value of θ if:

a $\cos\theta = \frac{1}{\sqrt{2}}$ **b** $\tan\theta - 1 = 0$ **c** $2\sin\theta = \sqrt{3}$

11 Find two possible values of θ, rounding to the nearest minute if necessary, if:

a $\sin\theta = 0.5$ **b** $\sin\theta = 0.7654$

PAGE 163

Answers **1 a** 70° **b** 25° **c** 45° **2 a** 70° **b** 15° **c** 48° **3 a** 0.940 **b** 0.940 **c** 0.616 **d** 0.616 **4 a** 60° **b** 72° **c** 45° **5** 0.75 **6 a** $\frac{1}{2}$ **b** 1 **c** $\frac{1}{2}$ **d** $\frac{\sqrt{3}}{2}$ **e** $\frac{1}{\sqrt{3}}$ **f** $\frac{1}{\sqrt{2}}$ **g** $\sqrt{3}$ **h** $\frac{1}{\sqrt{2}}$ **i** $\frac{\sqrt{3}}{2}$ **7 a** true **b** false **c** true **8 a** −4.7046 **b** 0.4067 **c** −0.9925 **9 a** 166° **b** 120° **c** 129° **10 a** 45° **b** 45° **c** 60° **11 a** 30°, 150° **b** 49°57′, 130°3′

COMPLEMENTARY RATIOS, EXACT VALUES & THE UNIT CIRCLE

Pythagoras and Trigonometry

25 MINUTES

INTERMEDIATE TEST

Part A Multiple Choice

1 $\cos 60° = ?$

A $\frac{1}{2}$ **B** $\frac{1}{\sqrt{2}}$ **C** $\frac{1}{\sqrt{3}}$ **D** $\frac{\sqrt{3}}{2}$ (1 mark)

2 $\sin 55° = ?$

Which statement is correct? I $\sin 125°$ II $\cos 35°$

A I only **B** II only **C** both I and II **D** neither I nor II (1 mark)

3 $\sin\theta = 0.5$

Which statement is correct? I $\theta = 30°$ II $\theta = 150°$

A I only **B** II only **C** either I or II **D** neither I nor II (1 mark)

4 $\tan 150° = ?$

A $\sqrt{3}$ **B** $\frac{1}{\sqrt{3}}$ **C** $-\sqrt{3}$ **D** $-\frac{1}{\sqrt{3}}$ (1 mark)

5 If $\sin\theta = 0.66$ and $\cos\theta = 0.75$, then $\tan\theta = ?$ *Hint 1*

A 0.41 **B** 0.495 **C** 0.88 **D** 1.41 (1 mark)

6 In which of the following quadrants are $\sin\theta$ and $\cos\theta$ both negative?

A 1st **B** 2nd **C** 3rd **D** 4th (1 mark)

Part B Written Response

7 Find the exact length of the side BC. *Hint 2* (3 marks)

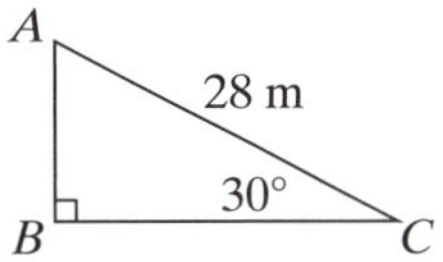

8 If $\sin A = \frac{7}{9}$ and $\angle A$ is obtuse find its size, to the nearest degree. *Hint 3* (3 marks)

9 **a** On the same diagram, draw the graphs of $y = \sin x°$ and $y = \cos x°$ for $0° \le x° \le 180°$. (4 marks)

b For what value(s) of x, $0° \le x \le 180°$, does $\sin x = \cos x$? *Hint 4* (1 mark)

10 **a** Find the length of AC. (2 marks)

b Show that $AB = 12\sqrt{2}$ cm. *Hint 5* (2 marks)

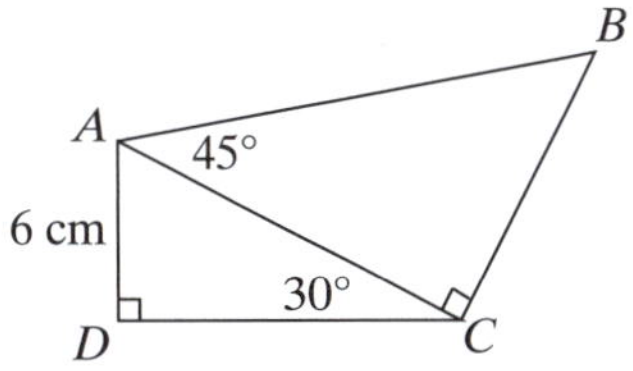

11 Solve $\sqrt{2}\sin x° = 1$ for $0° \le x° \le 180°$. *Hint 6* (2 marks)

12 The line $y = 5 - x$ makes an angle of θ with the positive direction of the x-axis. Find the size of θ. *Hint 7* (2 marks)

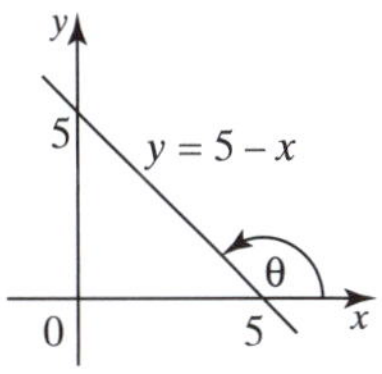

Hint 1: There is no need to find the value of θ.
Hint 2: Exact does not mean a decimal approximation.
Hint 3: First find the size of the acute angle.
Hint 4: There is no need to try and solve an equation. Read the value from the graph.
Hint 5: Use the answer from part **a** in your calculation.
Hint 6: There will be two answers.
Hint 7: The gradient $m = \tan\theta$.

Your Feedback

$\frac{\square}{25} \times 100\% = \square\ \%$

PAGES 163–164

PAGE 182

COMPLEMENTARY RATIOS, EXACT VALUES & THE UNIT CIRCLE

Pythagoras and Trigonometry

30 MINUTES

ADVANCED TEST

1 Find the exact value of cos 210°. (2 marks)

2 If $\sin A = 0.654$, find the size of $\angle A$ to the nearest minute, if:

- **a** $\angle A$ is acute (1 mark)
- **b** $\angle A$ is obtuse (1 mark)

3 If θ is acute and $\cos\theta = x$ and $\sin\theta = y$, write down the value of:

- **a** $\sin(180° - \theta)$ (1 mark)
- **b** $\cos(180° - \theta)$ (1 mark)
- **c** $\tan\theta$ (1 mark)
- **d** $\cos(90° - \theta)$ (1 mark)
- **e** $\sin(360° - \theta)$ (1 mark)
- **f** $\cos(180° + \theta)$ (1 mark)

4 If $\sin\theta = \dfrac{2}{\sqrt{7}}$ and θ is acute, find $\tan\theta$. (2 marks)

5 Find all values of θ, $0° \le \theta \le 360°$, for which $3\cos\theta = \cos\theta - 1$. (3 marks)

6 What is the approximate gradient of the line *m* that makes an angle of 76° with the positive direction of the *x*-axis? (1 mark)

7 Find the value of θ, to the nearest degree, if θ is obtuse and $\tan\theta = -\dfrac{1}{4}$. (1 mark)

8 $AB = 1$ m. $AC = DC$. $\angle ADC = 15°$.

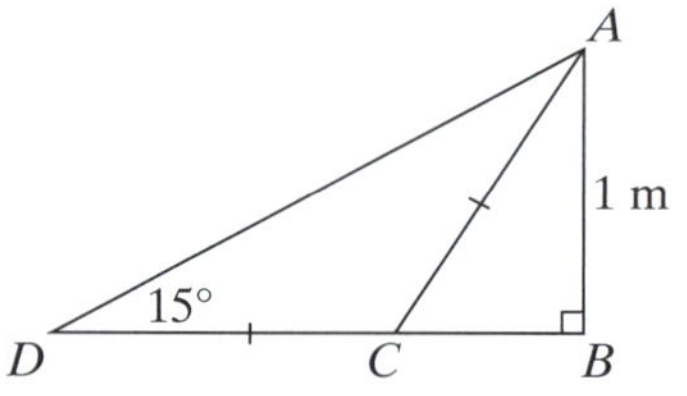

- **a** Find the size of $\angle ACB$. (1 mark)
- **b** Find the length of AC. (1 mark)
- **c** Find the exact length of CB. (1 mark)
- **d** What is the exact length of DB? (1 mark)
- **e** Hence show that $\tan 15° = 2 - \sqrt{3}$. (2 marks)
- **f** Find the exact value of $\tan 75°$. (1 mark)

9 The diagram shows the graph of $y = \sin 2x°$ for $0° \le x° \le 360°$.

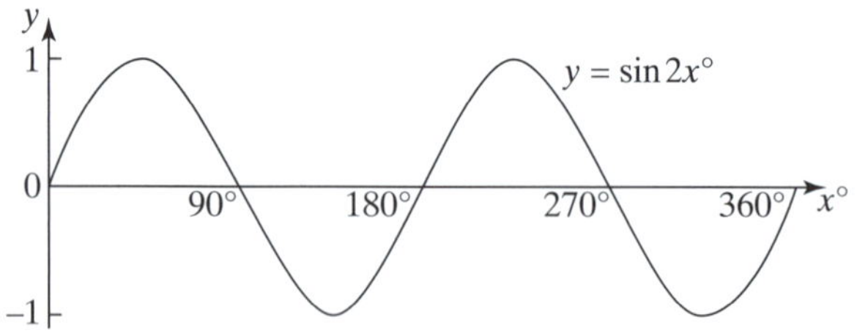

- **a** Show that, when $x = 60$, $\sin 2x° = \sin x°$. (1 mark)
- **b** Copy the diagram and sketch the graph of $y = \sin x°$ on the same diagram. (3 marks)
- **c** Write down all the values of x for which $\sin 2x° = \sin x°$, $0° \le x° \le 360°$. (2 marks)

Your Feedback

$\dfrac{\square}{30} \times 100\% = \square\%$

PAGES 164–165
PAGE 182

SINE RULE, COSINE RULE AND AREA OF A TRIANGLE

Trigonometry

STUDY NOTES

1 The ratios for sin, cos and tan have been used in reference to angles of right angled triangles, but it is also possible to find these **trigonometric ratios for angles larger than 90°**.

2 When **naming the angles and sides** of a triangle, upper case letters (capitals) are used for angles and lower case letters for sides. The sides are named for the angles they are opposite. i.e. side a is opposite angle A and angle B is opposite side b.

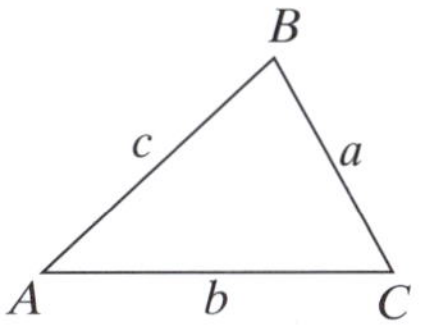

3 The **sine rule** can be used in any triangle.

The sine rule states: $\dfrac{a}{\sin A} = \dfrac{b}{\sin B} = \dfrac{c}{\sin C}$

The sine rule links all three sides and all three angles of a triangle, but in practice we use only two sides and two angles. If **finding a side**, given two angles and one side, we use $\dfrac{a}{\sin A} = \dfrac{b}{\sin B}$. If **finding an angle**, given two sides and one angle, we use $\dfrac{\sin A}{a} = \dfrac{\sin B}{b}$. The angles must match the sides. If two angles of a triangle are known the third can easily be found, because the angle sum of any triangle is 180°. It may be necessary to find an angle before using the sine rule, or to use the sine rule to find a different angle to the one we require.

4 When using the sine rule to find an angle there may be **two possible answers**, an **acute angle** (less than 90°) and an **obtuse angle** (between 90° and 180°). The acute angle will always be physically possible but the obtuse angle may not.

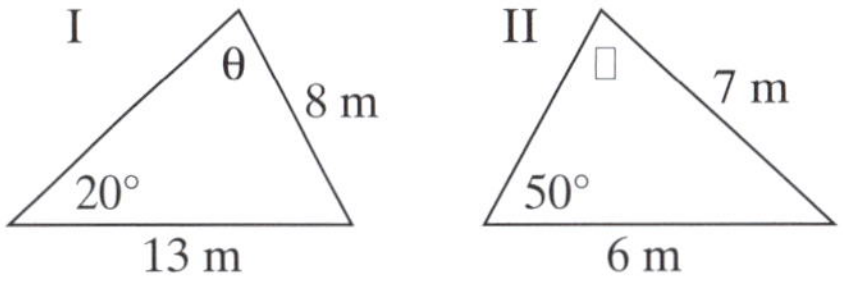

For example, in diagram I using the sine rule to find θ gives $\sin\theta = 0.555\,782\,732\ldots$ and hence $\theta = 34°$ or $146°$ (to the nearest degree), both of which are possible. In diagram II, $\sin\phi = 0.656\,609\,522\ldots$ and hence $\phi = 41°$ or $139°$. But if one angle of the triangle is 50°, the other two angles must add to 130° and so $\phi < 130°$. An angle of 139° is not possible. Unless told otherwise, always consider an obtuse angle as well as an acute angle.

5 The **cosine rule** can be used to **find a side** of a triangle when two sides and the included angle are known.

The cosine rule states: $a^2 = b^2 + c^2 - 2bc\cos A$

Applying this rule you must follow the correct order of operations. Take special care when entering the numbers into your calculator. Do not put an equals sign in the wrong place. Don't forget that the formula gives the value of a^2 and the square root must be taken to find the answer.

6 The **cosine rule** can also be used to **find an angle** when all three sides are known i.e. $\cos A = \dfrac{b^2 + c^2 - a^2}{2bc}$.

This formula is just a rearrangement of the one above. Again care must be taken to enter the numbers correctly.

7 The **area of any triangle** can be found, given two sides and the included angle, using the formula:

$$\text{Area} = \frac{1}{2}ab\sin C$$

Checklist

Can you:

1 *Name the angles and sides of a triangle?* ☐

2 *Use the sine rule to find sides and angles, remembering to consider an obtuse angle as well as an acute angle when finding an angle?* ☐

3 *Use the cosine rule to find sides and angles?* ☐

4 *Find the area of any triangle, given two sides and the included angle?* ☐

SINE RULE, COSINE RULE AND AREA OF A TRIANGLE

Trigonometry

SKILLS CHECK

1 Write down the length of the given side:

a p
b q
c r

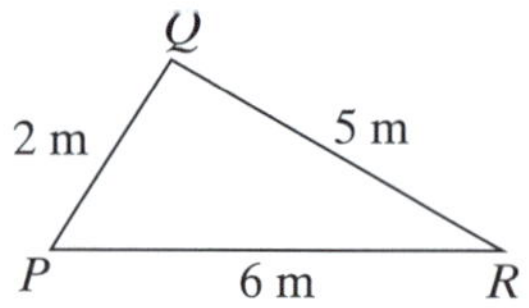

2 What is the size of:

a angle A?
b angle B?
c angle C?

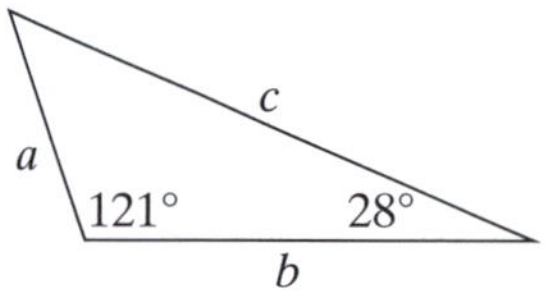

3 Given $\dfrac{a}{\sin A} = \dfrac{b}{\sin B}$ find a, correct to one decimal place, if $b = 5$, $\angle A = 70°$ and $\angle B = 56°$.

4 If $\sin C = 0.376$ find the size of $\angle C$, to the nearest minute, if $\angle C$ is obtuse.

5 If B is an acute angle, find its size given that $\dfrac{\sin B}{b} = \dfrac{\sin C}{c}$ and $b = 4.2$, $c = 5.7$ and $\angle C = 64°$.

6 Given $c^2 = a^2 + b^2 - 2ab\cos C$ find the value of c, correct to two decimal places, if:

a $a = 3.7$, $b = 2.8$ and $\angle C = 60°$
b $a = 65$, $b = 21.1$ and $\angle C = 146°$

7 If $\cos A = \dfrac{b^2 + c^2 - a^2}{2bc}$, find the size of angle A if $a = 7$, $b = 5$ and $c = 8$.

8 Which rule (sine rule or cosine rule) should be used to find the length of PQ?

a

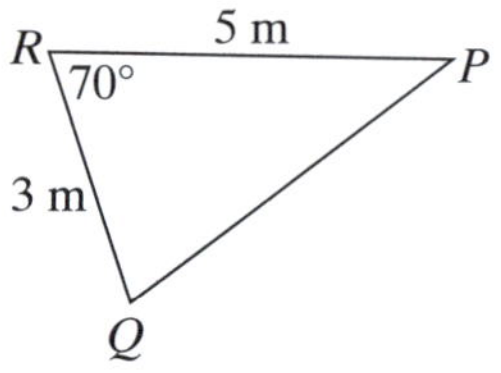

b

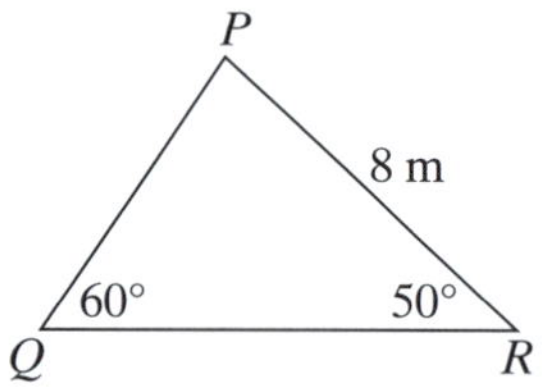

9 Given area $= \dfrac{1}{2}ab\sin C$ find the area of a triangle where $a = 8$, $b = 4$ and $\angle C = 30°$.

PAGES 165–166

Answers **1 a** 5 m **b** 6 m **c** 2 m **2 a** 28° **b** 31° **c** 121° **3** 5.7 **4** 157° 55' **5** 41° **6 a** 3.34 **b** 83.33 **7** 60° **8 a** cosine rule **b** sine rule **9** 8 units²

SINE RULE, COSINE RULE AND AREA OF A TRIANGLE

Trigonometry

25 MINUTES

INTERMEDIATE TEST

Part A Multiple Choice

1

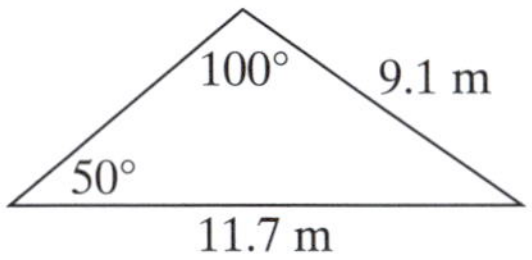

The area of this triangle is closest to:

A 27 m^2 **B** 52 m^2

C 41 m^2 **D** 46 m^2 (1 mark)

2

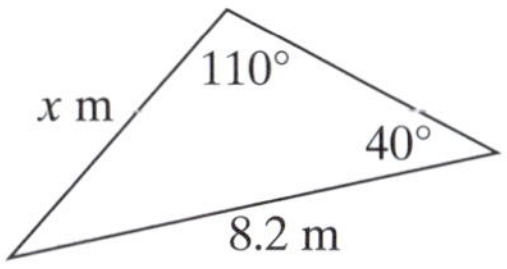

To one decimal place, x = ?

A 5.0 **B** 5.6

C 6.4 **D** 12.0 (1 mark)

3

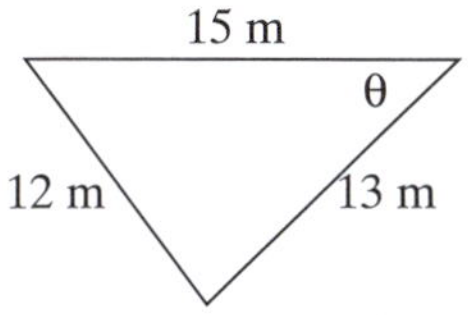

To the nearest degree, θ = ?

A 50° **B** 56°

C 44° **D** 74° (1 mark)

Part B Written Response

4 Use the sine rule to find the length of AB.

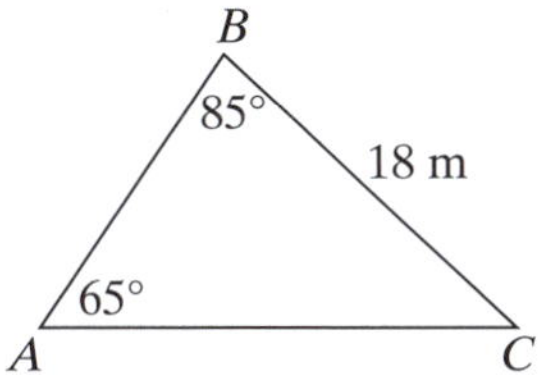

Give the answer in metres, correct to one decimal place.

Hint 1 (3 marks)

5 Use the cosine rule to find the size of the largest angle (to the nearest whole degree) of this triangle. *Hint 2*

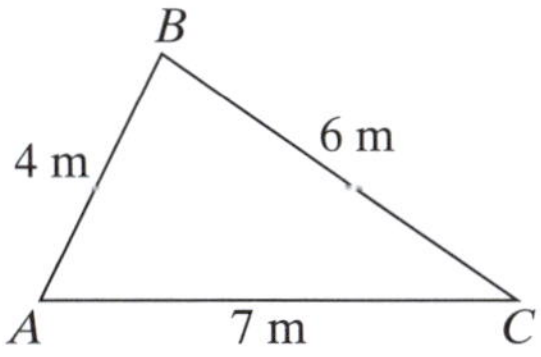

(3 marks)

6

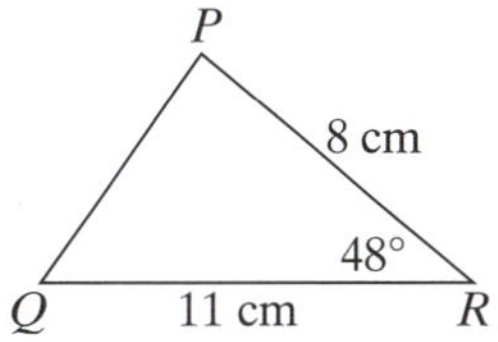

a Find the length of PQ to the nearest mm. *Hint 3* (3 marks)

b Find the area of triangle PQR. (2 marks)

7 Find the size of θ, to the nearest degree. *Hint 4*

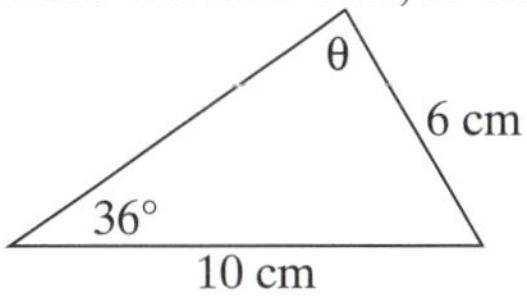

(3 marks)

8 A rescue helicopter has its base 85 km due North of a major hospital. The helicopter receives a call to retrieve a patient and flies 175 km from its base on a bearing of 072°. The helicopter flies the patient to the hospital and then returns to its base.

a Draw a diagram to show the path of the helicopter. (1 mark)

b How far did the helicopter fly altogether? *Hint 5* (4 marks)

9 A block of land is triangular in shape, as shown in the diagram.

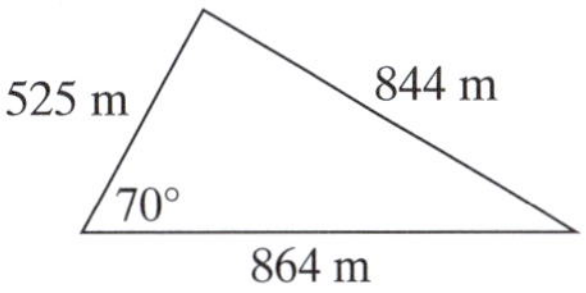

Find the area of the land, giving the answer in hectares, correct to 2 decimal places. *Hint 6* (3 marks)

Hint 1: You will need to find the size of ∠ACB first.
Hint 2: The largest angle will be opposite the largest side.
Hint 3: Use the cosine rule.
Hint 4: There are two possible answers.
Hint 5: First use the cosine rule to find the distance from the patient to the hospital.
Hint 6: You will not need to use all the given information.

Your Feedback

$\frac{\square}{25} \times 100\% = \square\%$

PAGES 166–167

PAGE 182

SINE RULE, COSINE RULE AND AREA OF A TRIANGLE

Trigonometry

40 MINUTES

ADVANCED TEST

1

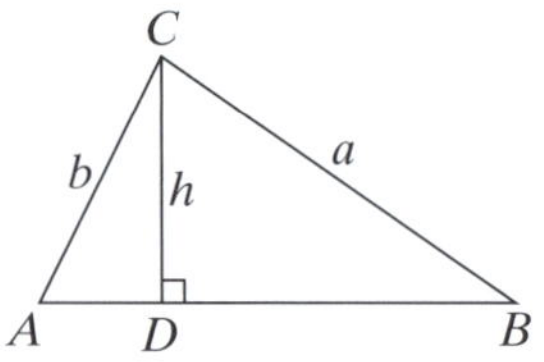

a Find an expression for h in terms of angle A and side b. (1 mark)

b Find another expression for h in terms of angle B and side a. (1 mark)

c Show that $\dfrac{a}{\sin A} = \dfrac{b}{\sin B}$ (1 mark)

2

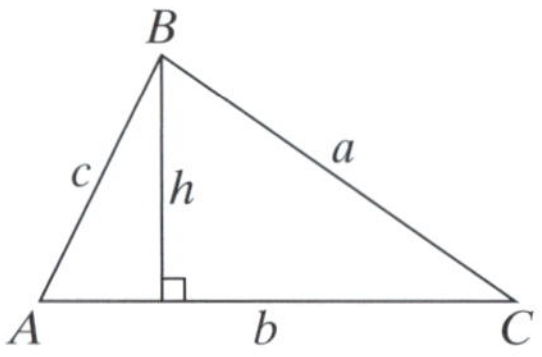

a Find an expression for h in terms of angle C and side a. (1 mark)

b Show that the area of the triangle is given by $\frac{1}{2}ab\sin C$. (1 mark)

3

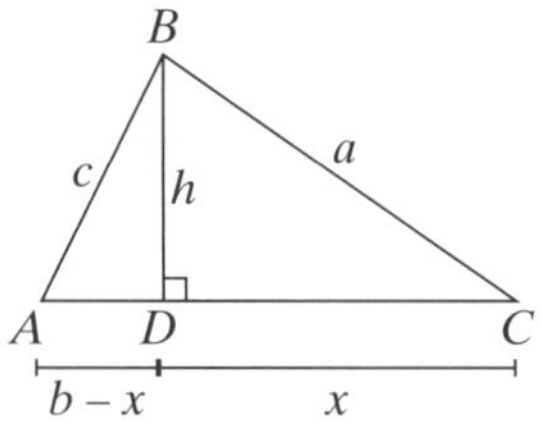

a Using Pythagoras' theorem in both triangles ABD and BCD, show that $c^2 = a^2 + b^2 - 2bx$. (3 marks)

b Show that $c^2 = a^2 + b^2 - 2ab\cos C$. (1 mark)

4

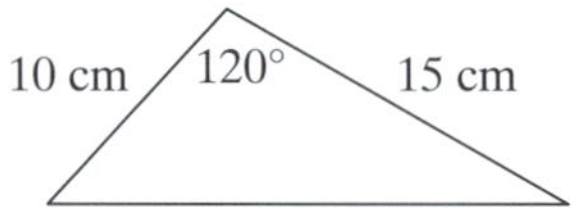

a Find the exact length of the third side of the triangle. (2 marks)

b Find the exact area of the triangle. (2 marks)

5 In an air-race, contestants fly 220 km from point P to point Q on a bearing of 154°. From Q they fly 176 km on a bearing of 280° to R. They then fly straight back to P.

a Show this information on a diagram. (1 mark)

b Find the total distance that the contestants must fly. (4 marks)

c Find the bearing on which they must fly as they return to P. (3 marks)

6

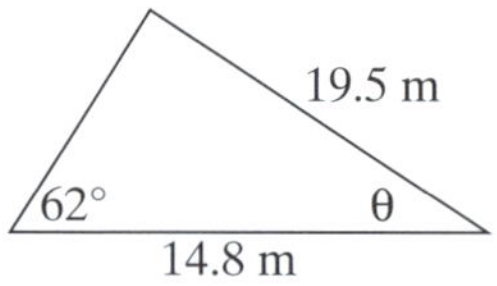

Find the size of θ. Give the answer to the nearest minute. (3 marks)

7 In ΔABC, $A = 26°$, $a = 13.7$ m and $b = 24.1$ m.

a Sketch two different possible triangles, one where B is acute and one where B is obtuse, using this information. (2 marks)

b Find the size of B in each triangle. Give the answers to the nearest minute. (3 marks)

c Hence find the approximate difference in area between the two triangles. (5 marks)

8 $BC = 520$ m and $BD = 740$ m. The area of ΔABC is 10.4 ha.

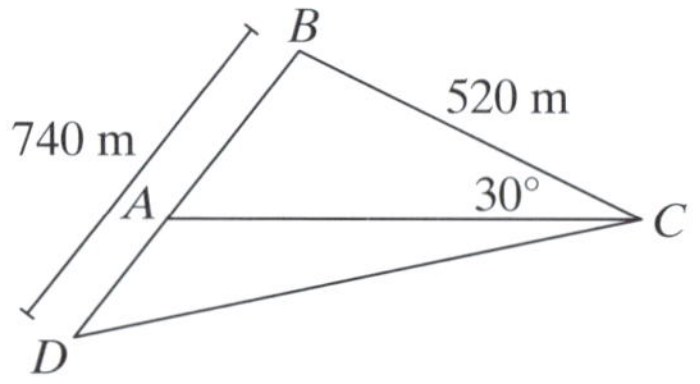

a Find the length of AC. (2 marks)

b Find the length of CD. (4 marks)

Your Feedback

$\dfrac{\square}{40} \times 100\% = \square\%$

PAGES 167–168

PAGE 182

MEAN AND STANDARD DEVIATION

Statistics and Probability

STUDY NOTES

1 The **mean** is commonly known as the average. To find the mean, add all the scores together and divide by the number of scores. The **mode** is the score with the highest frequency. It is the score that happens the most. The **range** is the difference between the highest score and the lowest score.

2 The **median** is the middle value when the scores are arranged in order. If there is an odd number of scores, there is one middle score. When there is an even number of scores the median is the average of the two middle scores.

3 **Quartiles** divide the scores into four equal parts. The **upper quartile** is the value for which one quarter of the scores are above. The **lower quartile** is the value for which one quarter of the scores are below. The median is the middle quartile. To find the upper or lower quartile, first find the median. The lower quartile is then the median of all of the scores that are less than the median. The upper quartile is the median of all of the scores that are greater than the median.

4 The **interquartile range** is the difference between the upper quartile and the lower quartile.

5 The **standard deviation** of a set of scores is a measure of the spread of the scores. It tells us whether the scores are generally close to the mean or are spread out.

6 Using the **statistics mode** of a calculator, we can easily find the mean $\boxed{\bar{x}}$ and standard deviation $\boxed{\sigma_n}$. Different calculators work with statistics in different ways. Make sure you know how to use your calculator. In particular make sure that all previous scores have been deleted before using your calculator with a new set of data. If your calculator has an $\boxed{n}$ button this will tell you the number of scores. It is a good idea to check with this button that you are working with the right number of scores.

7 The standard deviation of a set of scores can also be found using the formula $\sqrt{\frac{\sum(x-\bar{x})^2}{n}}$. This formula gives the standard deviation of a **population**. If only a sample of the data set is considered then the sample standard deviation is used. The sample standard deviation is σ_{n-1} and in the formula n is replaced by $n - 1$.

8 The lower the standard deviation the more **consistent scores** are. This is because the scores are closer to the mean.

9 **Comparisons** can be made between different **data sets**, by considering both the mean and standard deviation. For example, suppose you received 75% in two different tests and also suppose that the mean was 68% for both tests. If the standard deviation was 5.2 for the first test and 8.1 for the second test, then the first test is the better result because it is more than one standard deviation above the mean while the second test result is less than one standard deviation above the mean.

Checklist

Can you:

1 *Find the mean, mode, median and range of a set of scores?* ☐
2 *Determine the upper and lower quartiles and find the interquartile range?* ☐
3 *Find the mean and standard deviation of a set of scores using a calculator?* ☐
4 *Find the standard deviation of a small set of scores using the formula for standard deviation?* ☐
5 *Understand that standard deviation gives information about the spread of scores from the mean?* ☐
6 *Understand the difference between the sample standard deviation and the population standard deviation?* ☐
7 *Compare sets of data by referring to the mean and standard deviation?* ☐

MEAN AND STANDARD DEVIATION

Statistics and Probability

SKILLS CHECK

1 For the scores 1, 2, 2, 5, 6, 7, 7, 7, 7, 8 find the:

a mean **b** mode **c** median **d** range

2 For the scores 11, 11, 12, 13, 13, 13, 14, 14, 14, 14, 15, 15, 16, 16, 16, 16, 16, 17, 17, 19 find the:

a mean **b** median **c** range
d lower quartile **e** upper quartile **f** interquartile range

3 Use a calculator to find the mean ($\bar{x}$) and standard deviation (σ_n), to 1 decimal place, of:

a 5, 7, 9, 12, 12, 12, 15, 17, 18, 19
b 56, 49, 32, 48, 91, 66, 78, 55
c

Score	5	6	7	8	9	10
Frequency	4	6	8	9	9	7

4 Consider these scores: 18, 28, 34, 35, 39, 41, 44, 46, 46, 49

a Determine the mean of the scores.
b Find $\Sigma(x - \bar{x})^2$.
c Find $\sqrt{\dfrac{\sum(x-\bar{x})^2}{n}}$, giving the answer correct to two decimal places.

5 Consider this data set:
59, 64, 67, 69, 71, 72, 73, 75, 76, 76, 79, 83

a Find the mean.
b Find the standard deviation using a calculator. Give the answer correct to two decimal places.
c A score of 72 is added to the scores. Find the new mean and the new standard deviation. Briefly comment on the results compared to the original data set.
d Five is added to each of the original scores. Find the new mean and the new standard deviation. Briefly comment on the results compared to the original data set.
e Each of the original scores is doubled. Find the new mean and the new standard deviation. Briefly comment on the results compared to the original data set.

6 Consider these two sets of data:
A: 17, 19, 20, 23, 24, 24, 25, 26, 28, 29
B: 18, 19, 20, 20, 21, 24, 27, 27, 29, 30

a Which data set has the greater mean?
b Which data set has the greater standard deviation?
c Which data set has the greater range?
d Which data set has more consistent scores? Justify your answer.

PAGES 168–169

Answers **1 a** 5 **b** 7 **c** 6 **d** 7 **2 a** 14.6 **b** 14.5 **c** 8 **d** 13 **e** 16 **f** 3 **3 a** 12.6, 4.5 **b** 59.4, 17.3 **c** 7.8, 1.5 **4 a** 38 **b** 820 **c** 9.06 **5 a** 72 **b** 6.32 **c** 72, 6.08, the mean did not change but the standard deviation decreased **d** 77, 6.32, the mean increased by 5 but the standard deviation did not change **e** 144, 12.65, both the mean and standard deviation doubled **6 a** both the same **b** B **c** both the same **d** A, it has the smaller standard deviation.

MEAN AND STANDARD DEVIATION

Statistics and Probability

25 MINUTES

INTERMEDIATE TEST

Part A Multiple Choice

1 The frequency histograms for four tests are shown below. All of the tests had the same mean. Which test had the greater standard deviation? *Hint 1*

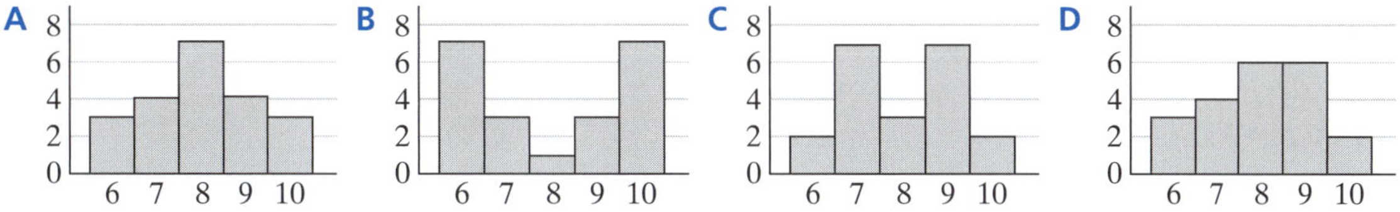

(1 mark)

2 For the set of scores 4, 5, 5, 5, 6, 7, 7, 8, 9, 9, 10, 10 which is largest? *Hint 2*

A mean **B** median **C** standard deviation **D** interquartile range (1 mark)

Part B Written Response

3 The results of an exam, out of 20, are shown in the table.

Score	11	12	13	14	15	16	17	18	19	20
Frequency	15	17	12	8	13	10	7	4	2	1

a How many people sat for the exam? (1 mark)

b Find the mean, correct to one decimal place. (1 mark)

c Find the standard deviation, correct to one decimal place. (1 mark)

d If a score of 15 was added to the results, what would happen to the mean? Justify your answer without any further calculations. (2 marks)

4 Josh received his exam marks for English and History. In English the mean mark was 65%, the standard deviation was 3.5 and Josh scored 72%. In History Josh scored 81%, while the mean mark was 77% and the standard deviation 4. Which was the better result? Justify your answer. *Hint 3* (2 marks)

5 The twelve students in a university course all sat for theory and practical exams, both marked out of 50.
The scores are: Theory: 35, 28, 30, 41, 32, 44, 36, 42, 37, 32, 38, 33
Practical: 33, 39, 29, 42, 37, 44, 33, 49, 38, 32, 37, 40

a What is the mean and standard deviation of the theory exam marks? (2 marks)

b What is the mean and standard deviation of the practical exam marks? (2 marks)

c Which test had the most consistent results? Justify your answer. *Hint 4* (2 marks)

d On which test did the students perform better? (1 mark)

e Elise is one of the students. She scored 41 in theory and 42 in the practical exam. Which was her better result? Justify your answer. (2 marks)

6 The numbers of vehicles that a dealer sold each month last year were:
24, 18, 21, 22, 29, 46, 17, 20, 25, 35, 32, 38

a Find the mean, median, standard deviation, range and interquartile range. (5 marks)

b Compare the relative merits of the measures of location and the measures of spread for this data. *Hint 5* (2 marks)

Hint 1: There is no need to do any calculations. Standard deviation measures the spread of scores. In which histogram are the scores the most spread out?

Hint 2: You will need to do some calculations to determine the answer.

Hint 3: Consider the mean and the standard deviation as well as the mark.

Hint 4: Refer to the standard deviation.

Hint 5: Consider answers to part **a** as well as the raw scores.

Your Feedback

$\frac{\square}{25} \times 100\% = \square\%$

PAGES 169–170

PAGE 182

MEAN AND STANDARD DEVIATION

Statistics and Probability

30 MINUTES

ADVANCED TEST

1 These are the percentage marks Marias scored in 14 tests this year.

59 68 73 69 82 75 70 87 67 78 73 74 66 81

a Find the mean. (1 mark)

b Find the standard deviation (σ_n) to two decimal places. (1 mark)

In his 15th test Marias scored 73%.

c What happens to the mean? Why? (1 mark)

d What happens to the standard deviation? Why? (2 marks)

In his 16th test, Marias scored 89%.

e What happens to the mean? Why? (1 mark)

f What happens to the standard deviation? Why? (2 marks)

2 **a** Find the mean of these scores 3, 5, 8, 9, 12, 13, 15, 17, 18, 20. (1 mark)

b Draw up a table with the headings shown below and complete it for the above scores. (2 marks)

x	$x-\bar{x}$	$(x-\bar{x})^2$

c Find the sum of the second column. (1 mark)

d Find the sum of the third column. (1 mark)

e Find $\sqrt{\dfrac{\sum(x-\bar{x})^2}{n}}$, giving the answer correct to three decimal places. (1 mark)

f Is the answer to part **e** the same as found using a calculator to find σ_n? Explain why or why not. (1 mark)

g Gwen said 'Because taking the square root is the opposite operation to squaring there is no need to do both. I could have just found $\dfrac{\sum(x-\bar{x})}{\sqrt{n}}$ in part **e**.'

Is Gwen correct?
Justify your answer. (2 marks)

3 **a** If every score in a data set increases by 4 what will happen to the mean and standard deviation?
Justify your answer. (2 marks)

b If every score in a data set doubles, what will happen to the mean and standard deviation?
Justify your answer. (2 marks)

4 Emma recorded the times (in minutes) to travel to and from school for a fortnight.

Day	Morning	Afternoon
Monday	26	19
Tuesday	39	29
Wednesday	41	25
Thursday	32	29
Friday	44	34
Monday	28	21
Tuesday	53	30
Wednesday	40	27
Thursday	37	24
Friday	35	32

a Find the mean and standard deviation of the morning times. (2 marks)

b Find the mean and standard deviation of the afternoon times. (2 marks)

c What observations can be made about the travelling times from these results? (2 marks)

5 Two data sets of exam marks have the same mean and the same number of scores. The standard deviation of the first set is 1.5 times the standard deviation of the second.

a What information can be gained from the above statement? (1 mark)

b When would a score of 72 be equivalent in both data sets? (1 mark)

c Assuming that it is not equivalent, would a score of 72 be better in the first set or in the second? (1 mark)

Your Feedback

$\dfrac{\square}{30} \times 100\% = \square\%$

PAGES 170–171

PAGE 182

INVESTIGATIONS OF DATA SETS AND REPORTS

Statistics and Probability

STUDY NOTES

1 If data has only one variable it is **univariate**. **Bivariate data** shows the relationship between two related sets of variables. For example, information about the age and value of cars could be shown in a table or graph.

2 If two related sets of data are **plotted** on a graph, **patterns** can be seen illustrating the relationship. For example, the points might tend to form a straight line or curves such as parabolas or hyperbolas.

3 A **line of fit** is a straight line drawn on a graph to show the pattern between two variables. If asked to **draw a line** of fit, there is no unique answer. Try to draw the line so that it goes roughly through the middle of the set of points.

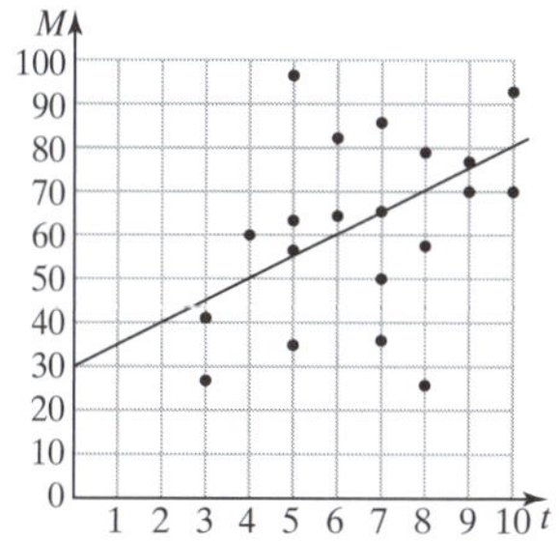

4 Because a line of fit is a straight line, its **equation** can be found. The equation will be of the form $y = mx + b$ where m is the gradient of the line and b is the y-intercept. (The equation for the line of fit might involve different variables, not always x and y.)

The gradient can be found by the gradient formula $m = \frac{y_2 - y_1}{x_2 - x_1}$ or by using $\frac{\text{vertical change in position}}{\text{horizontal change in position}}$. For example, the line shown has $m = \frac{50}{10} = 5$ and vertical intercept 30. Its equation is $M = 5t + 30$.

5 The equation of a line of fit can be used to make **predictions** about the data. This is done by substituting values into the equation. Such predictions might involve **interpolation** (finding values within the range of values in the data set) or extrapolation (finding values outside the range of values in the data set).

6 If there is a **strong relationship**, or correlation, between two data sets it doesn't mean that the strong relationship must exist for every member of the data set. If there is a strong relationship between two variables, it does not mean that one causes the other. For example, there might be a strong relationship between the weight and height of a group of students, but the fact that someone might have considerable weight doesn't cause them to be tall.

7 You must be able to **critically evaluate statistical reports** found in the media and elsewhere. Things to consider are where the data came from; is it a reputable source, is it from an independent source. What sort of **sample** was the information drawn from; was it large enough, was it comprehensive enough, was it typical of the general population? Is there any **bias** in the report? Are any conclusions drawn reasonable? Are there any other factors that could have influenced the data that have not been taken into account? Is the material presented accurately or is some misleading?

8 When **answering questions** about statistical reports be clear and concise. Don't waffle. Make sure you answer the question that has been asked, particularly if there are several parts to be addressed. If asked to give reasons or examples, give more than one if required.

Checklist

Can you:

1 *Understand what is meant by bivariate data?* ☐
2 *Read and understand graphs and draw lines of fit?* ☐
3 *Find the equation of a line of fit?* ☐
4 *Make predictions, interpolate and extrapolate?* ☐
5 *Critically evaluate and answer questions concerning statistical reports?* ☐

INVESTIGATIONS OF DATA SETS AND REPORTS

Statistics and Probability

SKILLS CHECK

1 A scatterplot was drawn to show the test results of a class in both spelling and mental arithmetic.

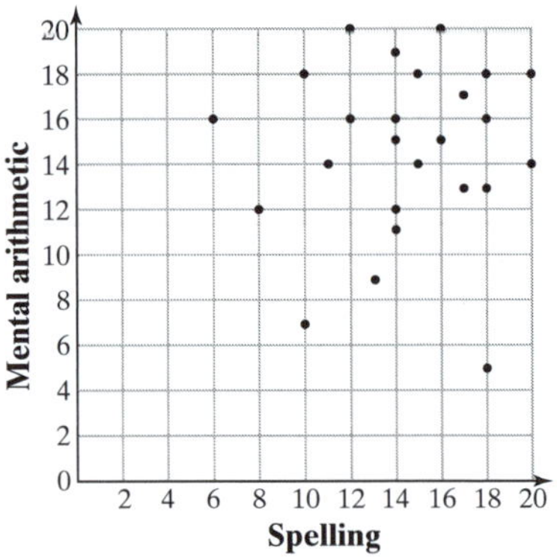

- **a** How many students are in the class?
- **b** One student scored 13 in spelling. What did she score in mental arithmetic?
- **c** One student scored 17 in mental arithmetic. What did he score in spelling?
- **d** What was the difference in mental arithmetic scores of the two students who scored 10 in spelling?
- **e** Is there any relationship between the spelling and mental arithmetic scores?
- **f** Harry said: 'Everyone who did well in spelling also did well in mental arithmetic.' Give one example that shows that Harry is wrong.

2 What is the equation of each line of fit?

a

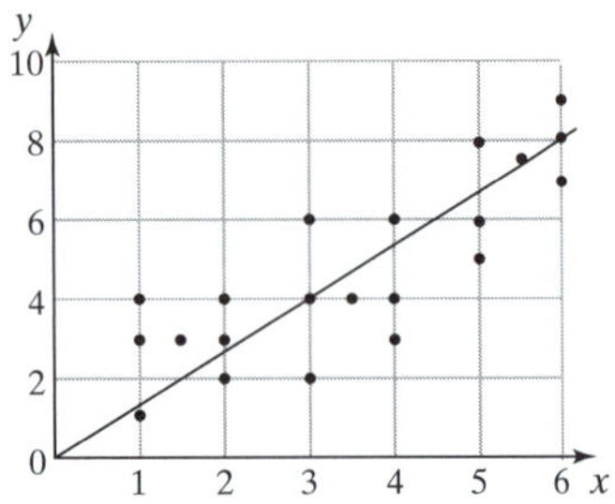

b

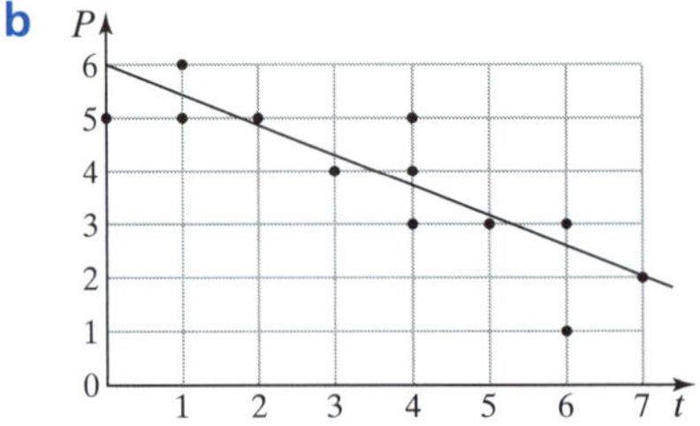

c

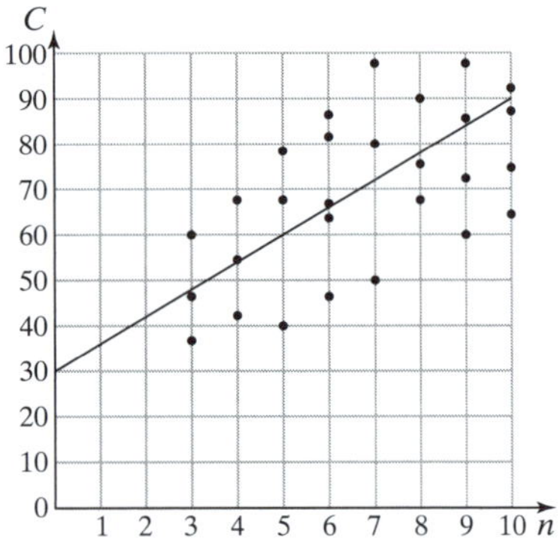

3 The equation of a line of fit was $h = 9 - \frac{3}{4}t$.

- **a** What is the value of h when $t = 6$?
- **b** What is the value of t when $h = 3$?
- **c** What is the value of h when $t = 20$?
- **d** If h represents height, why does the answer to part **c** not make sense?
- **e** What is the largest possible value of t for which h makes sense if it represents height?

4 A survey of 40 people was conducted at a Darwin hotel. Give reasons why this sample would not be a true representation of the Australian population.

5 Determine whether the following statements are true or false:

- **a** For any set of plotted data there is only one possible line of fit.
- **b** For any line of fit there is only one possible equation.
- **c** The size of a sample is not important when assessing data.
- **d** Bias should be avoided when gathering data.

PAGES 171–172

Answers **1 a** 25 **b** 9 **c** 17 **d** 11 **e** slight relationship **f** a student scored 18 and 5 **2 a** $y = \frac{4}{3}x$ **b** $P = \frac{-4}{7}t + 6$ **c** $C = 6n + 30$ **3 a** 4.5 **b** 8 **c** −6 **d** it cannot be negative **e** 12 **4 a** see worked solutions **5 a** false **b** true **c** false **d** true

INVESTIGATIONS OF DATA SETS AND REPORTS

Statistics and Probability

20 MINUTES

INTERMEDIATE TEST

Part A Multiple Choice

Questions 1–3 refer to this scatterplot which shows the results some design students achieved in both their theory and practical tests.

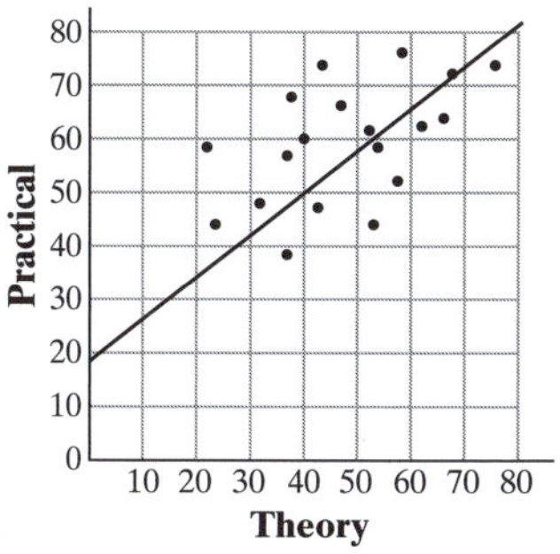

1 Josh scored 66% on his theory test but missed the practical test. His teacher gave him an estimate, using the line of fit. What might Josh's estimated mark have been?

A 57 B 62 C 70 D 73 (1 mark)

2 Jess wasn't well on the day of the theory test. She scored the lowest mark. The teacher decided to increase her mark by averaging Jess' actual theory mark with the mark predicted by the line of fit. What was Jess's adjusted mark?

A 36 B 43 C 47 D 53 (1 mark)

3 Juno only did the theory test and scored 90. What does the line of fit suggest his practical mark might have been?

A 90 B 94 C 98 D 100 (1 mark)

4 A media release reported that only 33% of people surveyed supported Y while the remainder supported X. Which statement must be correct?

A The majority of the population supported X.
B X must be better than Y.
C Around twice as many of those surveyed preferred X to Y.
D All of the statements are correct.
(1 mark)

Part B Written Response

5 A survey was taken of the length of time (in hours) that students spent learning a procedure and the time (in minutes) each took to complete the procedure. The results are shown in the table.

Time learning (hours)	7	15	10	3	12	14	6	12	4	18	9	5
Time to complete (minutes)	32	25	29	37	30	27	35	26	35	22	30	35

a Draw a graph and plot the information. Draw a line of fit on your graph. *Hint 1* (3 marks)

b Find the equation of the line of fit. (2 marks)

c Use the equation of the line of fit to predict the length of time it would take someone who spent 24 hours learning the procedure to complete it. (1 mark)

d If a student took 8 minutes to complete the procedure, for how long would you expect she spent learning it? (1 mark)

e Briefly explain why the equation of the line of fit could not be used to predict the amount of time needed to complete the procedure by a student who studied it for 48 hours. (1 mark)

6 A survey was taken at a Melbourne railway station about plans being put forward by a federal government committee. People were asked which of the two plans they preferred. The results are shown in the table.

	Male	Female
Plan A	84	68
Plan B	57	51
Undecided	19	21

a What percentage of respondents were male? (1 mark)

b What percentage of people were not in favour of B? (1 mark)

c Is this survey based on a good sample of all Australians? Justify your answer. (2 marks)

d Which is more likely: A male will prefer plan A or a person who prefers plan A is male? Justify your answer. (2 marks)

e A report in a newspaper stated that the majority of both males and females favoured plan A. Is that correct? Justify your answer. (2 marks)

Hint 1: Show the time learning on the *x*-axis and the time to complete on the *y*-axis. A sensible approach to drawing the line will make answering the next parts of the question easier.

Your Feedback

$\frac{\square}{20} \times 100\% = \square\%$

PAGE 172
PAGE 182

INVESTIGATIONS OF DATA SETS AND REPORTS

Statistics and Probability

20 MINUTES

ADVANCED TEST

1 Charlotte collected some information about the maximum and minimum daily temperatures in both January and July and used a spreadsheet program to produce these two scatterplots.

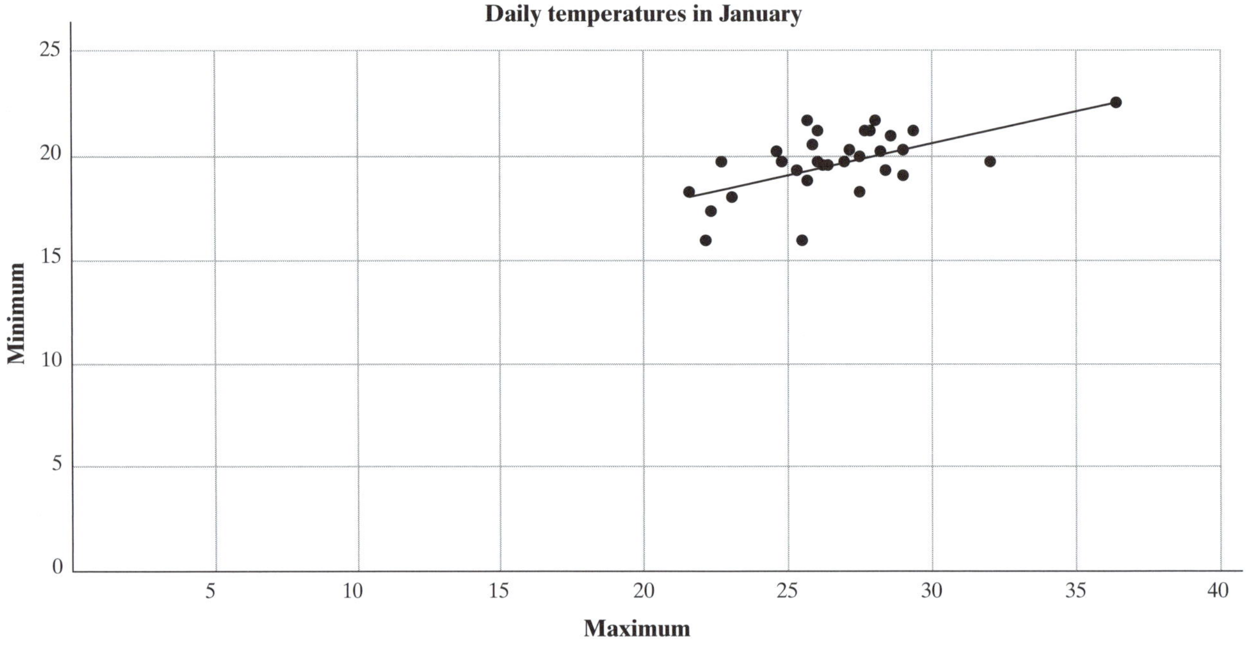

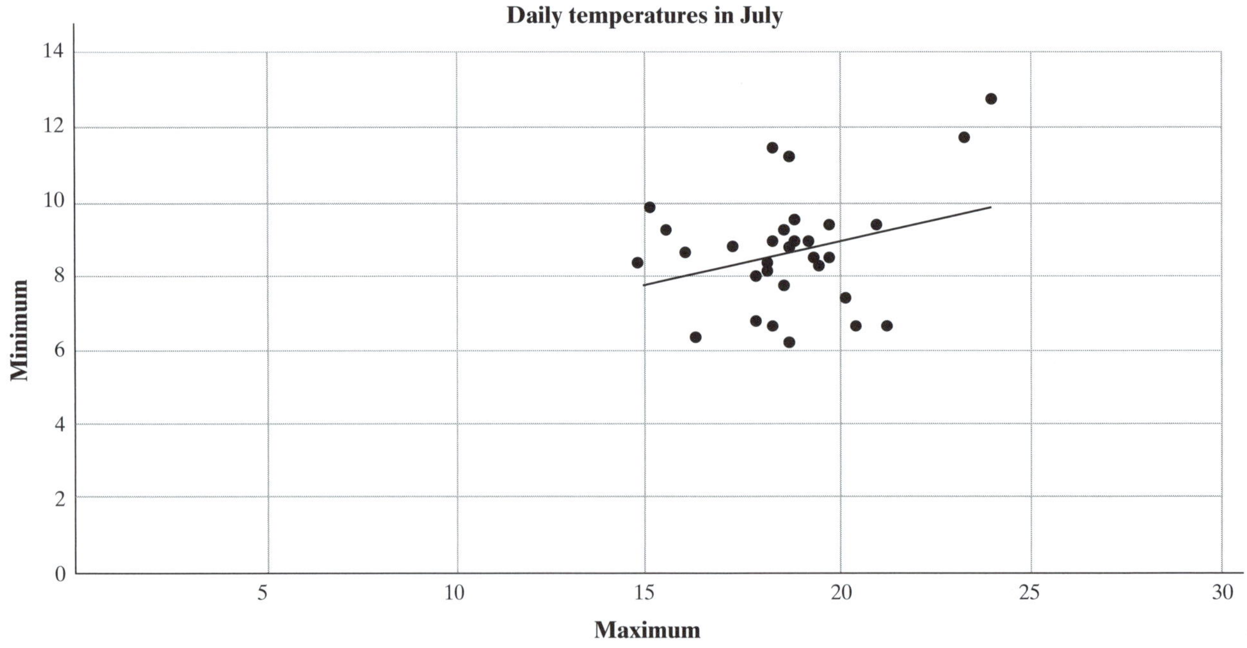

a Charlotte said that the line of fit had the same gradient on both plots. Is that correct? Justify your answer. (2 marks)

b In which month is the correlation between the maximum and minimum temperatures strongest? Justify your answer. (2 marks)

c What is the difference between the minimum temperatures, as predicted by the lines of fit, on a day when the maximum is 23 °C? (1 mark)

d On about how many days in July was the maximum temperature hotter than the lowest maximum in January? (1 mark)

e If in July there was a day with a maximum temperature of 27 °C, what would you predict the minimum would be? Justify your answer. (2 marks)

2 A report in the media quoted research that showed that 'drivers of smaller cars are more likely to use aggressive or reckless tactics on the road.' A parent, on reading this article, decided to buy a larger car for his daughter as her first car. Do you think this was a valid response? Discuss. (2 marks)

3 A survey was conducted about a school's discipline policy. The first question asked: 'What is the biggest problem with this policy?' Is this a good question? Justify your answer. (2 marks)

4 A local council bases its road maintenance program on guidelines issued by their insurance company. The insurance company developed the guidelines using statistics on road accidents and probability. Do you think this is a sensible approach by council? What issues do you think the council needs to take into consideration? (2 marks)

5 The media reported the conclusions of a controlled experiment conducted in Britain on people giving up smoking. It found that most people put on weight and considerably more weight than previously thought. What implications, if any, might this report have for people intending to give up smoking in Australia? Discuss. (2 marks)

6 The headline in a local newspaper stated that a certain road was rated the worst in the district in an online survey.

a Do you think that that road would be the worst in the district? What further information would be helpful? (2 marks)

b What is the significance of it being an online survey? How might it be possible to influence the result of such a survey? (2 marks)

Your Feedback

$\frac{\square}{20} \times 100\% = \square\%$

PAGE 173

PAGE 182

TIPS FOR THE SAMPLE EXAM PAPERS

STUDY TIPS

Know what to expect

- **Find out** from your teacher exactly what topics will be assessed in the exam.
- Yearly exams normally test **all work covered** throughout the year.
- **Determine what format** is being used for the exam. Are there multiple-choice questions? Will calculators be allowed in any part of the examination?

Show all working

- **Read the instructions** on the exam paper—marks may be allocated for working.
- Even when **marks are not allocated,** working is important and it is rewarded, even though your answer may be incorrect.
- **Never use correction fluid**—just put a line through any incorrect working. Let your teacher see all your work.

Allocate your time

- Ensure you are **working through the paper efficiently** and not spending too much time on each question, only to find you run out of time at the end of the exam.

Understand mark allocations

- **Questions worth more marks** are often more complex and difficult. Working is crucial in these types of questions.

Reread and check

- Once you have **completed a question**, rather than moving on to the next, reread that question to make sure you have in fact answered the correct question. This will only take a second or two.
- If you **finish the exam** with time still remaining, check your answers—often mistakes are found and can be corrected.

Use quality diagrams

- Diagrams should be of **good quality, large and drawn with a lead pencil**. Use an eraser, not correction fluid, to delete mistakes.

Be ready

- Finally, you need to be **prepared for the exam**.
- Always **study mathematics actively**. Active study means using pen and paper to make notes, writing down difficult questions and their solutions, and recording rules to learn.

SAMPLE EXAM PAPERS
Level of difficulty—Average

60 MINUTES

PAPER 1

Part A Multiple Choice (25 marks)

1 $\sqrt{36x^{36}} = ?$

A $6x^6$ **B** $6x^{18}$ **C** $18x^6$ **D** $18x^{18}$ (1 mark)

2 Which diagram shows Q on a bearing of 075° from P?

A

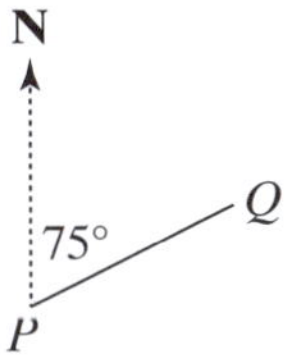

B

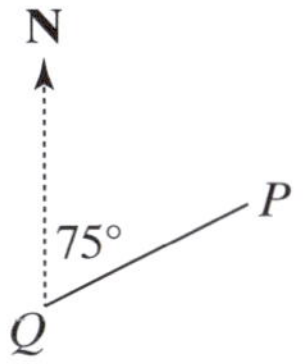

C

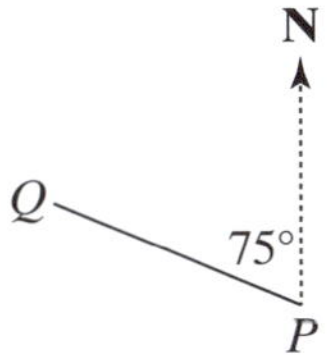

D

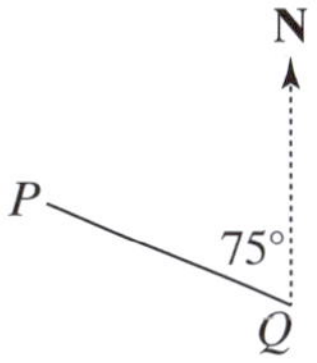

(1 mark)

3 The volume of a cone with diameter 6 cm and height 4 cm is closest to:

A 38 cm^3 **B** 113 cm^3 **C** 151 cm^3 **D** 452 cm^3 (1 mark)

4 $\frac{x^2 - x}{x^2 - 1} = ?$

A 1 **B** x **C** $\frac{x}{x + 1}$ **D** cannot be simplified (1 mark)

5 $3n^0 - n^{-1} = ?$

A $-\frac{1}{n}$ **B** $1 - \frac{1}{n}$ **C** $3 - \frac{1}{n}$ **D** none of these (1 mark)

6 $\frac{2}{\sqrt{2}} + \frac{\sqrt{2}}{2} = ?$

A $\frac{2 + \sqrt{2}}{2\sqrt{2}}$ **B** $\frac{3\sqrt{2}}{2}$ **C** $\frac{6}{\sqrt{2}}$ **D** 1 (1 mark)

7 The equation of the axis of symmetry of the parabola $y = x^2 - 6x + 8$ is:

A $x = -4$ **B** $x = 4$ **C** $x = -3$ **D** $x = 3$ (1 mark)

8 $\log_a 6 + \log_a 8 - \log_a 12 = ?$

A $\log_a 2$ **B** $\log_a 4$ **C** $\log_a 10$ **D** $\log_a 16$ (1 mark)

9 O is the centre of the circle.
$x = ?$

A 40
B 70
C 80
D 100

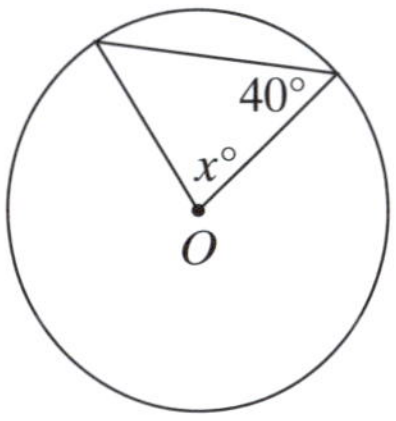

(1 mark)

10 $xy - x - y + 1 = ?$

A $(x - 1)(y - 1)$ **B** $(x - 1)(y + 1)$ **C** $x(y - 1)$ **D** $x(y + 1)$ (1 mark)

11 Which line is perpendicular to $y = \frac{1}{3}x - 4$?

A $3x + y - 5 = 0$ **B** $3x - y + 2 = 0$ **C** $x + 3y - 1 = 0$ **D** $x - 3y = 0$ (1 mark)

12 The curved surface area of this cone is closest to:

A 427 cm^2
B 801 cm^2
C 855 cm^2
D 1005 cm^2

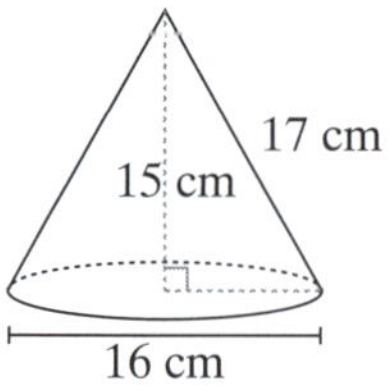

(1 mark)

13 $3m^2 + m - 10$ when expressed as a product of its factors is:

A $(3m - 2)(m + 5)$ **B** $(3m + 2)(m - 5)$ **C** $(3m + 5)(m - 2)$ **D** $(3m - 5)(m + 2)$ (1 mark)

14 The dot plots have been drawn on the same axis to show the scores of the same class of students over two different tests.

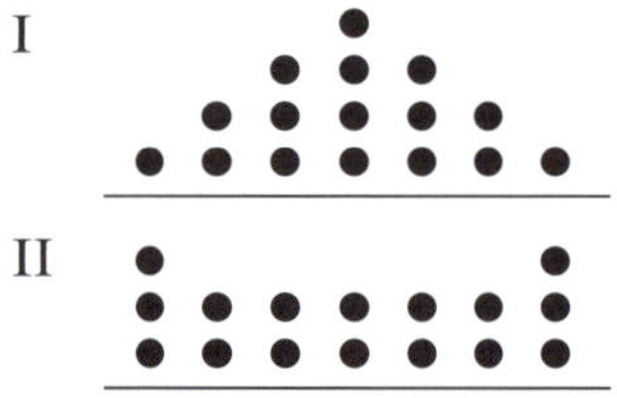

Which statement is correct?

A Both tests have the same mean and the same standard deviation.
B Both tests have the same mean but test I has a larger standard deviation.
C Both tests have the same mean but test II has a larger standard deviation.
D Both tests have the same standard deviation but different means. (1 mark)

15 The remainder when $P(x) = x^3 - 3x^2 + 6x - 9$ is divided by $(x + 1)$ is:

A –9 **B** 9 **C** –5 **D** –19 (1 mark)

16 $x = ?$

A 79
B 99
C 101
D There is not enough information to determine the value of x.

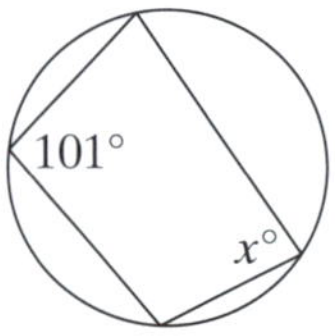

(1 mark)

17 The length of QR is closest to:

A 13 m
B 14 m
C 15 m
D 16 m

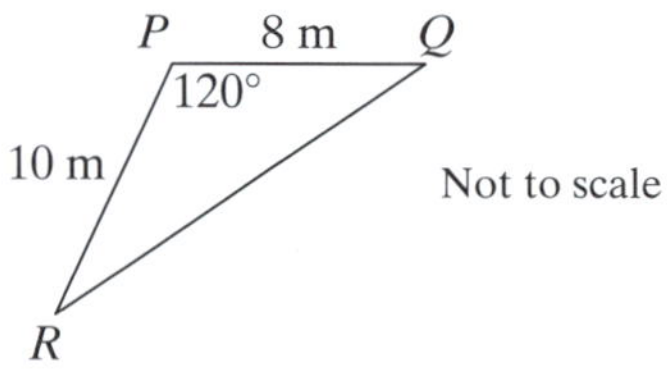

(1 mark)

18 The lengths of the sides of two similar figures are in the ratio 2:3. The volume of the larger figure is 81 m^3. The volume of the smaller figure is:

A 16 m^3 **B** 24 m^3 **C** 36 m^3 **D** 54 m^3 (1 mark)

19 Which diagram could show the graph of $y = 2^{-x}$?

A
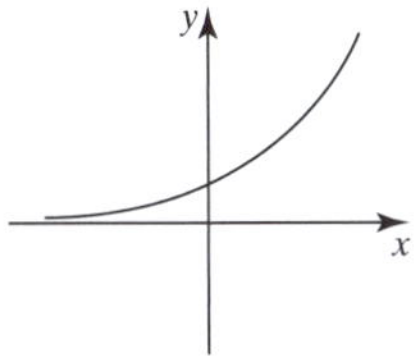

B
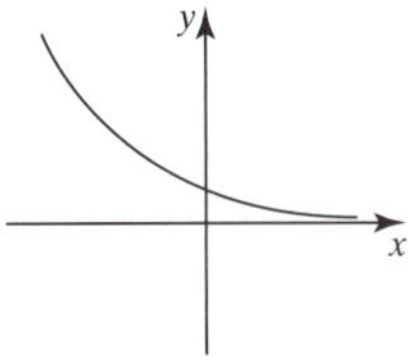

C
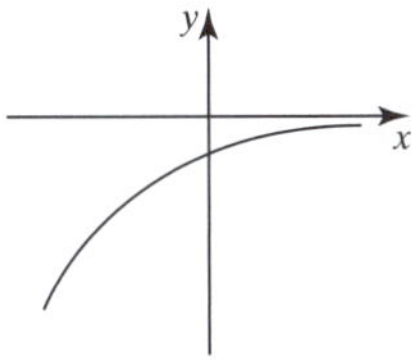

D
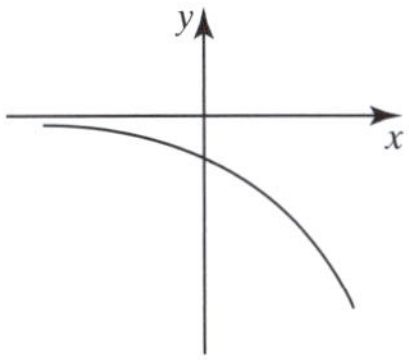

(1 mark)

20 Which is the equation of a circle with centre (–1, 3) and radius 2 units?

A $(x - 1)^2 + (y + 3)^2 = 2$ **B** $(x + 1)^2 + (y - 3)^2 = 2$
C $(x - 1)^2 + (y + 3)^2 = 4$ **D** $(x + 1)^2 + (y - 3)^2 = 4$ (1 mark)

21 If $f(x) = x^2 - x + 1$ then $f(-1)$ equals:

A –1 **B** 1 **C** 2 **D** 3 (1 mark)

22 Which expression will give $\sin P$?

A $\dfrac{5\sin 60^\circ}{9.5}$ **B** $\dfrac{9.5\sin 60^\circ}{5}$

C $\dfrac{5\sin 60^\circ}{11}$ **D** $\dfrac{11\sin 60^\circ}{5}$

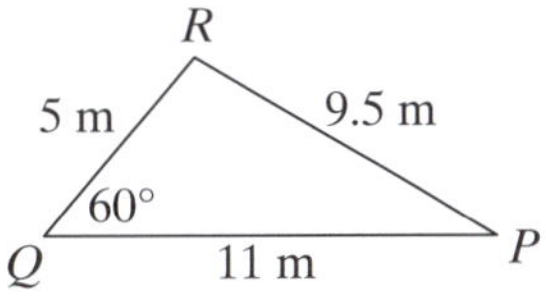

(1 mark)

23 The graph shows the mass of a chemical element over time.
The mass is:

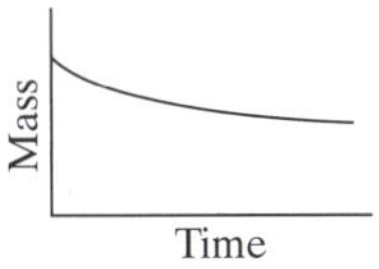

A increasing at an increasing rate
B increasing at a decreasing rate
C decreasing at an increasing rate
D decreasing at a decreasing rate (1 mark)

24 $(3m - 2)(m - 4) = ?$

A $3m^2 - 6m + 8$ B $3m^2 - 10m + 8$ **C** $3m^2 - 14m + 8$ **D** $3m^2 - 2m + 8$ (1 mark)

25 What must be added to the expression $9x^2 + 6x$ to complete the square?

A 4 **B** 9 **C** 1 **D** 144 (1 mark)

Your Score

/25

PAGES 173–175

Part B Written Response (25 marks)

26 Simplify $\dfrac{8}{x^2 - 4} - \dfrac{1}{2x - 4}$ (2 marks)

27 Solve $8^x = 32$. (2 marks)

28 Simplify $\sqrt{28} + \sqrt{63}$. (1 mark)

29 Use the quadratic formula to find the values of x, correct to two decimal places, for which $3x^2 - 7x - 1 = 0$. (2 marks)

30 Find the exact value of x. (2 marks)

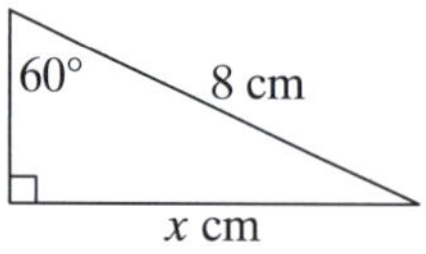

31 **a** Use the cosine rule to find the size of angle A to the nearest degree. (2 marks)

b Find the area of triangle ABC. (1 mark)

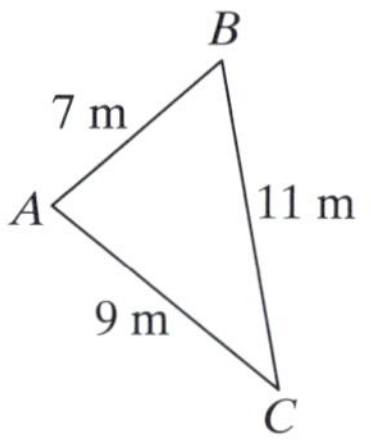

32 O is the centre of the circle. M is the midpoint of chord AB. By proving triangles AOM and BOM congruent, prove that OM is perpendicular to AB. (3 marks)

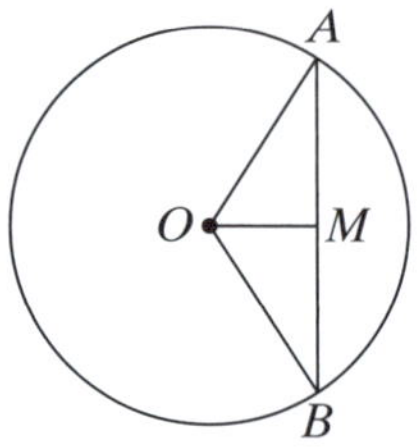

33 The rectangular base of this pyramid is 72 cm long and 16 cm wide. The perpendicular height of the shaded face is 17 cm.

a Find the perpendicular height of the pyramid. (1 mark)

b Find the volume of the pyramid. (1 mark)

c Find the total surface area of the 5 faces of the pyramid. (4 marks)

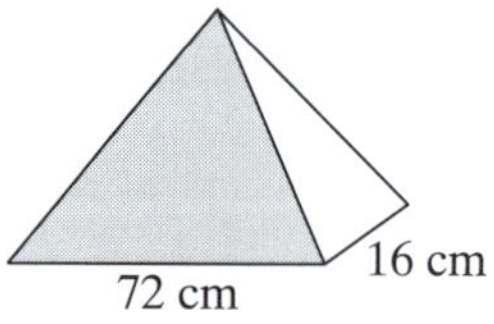

34 **a** Factorise $x^3 - 3x^2 - 22x + 24$ (3 marks)

b Without sketching, briefly describe the features of the graph of $y = x^3 - 3x^2 - 22x + 24$. (1 mark)

Your Score

____ / 25

PAGES 175–176

Your Feedback

Part A + Part B = ____ / 50 = ____ %

PAGE 182

SAMPLE EXAM PAPERS

Level of difficulty—Above average

60 MINUTES

PAPER 2

Part A Multiple Choice (25 marks)

1 $a^2 - 25 = ?$

A $(a-5)^2$ **B** $(a+5)^2$ **C** $(a+5)(a-5)$ **D** $a(a-25)$ (1 mark)

2 Which curve passes through the point $(-2, -6)$?

A $y = x^2 + 5x + 13$ **B** $y = \frac{1}{2}x^3$ **C** $y = \frac{6}{x+1}$ **D** $y = -2^{-x}$ (1 mark)

3 $P(x) = (2x-7)(4x^3 + 5x^2 - 7x + 3)$. What is the leading coefficient?

A 2 **B** 4 **C** 6 **D** 8 (1 mark)

4 Which is **not** equal to $\frac{\sqrt{3}}{2}$?

A $\sin 60°$ **B** $\sin 120°$ **C** $\cos 30°$ **D** $\cos 150°$ (1 mark)

5 $x = ?$

A 40
B 50
C 60
D 70

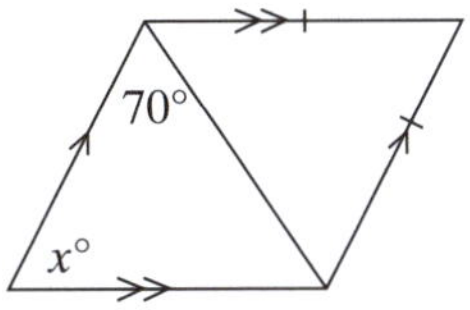

(1 mark)

6 The vertex of the parabola $y = x^2 - 6x + 11$ is at:

A (6, 11) **B** (3, 2) **C** (–3, –2) **D** (3, –2) (1 mark)

7 Which diagram could be a sketch of $y = \frac{-2}{x}$?

A

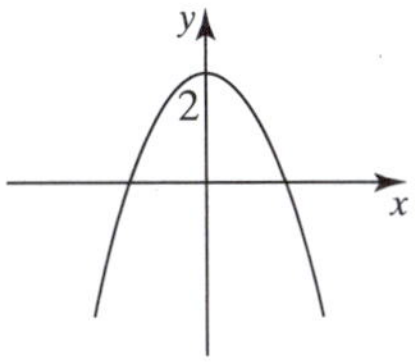

B

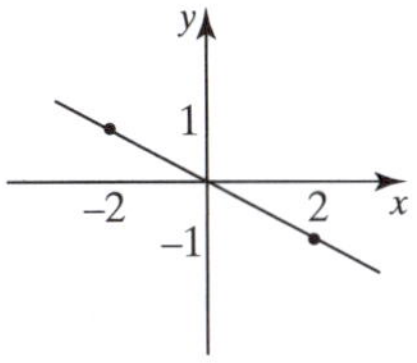

C

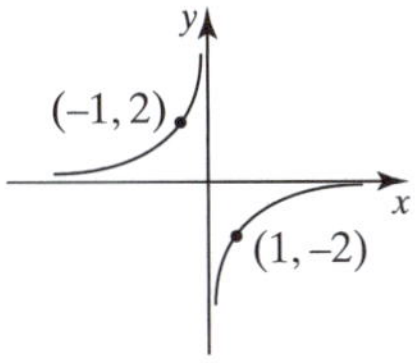

D

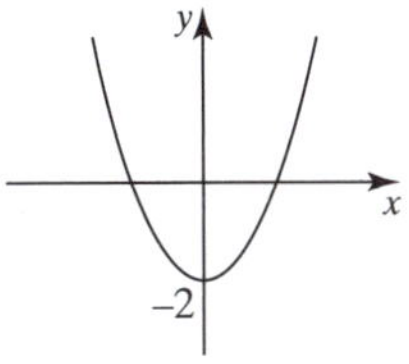

(1 mark)

8 $2x^{\frac{1}{2}} \times 4x^{\frac{1}{2}} = ?$

A $8x$ **B** $8x^2$ **C** $8x^{\frac{1}{4}}$ **D** $8x^{\frac{1}{2}}$ (1 mark)

9 A triangular pyramid has perpendicular height 8 cm and volume 36 cm³. What is the area of its base?

A 1.5 cm² **B** 4.5 cm² **C** 6.75 cm² **D** 13.5 cm² (1 mark)

10 $\tan 30° = ?$

A 1 **B** $\frac{1}{\sqrt{3}}$ **C** $\frac{\sqrt{3}}{2}$ **D** $\sqrt{3}$ (1 mark)

11 When simplified, $\sqrt{8} + 3\sqrt{2} = ?$

A $5\sqrt{2}$ **B** 12 **C** $3\sqrt{10}$ **D** It cannot be simplified (1 mark)

12 Which is the correct formula to find the value of x if $ax^2 + bx + c = 0$?

A $x = \dfrac{-b \pm \sqrt{b^2 - 4ac}}{2a}$ **B** $x = \dfrac{-b \pm \sqrt{b^2 + 4ac}}{2a}$

C $x = \dfrac{b \pm \sqrt{b^2 - 4ac}}{2a}$ **D** $x = \dfrac{b \pm \sqrt{b^2 + 4ac}}{2a}$ (1 mark)

13 Quentin found the surface area and volume of this object.
He found the surface area was 810 cm^2 and the volume was 900 cm^3.
Which was correct?

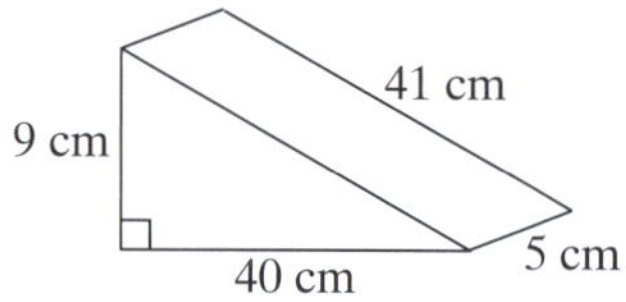

A both surface area and volume
B surface area but not volume
C volume but not surface area
D neither surface area nor volume (1 mark)

14. The area of this triangle is closest to:

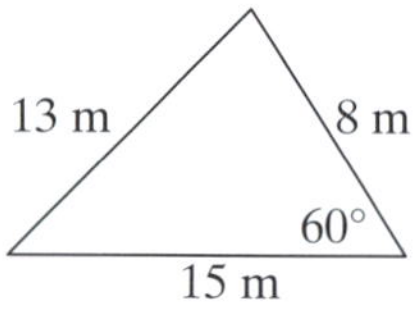

A 30 m^2
B 45 m^2
C 52 m^2
D 84 m^2 (1 mark)

15 $(2x + 5)^2 = 4x^2 +$ ______ $+ 25$
Which term belongs in the space?
A $9x$ **B** $10x$ **C** $20x$ **D** $40x$ (1 mark)

16 Students in a particular course had to write an essay and sit for an exam. The teacher placed the marks the students scored on a scatterplot and drew a line of fit as shown. Olivia was one of the students. She was sick on the day of the exam and scored the lowest mark. The teacher increased her mark using the equation of the line of fit. About how many marks higher was Olivia's new exam mark than the actual one?

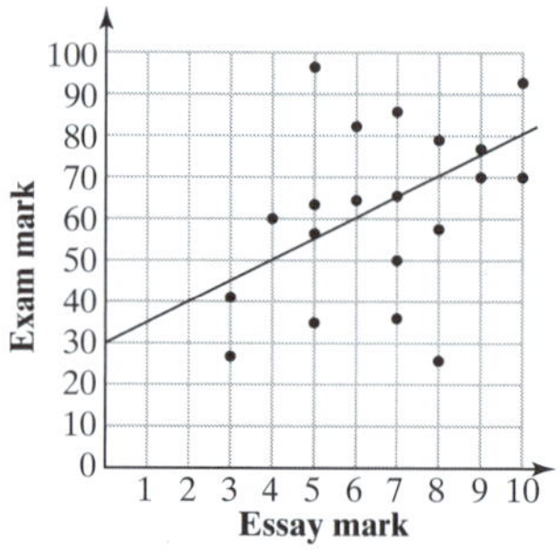

A 20
B 25
C 35
D 45 (1 mark)

17 The length AG is:

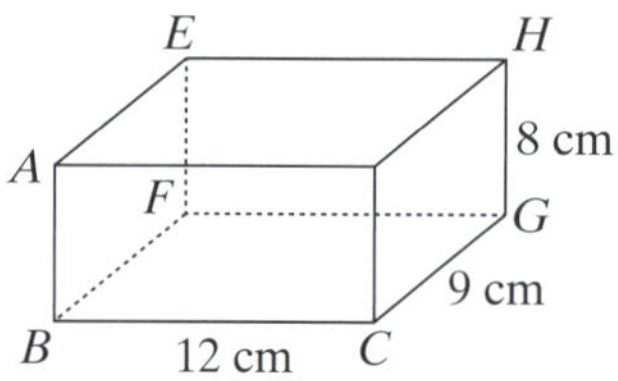

A 15 cm
B 17 cm
C 25 cm
D 29 cm (1 mark)

18 $x^2 + 5x + 6 = 0$
Which is correct?
A $x = -2$ or $x = -3$ **B** $x = 2$ or $x = 3$ **C** $x = -5$ or $x = -1$ **D** $x = 1$ or $x = 5$ (1 mark)

19 A line has gradient 2 and y-intercept –1. The equation of the line is:

A $2x - y - 1 = 0$ **B** $2x - y + 1 = 0$ **C** $2x + y - 1 = 0$ **D** $2x + y + 1 = 0$ (1 mark)

20 If $\log_8 2 = x$, x equals:

A $\frac{1}{3}$ **B** $\frac{1}{4}$ **C** 3 **D** 4 (1 mark)

21 The centre of the circle $(x + 4)^2 + (y - 2)^2 = 16$ is at:

A $(-4, -2)$ **B** $(-4, 2)$ **C** $(4, -2)$ **D** $(4, 2)$ (1 mark)

22 Which is a factor of the polynomial $x^4 - 6x^3 - 4x^2 + 54x - 45$?

A $x + 5$ **B** $x + 3$ **C** $x + 2$ **D** $x + 1$ (1 mark)

23 O is the centre of the circle.
$x = ?$

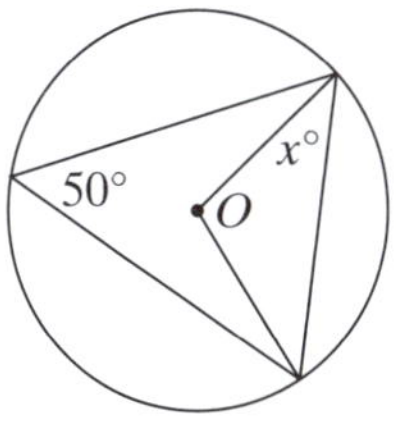

A 20
B 25
C 40
D 50 (1 mark)

24 Which graph best shows the depth of water over time as this contained is filled by water flowing into it at a constant rate?

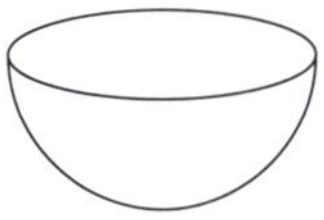

A
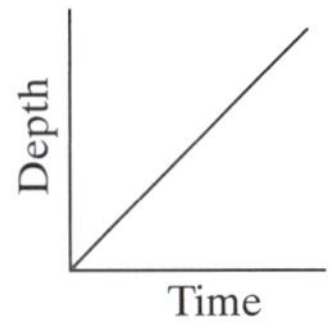

B
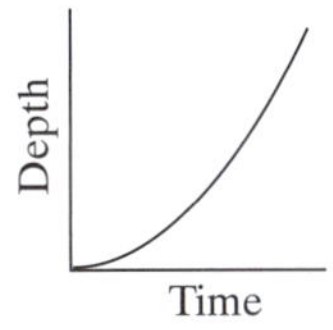

C
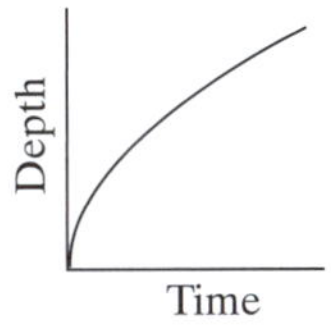

D
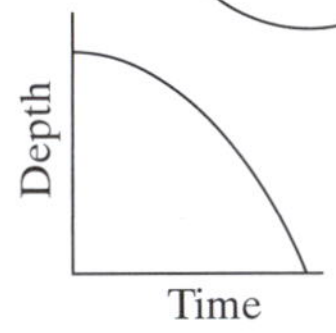

(1 mark)

25 Which could be the graph of $y = x^2(x - 3)$?

A
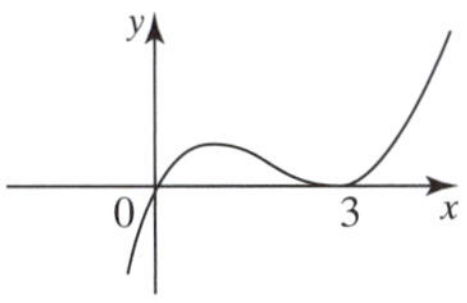

B
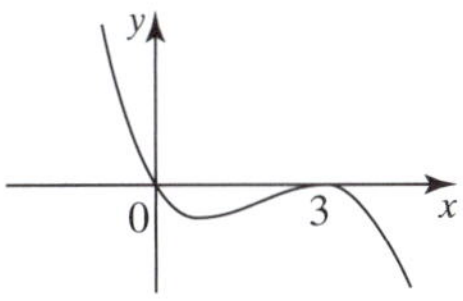

C
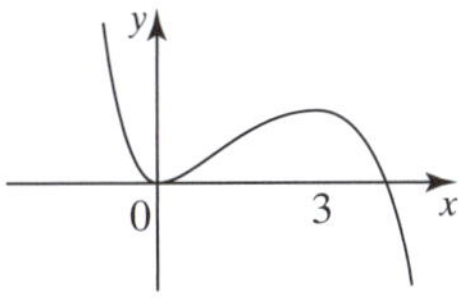

D
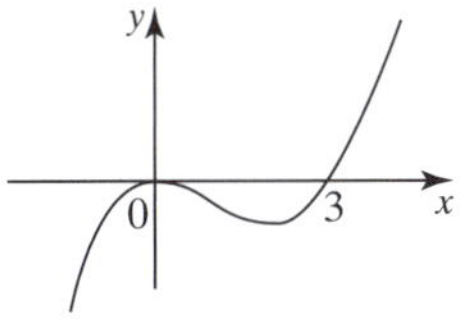

(1 mark)

Your Score

/ 25

PAGES 176–178

Part B Written Response (25 marks)

26 Factorise fully:

a $4x^2 + 10x + 6$ (1 mark)

b $x^2 + xy - x - y$ (2 marks)

27 Solve $n^2 - 6n = 18$ by completing the square, leaving the answer in simplest surd form. (2 marks)

28 A boat sails 16 km from P to Q on a bearing of 052°. From Q it sails 28 km on a bearing of 118° to R. (Diagram is not to scale)

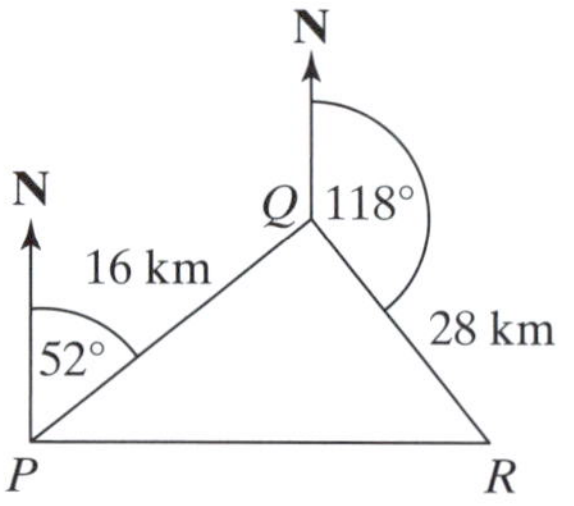

a Find the size of $\angle PQR$ (1 mark)

b Find the distance from P to R. (2 marks)

c Find the bearing of P from R. (2 marks)

29 Simplify $3\log_3 6 - (\log_3 12 + \log_3 2)$. (2 marks)

30 O is the centre of the circle of radius 8 m. P is the point of contact of the tangent drawn from Q to the circle. OQ meets the circle at R. $RQ = 9$ m. Find the length of PQ. (2 marks)

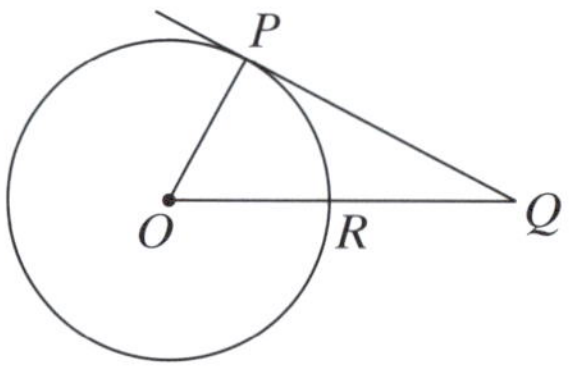

31 Find the values of a and b such that $\frac{4}{3 + \sqrt{7}} = a + b\sqrt{7}$. (2 marks)

32 Find the points of intersection of the line $y = 3x + 1$ with the parabola $y = x^2 + 2x - 5$. (2 marks)

33 The surface area of a sphere is 2124 cm^2. Find the volume of the sphere. (3 marks)

34 The marks in a music test were 78, 86, 52, 67, 74, 82, 91, 66, 81, 60, 78, 75.

a Determine the mean mark. (1 mark)

b What is the standard deviation? (1 mark)

c Ryan was one of the students who sat for the music test. He scored 81. In an art exam Ryan scored 73, the mean mark was 66 and the standard deviation was 3.5. Did Ryan perform better in the art exam or the music test? Justify your answer. (2 marks)

Your Score

$\frac{\square}{25}$

PAGES 178–179

Your Feedback

$\frac{\square}{\text{Part A}} + \frac{\square}{\text{Part B}} = \frac{\square}{50} = \square\%$

PAGE 182

SAMPLE EXAM PAPERS

Level of difficulty—Difficult

60 MINUTES

PAPER 3

Part A Multiple Choice (20 marks)

1 The parabola $y = x^2 - 6x + 14$ will meet the x-axis in how many places?

A 0 **B** 1 **C** 2 **D** more than 2 (1 mark)

2 $(16x^{-\frac{1}{2}})^{\frac{1}{2}} = ?$

A $\dfrac{4}{\sqrt[4]{x}}$ **B** $\dfrac{2}{\sqrt[4]{x}}$ **C** $\dfrac{1}{4\sqrt[4]{x}}$ **D** $\dfrac{1}{2\sqrt[4]{x}}$ (1 mark)

3 The equations of the asymptotes of the curve $y = \dfrac{4}{x+2}$ are:

A $x = 2$ and $y = 0$ **B** $x = -2$ and $y = 0$ **C** $x = 2$ and $y = 4$ **D** $x = -2$ and $y = 4$ (1 mark)

4 If $x = 2at$, then $x^2 - tx = ?$

A 0 **B** $2at^2$ **C** $2a^2t^2 - 2at^2$ **D** $4a^2t^2 - 2at^2$ (1 mark)

5 Which set of scores is the most consistent?

A 10, 12, 13, 13, 13, 15, 17, 18, 18, 20
B 10, 11, 12, 14, 15, 15, 16, 18, 19, 20
C 10, 10, 12, 12, 14, 16, 18, 18, 20, 20
D 10, 11, 12, 13, 15, 15, 17, 18, 19, 20 (1 mark)

6 Chords AB and PQ intersect at X.
$AX = 18$ cm. $BX = 24$ cm and $QX = 12$ cm.
What is the length of PQ?

A 21 cm
B 36 cm
C 42 cm
D 48 cm (1 mark)

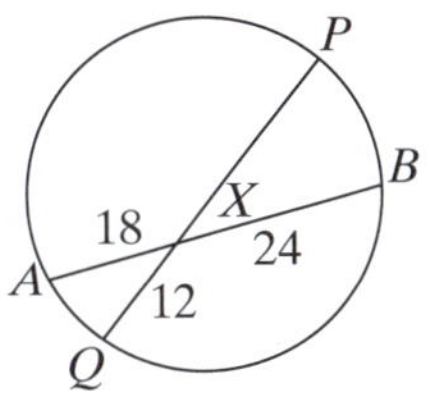

7 To the nearest degree, θ = ?

A 18°
B 27°
C 34°
D 56° (1 mark)

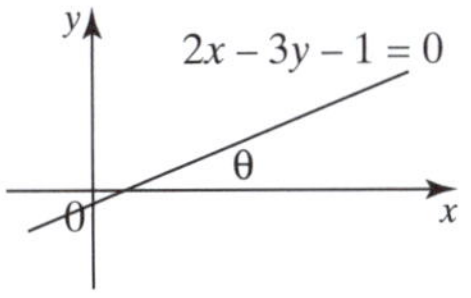

8 $ABCD$ is a parallelogram. E is a point on diagonal BD such that $AD = ED = CD$.
$\angle BAE = 42°$. What is the size of $\angle BCD$?

A 106°
B 111°
C 116°
D 138° (1 mark)

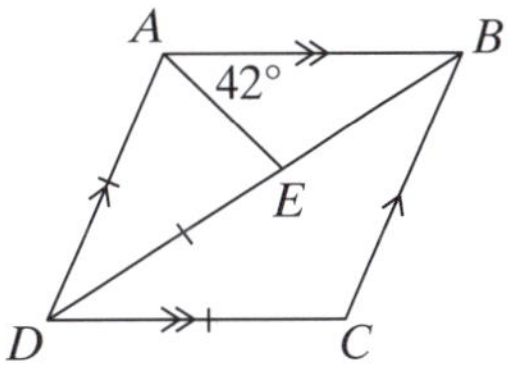

9 $\sin 40° = ?$
Which statement is correct?
I $\cos 50°$ II $\sin 140°$

A I only **B** II only **C** neither I nor II **D** both I and II (1 mark)

10 The inverse function of $y = 3x - 6$ is:

A $y = \frac{1}{3}x + 6$ **B** $y = \frac{1}{3}x + 2$ **C** $y = \frac{1}{3}x - 2$ **D** $y = \frac{1}{3}x - 6$ (1 mark)

11 Which expression will give the value of x?

A $\frac{25}{2\sin 70°}$ **B** $50\sin 70°$

C $\frac{1}{50\sin 70°}$ **D** $\frac{2\sin 70°}{25}$

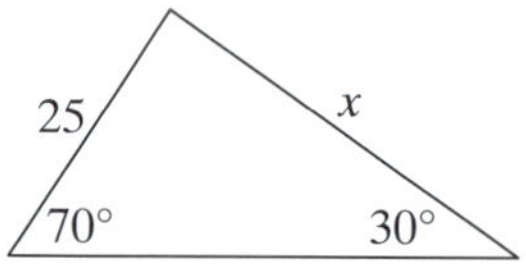

(1 mark)

12 If θ is an acute angle and $\cos\theta = x$ and $\sin\theta = y$, what is $\tan\theta$?

A $\frac{x}{y}$ **B** $\frac{y}{x}$ **C** $\sqrt{x^2 + y^2}$ **D** 1 (1 mark)

13 $\frac{x^2 + 7x + 10}{x^2 - x - 6} \times \frac{x^2 - 7x + 6}{x^2 + 4x - 5} = ?$

A $\frac{x-6}{x-3}$ **B** $\frac{x+2}{x-1}$ **C** $\frac{x-3}{x+2}$ **D** $\frac{x-1}{x-2}$ (1 mark)

14 The solution to the equation $3x^2 - 6x - 1 = 0$ is:

A $x = \frac{3 \pm \sqrt{6}}{3}$ **B** $x = \frac{3 \pm 2\sqrt{3}}{3}$ **C** $x = 1 \pm 4\sqrt{3}$ **D** $x = 1 \pm 2\sqrt{3}$ (1 mark)

15 What is the radius of the circle $x^2 - 6x + y^2 + 4y = 36$?

A 4 **B** 5 **C** 6 **D** 7 (1 mark)

16 O is the centre of the circle.
PQ is parallel to RO. $\angle PQR = 24°$.
What is the size of $\angle OPQ$?

A 12°
B 24°
C 36°
D 48°

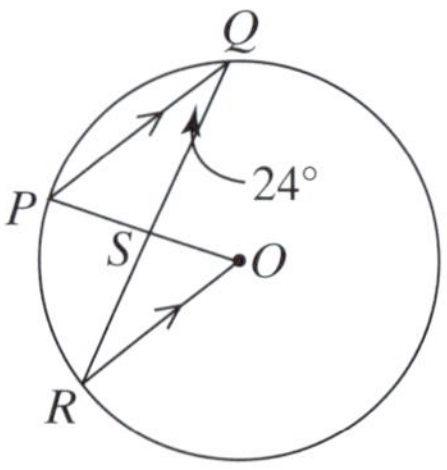

(1 mark)

17 If $\tan A = -0.875$ and $\angle A$ is an obtuse angle, what is the size of $\angle A$ to the nearest whole degree?

A 119° **B** 139° **C** 151° **D** 153° (1 mark)

18 A cylinder has volume 108 cm³. What is the volume of a cone with twice the radius and half the height of the cylinder?

A 18 cm³ **B** 36 cm³ **C** 72 cm³ **D** 144 cm³ (1 mark)

19 Two similar figures have volumes in the ratio 1 : 8. The surface area of the larger figure is 72 m². What is the surface area of the smaller figure?

A 9 m² **B** 16 m² **C** 18 m² **D** 36 m² (1 mark)

20 $P(x) = x^3 - 4x^2 + mx + n$. When $P(x)$ is divided by $(x - 4)$ the remainder is 7 and when $P(x)$ is divided by x the remainder is -5. What is the remainder when $P(x)$ is divided by $(x + 2)$?

A 2 **B** -12 **C** -35 **D** 70 (1 mark)

Your Score

/20

PAGES 179–180

Part B Written Response (30 marks)

21 By letting $u = x^2$, solve $x^4 - 10x^2 + 9 = 0$. (2 marks)

22 Find $\frac{1}{\sqrt{3}-2} + \frac{2}{\sqrt{5}-\sqrt{3}}$ in simplest form. (2 marks)

23 Solve $2\sin\theta - \sqrt{2} = 0$ for $0° \le \theta \le 360°$. (2 marks)

24 Simplify $\frac{x}{x^2-1} - \frac{x+1}{x-1}$. (2 marks)

25 Find m if $\log_m 4 = \log_8 2$. (1 mark)

26 If $\log_7(x + y) = 0$ and $\log_7(x - y) = 1$, find x and y. (2 marks)

27 The mass y kg and height h cm, was recorded on a scatterplot for a group of people. A line of fit was then drawn on the graph. What is the equation of the line of fit?

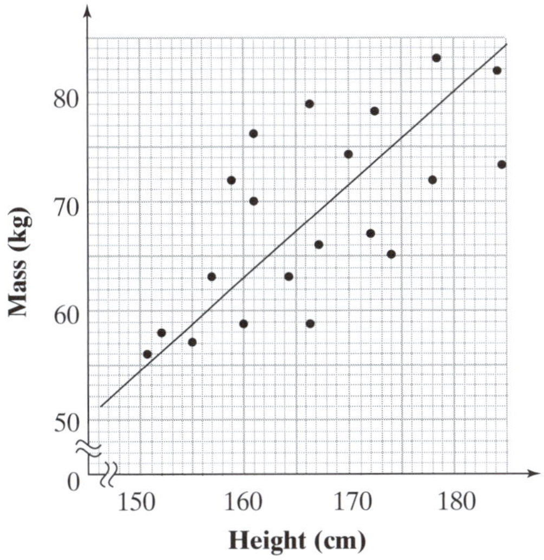

(2 marks)

28 a Show that $x - 3$ is a factor of $x^3 - 4x^2 + x + 6$. (1 mark)

b Hence factorise $x^3 - 4x^2 + x + 6$. (2 marks)

c Draw a sketch of the graph of $y = x^3 - 4x^2 + x + 6$. (2 marks)

29 P is the top and Q is the base of a tower that is 120 m high. Points A, B and Q are all on level ground. From A, which is due south of Q the angle of elevation of P is 45°. From B, due east of Q, the angle of elevation of P is 28°.

a Show the information on a diagram. (1 mark)
b Briefly explain why AQ is 120 m. (1 mark)
c Find the length of BQ to the nearest metre. (2 marks)
d Find the length of AB to the nearest metre. (1 mark)

30 AB is a chord of a circle of radius 89 cm. The perpendicular distance from the centre of the circle, O, to the chord is 39 cm.

a Show this information on a diagram and shade the minor segment. (1 mark)
b Find the length of the chord AB. (2 marks)
c Find the size of $\angle AOB$ to the nearest degree. (2 marks)
d Find the shaded area to the nearest square centimetre. (2 marks)

Your Score

$\frac{\square}{30}$

PAGES 180–181

Your Feedback

$\frac{\square}{\text{Part A}} + \frac{\square}{\text{Part B}} = \frac{\square}{50} = \square\%$

PAGE 182

WORKED SOLUTIONS

CHECK YOUR SOLUTIONS

RATIONAL AND IRRATIONAL NUMBERS
SKILLS CHECK PAGE 2

1 a $\frac{2}{3}$ is rational

b -7 is rational $[\frac{-7}{1}]$

c $\sqrt{3}$ is irrational

d 0 is rational

e π is irrational

[Sometimes the value $\frac{22}{7}$ is given as an approximate value of π, but it is only an approximate value. The exact value of π cannot be written as a fraction.]

2 The value of x is the same as the length of the hypotenuse of the triangle.

$x^2 = 1^2 + 1^2$
$= 2$
$x = \sqrt{2}$

3 a $(\sqrt{4})^2 = 2^2 = 4$

b $(\sqrt{9})^2 = 3^2 = 9$

c $(\sqrt{36})^2 = 36$

d $(\sqrt{7})^2 = 7$

e $(\sqrt{5})^2 = 5$

f $(2\sqrt{2})^2 = 4 \times 2$
$= 8$
$[2^2 = 4,\ (\sqrt{2})^2 = 2]$

4 a $4^{\frac{1}{2}} = 2$

b $8^{\frac{1}{3}} = \sqrt[3]{8} = 2$

c $25^{\frac{1}{2}} = 5$

d $1000^{\frac{1}{3}} = \sqrt[3]{1000} = 10$

e $32^{\frac{2}{5}} = \left(\sqrt[5]{32}\right)^2 = 4$

f $9^{\frac{3}{2}} = (\sqrt{9})^3 = 27$

g $\sqrt[8]{6561} = 3$

h $\sqrt[6]{11\,390\,625} = 15$

5 a $\sqrt{x} = x^{\frac{1}{2}}$

b $\sqrt[3]{x} = x^{\frac{1}{3}}$

c $\frac{\sqrt{a}}{2} = \frac{1}{2}a^{\frac{1}{2}}$ $\left[\text{or } \frac{a^{\frac{1}{2}}}{2}\right]$

6 a $m^{\frac{1}{2}} = \sqrt{m}$

b $t^{\frac{1}{3}} = \sqrt[3]{t}$

c $4x^{\frac{1}{2}} = 4 \times x^{\frac{1}{2}} = 4\sqrt{x}$

d $a^{\frac{2}{3}} = (\sqrt[3]{a})^2 = \sqrt[3]{a^2}$

e $x^{\frac{3}{2}} = (\sqrt{x})^3 = \sqrt{x^3}$

f $(9n)^{\frac{1}{2}} = \sqrt{9n} = 3\sqrt{n}$

7 a $\frac{2}{9} = 0.222\,2222\ldots$ $[2 \div 9]$
$= 0.\dot{2}$

b $\frac{5}{12} = 0.416\,666\,666\ldots$ $[5 \div 12]$
$= 0.41\dot{6}$

c $1\frac{1}{6} = 1.166\,666\,666\ldots$
$= 1.1\dot{6}$

d $\frac{4}{7} = 0.571\,428\,571\ldots$
$= 0.\dot{5}7142\dot{8}$

8 a $\frac{2}{5} = 0.4$
terminating

b $\frac{2}{3} = 0.666\,666\,66\ldots$
recurring

c $\frac{3}{7} = 0.428\,571\,428\ldots$
recurring

d $\frac{7}{8} = 0.875$
terminating

9 a $\sqrt{2a} = (2a)^{\frac{1}{2}} \neq 2a^{\frac{1}{2}}$ False

b $0.\dot{3} = 0.333\,33\ldots = \frac{1}{3}$ True

c $0.\dot{5} = 0.555\,555\ldots$
$\frac{1}{5} = 0.2$ False

10 a $2x^{\frac{1}{2}} \times 4x^{\frac{1}{2}} = 8x$

b $20p^{\frac{1}{2}} \div 2p^{\frac{1}{2}} = 10$

c $6y^{\frac{1}{3}} \times 3y^{\frac{1}{3}} = 18y^{\frac{2}{3}}$

RATIONAL AND IRRATIONAL NUMBERS
INTERMEDIATE TEST PAGE 3

1 $\sqrt{6}$ is irrational
[It cannot be expressed as a fraction.]
$\therefore$ [C] ✓ (1 mark)

2 $8^{\frac{4}{3}} = (\sqrt[3]{8})^4$
$= 2^4$
$= 16$
$\therefore$ [D] ✓ (1 mark)

3 $\sqrt{64a^{64}} = 8 \times \sqrt{a^{64}}$
$= 8 \times (a^{64})^{\frac{1}{2}}$
$= 8a^{32}$
$\therefore$ [B] ✓ (1 mark)

4 $\frac{1}{3} = 0.333\,333\ldots$
$\frac{5}{6} = 0.833\,3333\ldots$
$\frac{7}{8} = 0.875$
$\frac{7}{9} = 0.777\,777\,77\ldots$
[Divide the numerator by the denominator to change a fraction to a decimal.]
$\therefore$ [C] ✓ (1 mark)

5 $8n^{\frac{1}{3}} = 8 \times n^{\frac{1}{3}}$
$= 8 \times \sqrt[3]{n}$
$= 8\sqrt[3]{n}$
$\therefore$ [B] ✓ (1 mark)

6 $\frac{2}{3} = 0.666\,6666\ldots$
$= 0.7$ (1 d.p.)
$\therefore$ [D] ✓ (1 mark)

7 $a^{\frac{m}{n}} = \sqrt[n]{a^m}$
$\therefore$ [A] ✓ (1 mark)

8 $\sqrt[3]{1000^5} = (\sqrt[3]{1000})^5$
$= 10^5$
$= 100\,000$ ✓ (1 mark)

9 $\left(2x^{\frac{1}{3}}\right)^6 \div 8x^4 = 2^6 \times \left(x^{\frac{1}{3}}\right)^6 \div 8x^4$
$= 64 \times x^{\frac{1}{3}\times 6} \div 8x^4$ ✓
$= 64x^2 \div 8x^4$
$= 8x^{-2}$ ✓
(2 marks)

10 $(9x^2y^4)^{\frac{1}{2}} = 9^{\frac{1}{2}} \times (x^2)^{\frac{1}{2}} \times (y^4)^{\frac{1}{2}}$
$= \sqrt{9} \times x^{2\times\frac{1}{2}} \times y^{4\times\frac{1}{2}}$
$= 3 \times x^1 \times y^2$
$= 3xy^2$ ✓ (1 mark)

11 $\sqrt{9x^{100}y^{16}} = 3x^{50}y^8$ ✓ (1 mark)

12 $\sqrt[3]{8n^{27}} = 2n^9$ ✓ (1 mark)

13 **a** $\sqrt[5]{32^7} = 32^{\frac{7}{5}}$ ✓ (1 mark)

b $\sqrt[5]{32^7} = 128$ ✓ (1 mark)

14 **a** $0.\dot{8} = 0.888\,888\,88\ldots$ ✓
Let $x = 0.888\,888\,88\ldots$
$10x = 8.888\,888\,88\ldots$
$9x = 8$
$x = \frac{8}{9}$
$\therefore\ 0.\dot{8} = \frac{8}{9}$ ✓ (2 marks)

b $0.1\dot{6} = 0.166\,666\,666\ldots$ ✓
Let $x = 0.166\,666\,666\ldots$
$10x = 1.666\,666\,666\ldots$
$9x = 1.5$
$x = \frac{1.5}{9}$ ✓
$= \frac{15}{90}$
$= \frac{1}{6}$
$\therefore\ 0.1\dot{6} = \frac{1}{6}$ ✓ (3 marks)

(Total: 20 marks)

RATIONAL AND IRRATIONAL NUMBERS
ADVANCED TEST PAGE 4

1 **a** $\frac{7}{12}$

$\quad\ \ 0.\ 5\ 8\ 3\ 3\ 3\ldots$
$12\overline{)7.{}^{7}0{}^{10}0{}^{4}0{}^{4}0{}^{4}0\ldots}$

$\therefore \frac{7}{12} = 0.58\dot{3}$ ✓ (1 mark)

b $\frac{37}{40}$

Now $37 \div 40 = 3.7 \div 4$

$\quad 0.\ 9\ 2\ 5$
$4\overline{)3.{}^{3}7{}^{1}0{}^{2}0}$

$\therefore \frac{37}{40} = 0.925$ ✓ (1 mark)

c $\frac{11}{14}$

Now $11 \div 14 = 5.5 \div 7$

$\quad 0.\ 7\ 8\ 5\ 7\ 1\ 4\ 2\ 8\ 5\ldots$
$7\overline{)5.{}^{5}5{}^{6}0{}^{4}0{}^{5}0{}^{1}0{}^{3}0{}^{2}0{}^{6}0{}^{4}0\ldots}$

$\therefore \frac{11}{14} = 0.7\dot{8}5714\dot{2}$ ✓ (1 mark)

2 **a** $0.34\dot{5}$
Let $x = 0.345\,555\,55\ldots$
$10x = 3.455\,555\,55\ldots$
$9x = 3.11$
$x = \frac{3.11}{9}$ ✓
$= \frac{311}{900}$
$\therefore 0.34\dot{5} = \frac{311}{900}$ ✓ (2 marks)

b $0.\overline{345}$
Let $x = 0.345\,345\,345\ldots$
$1000x = 345.345\,345\,345\ldots$
$999x = 345$
$x = \frac{345}{999}$ ✓
$= \frac{115}{333}$
$\therefore 0.\overline{345} = \frac{115}{333}$ ✓ (2 marks)

c $0.3\dot{4}\dot{5}$
Let $x = 0.345\,454\,545\ldots$
$100x = 34.545\,454\,545\ldots$
$99x = 34.2$
$x = \frac{34.2}{99}$ ✓
$= \frac{342}{990}$
$= \frac{19}{55}$
$\therefore 0.3\dot{4}\dot{5} = \frac{19}{55}$ ✓ (2 marks)

3 $3.2\dot{1}\dot{4}$ is rational. It is a repeating decimal and all repeating decimals are rational. ✓ (1 mark)

4 $x = \frac{2}{9}, y = \frac{3}{4}, z = \frac{1}{6}$

$\frac{x^2y^3}{z^4} = \frac{\left(\frac{2}{9}\right)^2 \times \left(\frac{3}{4}\right)^3}{\left(\frac{1}{6}\right)^4}$
$= \frac{2^2}{9^2} \times \frac{3^3}{4^3} \div \frac{1}{6^4}$ ✓
$= \frac{2^2}{9^2} \times \frac{3^3}{4^3} \times 6^4$
$= \frac{2^2 \times 3^3 \times 2^4 \times 3^4}{3^4 \times 2^6}$
$= 3^3$
$= 27$ ✓ (2 marks)

5 **a** Yes the number e is real. All irrational numbers are real so if it is irrational it must be real. ✓ (1 mark)

b The number 2.718 281 828 has some repeating digits but we would need to know for certain that those digits continued to repeat to say it must be rational. As it is an irrational number the digits must obviously not continue to repeat. ✓ (1 mark)

6 **a** $\frac{a}{b^4} = ab^{-4}$ ✓ (1 mark)

b $\frac{1}{\sqrt{a^3}} = a^{-\frac{3}{2}}$ ✓ (1 mark)

c $\sqrt[3]{a^2} = a^{\frac{2}{3}}$ ✓ (1 mark)

WORKED SOLUTIONS

7 a $x^{-\frac{1}{4}} = \frac{1}{\sqrt[4]{x}}$ ✓ (1 mark)

b $x^{\frac{5}{3}} = \sqrt[3]{x^5}$ ✓ (1 mark)

c $x^{-2}y^{\frac{1}{2}} = \frac{\sqrt{y}}{x^2}$ ✓ (1 mark)

8 a $16^{\frac{3}{4}} = \left(\sqrt[4]{16}\right)^3$
$= 2^3$
$= 8$ ✓ (1 mark)

b $64^{\frac{2}{3}} = \left(\sqrt[3]{64}\right)^2$
$= 4^2$
$= 16$ ✓ (1 mark)

c $100^{\frac{3}{2}} = \left(\sqrt{100}\right)^3$
$= 10^3$
$= 1000$ ✓ (1 mark)

d $\left(\frac{1}{2}\right)^{-3} = 2^3$
$= 8$ ✓ (1 mark)

e $\left(1\frac{1}{2}\right)^{-1} = \left(\frac{3}{2}\right)^{-1}$
$= \frac{2}{3}$ ✓ (1 mark)

f $\left(\frac{1}{4}\right)^{-\frac{1}{2}} = 4^{\frac{1}{2}}$
$= 2$ ✓ (1 mark)

g $25^{-\frac{1}{2}} \div 8^{\frac{2}{3}} = \frac{1}{5} \div 4$
$= \frac{1}{20}$ ✓ (1 mark)

h $64^{-\frac{3}{2}} \times \left(\frac{1}{2}\right)^{-5} = \frac{1}{8^3} \times 2^5$
$= \frac{2^5}{2^9}$
$= \frac{1}{2^4}$
$= \frac{1}{16}$ ✓ (1 mark)

i $\sqrt{\frac{3^4}{4^3}} = \frac{\sqrt{3^4}}{\sqrt{4^3}}$
$= \frac{3^2}{2^3}$
$= \frac{9}{8}$
$= 1\frac{1}{8}$ ✓ (1 mark)

9 $\sqrt[3]{a} = a^{\frac{1}{3}}$ ✓
$(a^{\frac{1}{3}})^3 = a^{\frac{1}{3}\times 3} = a^1 = a$
But $(\sqrt[3]{a})^3 = a$
So $a^{\frac{1}{3}} = \sqrt[3]{a}$ ✓ (2 marks)

10 a $4^x = 2^3$
$(2^2)^x = 2^3$
$2^{2x} = 2^3$ ✓
$2^x = 3$
$x = 1.5$ ✓ (2 marks)

b $5^{2x-1} = \frac{1}{125}$
$= 5^{-3}$
$\therefore 2x - 1 = -3$ ✓
$2x = -2$
$x = -1$ ✓ (2 marks)

c $9^{4-x} = 27$
$(3^2)^{4-x} = 3^3$
$3^{8-2x} = 3^3$ ✓
$8 - 2x = 3$
$-2x = -5$
$x = 2.5$ ✓ (2 marks)

11

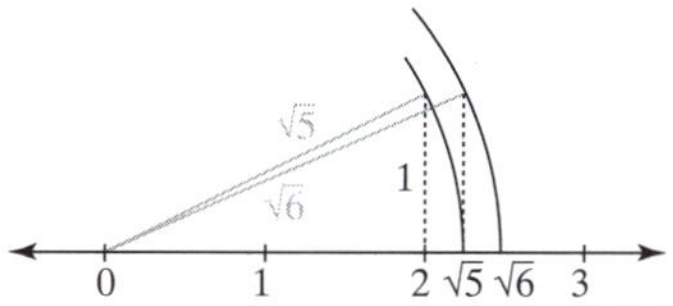

✓✓✓
(3 marks)

(Total: 40 marks)

SURDS
SKILLS CHECK PAGE 6

1 a $3\sqrt{7} + 5\sqrt{7} = 8\sqrt{7}$

b $5\sqrt{2} - \sqrt{2} = 4\sqrt{2}$
$[5\sqrt{2} - 1\sqrt{2}]$

c $\sqrt{6} + \sqrt{6} = 2\sqrt{6}$

d $7 - \sqrt{6} + 5\sqrt{6} - 2 + \sqrt{7}$
$= 5 + 4\sqrt{6} + \sqrt{7}$

e $8\sqrt{3} - 2\sqrt{5} = 8\sqrt{3} - 2\sqrt{5}$
[Unlike surds cannot be added or subtracted.]

f $3\sqrt{10} + 2\sqrt{3} + 5\sqrt{3} - 2\sqrt{10}$
$= \sqrt{10} + 7\sqrt{3}$

2 a $\sqrt{12} = \sqrt{4} \times \sqrt{3}$
$= 2\sqrt{3}$

b $\sqrt{45} = \sqrt{9} \times \sqrt{5}$
$= 3\sqrt{5}$

c $\sqrt{75} = \sqrt{25} \times \sqrt{3}$
$= 5\sqrt{3}$

d $\sqrt{600} = \sqrt{100} \times \sqrt{6}$
$= 10\sqrt{6}$

e $2\sqrt{20} = 2 \times \sqrt{4} \times \sqrt{5}$
$= 2 \times 2 \times \sqrt{5}$
$= 4\sqrt{5}$

f $3\sqrt{8} = 3 \times \sqrt{4} \times \sqrt{2}$
$= 3 \times 2 \times \sqrt{2}$
$= 6\sqrt{2}$

g $6\sqrt{300} = 6 \times \sqrt{100} \times \sqrt{3}$
$= 6 \times 10 \times \sqrt{3}$
$= 60\sqrt{3}$

h $5\sqrt{98} = 5 \times \sqrt{49} \times \sqrt{2}$
$= 5 \times 7 \times \sqrt{2}$
$= 35\sqrt{2}$

3 a $\sqrt{8} + 5\sqrt{2} = \sqrt{4} \times \sqrt{2} + 5\sqrt{2}$
$= 2\sqrt{2} + 5\sqrt{2}$
$= 7\sqrt{2}$

b $\sqrt{18} - \sqrt{98}$
$= \sqrt{9} \times \sqrt{2} - \sqrt{49} \times \sqrt{2}$
$= 3\sqrt{2} - 7\sqrt{2}$
$= -4\sqrt{2}$

c $2\sqrt{300} + 3\sqrt{200} - 4\sqrt{27}$
$= 2 \times \sqrt{100} \times \sqrt{3}$
$+ 3 \times \sqrt{100} \times \sqrt{2}$
$- 4 \times \sqrt{9} \times \sqrt{3}$
$= 20\sqrt{3} + 30\sqrt{2} - 12\sqrt{3}$
$= 8\sqrt{3} + 30\sqrt{2}$

4 a $\sqrt{3} \times \sqrt{5} = \sqrt{15}$

b $2\sqrt{6} \times \sqrt{7} = 2\sqrt{42}$

c $3\sqrt{2} \times 5\sqrt{3} = 15\sqrt{6}$

d $7\sqrt{2} \times \sqrt{2} = 7\sqrt{4}$
$= 7 \times 2$
$= 14$

e $3\sqrt{6} \times 4\sqrt{3} = 12\sqrt{18}$
$= 12 \times \sqrt{9} \times \sqrt{2}$
$= 36\sqrt{2}$

f $-2\sqrt{3} \times 5\sqrt{2} \times 7\sqrt{10}$
$= -70\sqrt{60}$
$= -70 \times \sqrt{4} \times \sqrt{15}$
$= -140\sqrt{15}$

5 a $\sqrt{10} \div \sqrt{2} = \sqrt{5}$

b $6\sqrt{15} \div 2\sqrt{3} = 3\sqrt{5}$

c $8\sqrt{5} \div 2 = 4\sqrt{5}$

d $4\sqrt{7} \div \sqrt{7} = 4$ $[\sqrt{7} \div \sqrt{7} = 1]$

WORKED SOLUTIONS

CHECK YOUR SOLUTIONS

6 a $2(3\sqrt{2}-5)=6\sqrt{2}-10$

b $\sqrt{3}(\sqrt{5}+2)=\sqrt{15}+2\sqrt{3}$

c $3\sqrt{7}(\sqrt{5}-2\sqrt{7})=3\sqrt{35}-6\times 7$
$=3\sqrt{35}-42$

7 a $(\sqrt{3}+2)(\sqrt{5}-3)$
$=\sqrt{15}-3\sqrt{3}+2\sqrt{5}-6$

b $(\sqrt{7}+\sqrt{2})(\sqrt{6}+\sqrt{5})$
$=\sqrt{42}+\sqrt{35}+\sqrt{12}+\sqrt{10}$
$=\sqrt{42}+\sqrt{35}+\sqrt{4}\times\sqrt{3}+\sqrt{10}$
$=\sqrt{42}+\sqrt{35}+2\sqrt{3}+\sqrt{10}$

c $(\sqrt{2}-1)(\sqrt{3}-1)$
$=\sqrt{6}-\sqrt{2}-\sqrt{3}+1$

d $(\sqrt{5}+\sqrt{3})(\sqrt{5}-\sqrt{3})$
$=\sqrt{25}-\sqrt{15}+\sqrt{15}-\sqrt{9}$
$=5-3$
$=2$

e $(2-\sqrt{10})(2+\sqrt{10})$
$=4+2\sqrt{10}-2\sqrt{10}-10$
$=-6$

f $(2\sqrt{3}-\sqrt{5})(2\sqrt{3}+\sqrt{5})$
$=4\sqrt{9}+2\sqrt{15}-2\sqrt{15}-\sqrt{25}$
$=12-5$
$=7$

8 a $\frac{2}{\sqrt{5}}\times\frac{\sqrt{5}}{\sqrt{5}}=\frac{2\sqrt{5}}{5}$

b $\frac{3}{\sqrt{3}}\times\frac{\sqrt{3}}{\sqrt{3}}=\frac{3\sqrt{3}}{3}$
$=\sqrt{3}$

c $\frac{\sqrt{5}}{\sqrt{3}+2}\times\frac{\sqrt{3}-2}{\sqrt{3}-2}=\frac{\sqrt{15}-2\sqrt{5}}{3-4}$
$=\frac{\sqrt{15}-2\sqrt{5}}{-1}$
$=2\sqrt{5}-\sqrt{15}$

SURDS INTERMEDIATE TEST PAGE 7

1 $5\sqrt{5}-4\sqrt{3}+3\sqrt{5}-\sqrt{3}$
$=8\sqrt{5}-5\sqrt{3}$
∴ [A] ✓ (1 mark)

2 $\sqrt{32}+\sqrt{8}$
$=\sqrt{16}\times\sqrt{2}+\sqrt{4}\times\sqrt{2}$
$=4\sqrt{2}+2\sqrt{2}$
$=6\sqrt{2}$
∴ [C] ✓ (1 mark)

3 $\sqrt{100x^{100}}=10x^{50}$
$[\sqrt{100}=10,$
$\sqrt{x^{100}}=(x^{100})^{\frac{1}{2}}=x^{50}]$
∴ [B] ✓ (1 mark)

4 $\sqrt{3}(\sqrt{2}+\sqrt{3})=\sqrt{6}+3$
$[\sqrt{3}\times\sqrt{3}=\sqrt{9}=3]$
∴ [A] ✓ (1 mark)

5 $\frac{14}{\sqrt{7}}=\frac{14}{\sqrt{7}}\times\frac{\sqrt{7}}{\sqrt{7}}$
$=\frac{14\sqrt{7}}{7}$
$=2\sqrt{7}$
∴ [C] ✓ (1 mark)

6 $3\sqrt{2}=4.242\,640\,687\ldots$
It lies between 4 and 5.
∴ [C] ✓ (1 mark)

7 $\sqrt{63}-\sqrt{28}=\sqrt{9}\times\sqrt{7}-\sqrt{4}\times\sqrt{7}$
$=3\sqrt{7}-2\sqrt{7}$
$=\sqrt{7}$
∴ [D] ✓ (1 mark)

8 $\frac{4\sqrt{12}}{2\sqrt{2}}=2\sqrt{6}$ ✓ (1 mark)

9 $3\sqrt{5}=\sqrt{9}\times\sqrt{5}$
$=\sqrt{45}$ ✓ (1 mark)

10 a $\frac{6}{\sqrt{2}}=\frac{6}{\sqrt{2}}\times\frac{\sqrt{2}}{\sqrt{2}}$
$=\frac{6\sqrt{2}}{2}$ ✓
$=3\sqrt{2}$ ✓ (2 marks)

b $\frac{\sqrt{2}}{5\sqrt{10}}=\frac{\sqrt{2}}{5\sqrt{10}}\times\frac{\sqrt{10}}{\sqrt{10}}$
$=\frac{\sqrt{20}}{5\times 10}$ ✓
$=\frac{\sqrt{4}\times\sqrt{5}}{50}$
$=\frac{2\sqrt{5}}{50}$
$=\frac{\sqrt{5}}{25}$ ✓ (2 marks)

11 $\sqrt{6}(\sqrt{2}+\sqrt{3})$
$=\sqrt{12}+\sqrt{18}$ ✓
$=\sqrt{4}\times\sqrt{3}+\sqrt{9}\times\sqrt{2}$
$=2\sqrt{3}+3\sqrt{2}$ ✓ (2 marks)

12 $\sqrt{75}+2\sqrt{50}-8\sqrt{27}$
$=\sqrt{25}\times\sqrt{3}+2\times\sqrt{25}\times\sqrt{2}-8\times\sqrt{9}\times\sqrt{3}$
$=5\sqrt{3}+10\sqrt{2}-24\sqrt{3}$ ✓
$=10\sqrt{2}-19\sqrt{3}$ ✓ (2 marks)

13 a $(5\sqrt{2}-2\sqrt{3})(3\sqrt{2}-4\sqrt{3})$
$=15\times 2-20\sqrt{6}-6\sqrt{6}+8\times 3$ ✓
$=30-26\sqrt{6}+24$
$=54-26\sqrt{6}$ ✓ (2 marks)

b $(2\sqrt{7}-3)(2\sqrt{7}+3)$
$=(2\sqrt{7})^2-3^2$ ✓
$=28-9$
$=19$ ✓ (2 marks)

14 $\sqrt{48}+\sqrt{108}=x\sqrt{3}$
$\sqrt{48}+\sqrt{108}$
$=\sqrt{16}\times\sqrt{3}+\sqrt{36}\times\sqrt{3}$ ✓
$=4\sqrt{3}+6\sqrt{3}$
$=10\sqrt{3}$
$\therefore x=10$ ✓ (2 marks)

15 $(3\sqrt{5}+10\sqrt{6})^2$
$=(3\sqrt{5})^2+2\times 3\sqrt{5}\times 10\sqrt{6}+(10\sqrt{6})^2$ ✓
$=45+60\sqrt{30}+600$
$=645+60\sqrt{30}$ ✓ (2 marks)

(Total: 25 marks)

SURDS ADVANCED TEST PAGE 8

1 a $(6\sqrt{2})^2=36\times 2$
$=72$ ✓ (1 mark)

b $\frac{2\sqrt{24}}{8\sqrt{6}}=\frac{\sqrt{4}}{4}$
$=\frac{1}{2}$ ✓ (1 mark)

c $\frac{\sqrt{56}}{\sqrt{7}}=\sqrt{8}$
$=2\sqrt{2}$ ✓ (1 mark)

2 a $\sqrt{700}+\sqrt{200}-\sqrt{98}$
$=\sqrt{100}\times\sqrt{7}+\sqrt{100}\times\sqrt{2}-\sqrt{49}\times\sqrt{2}$ ✓
$=10\sqrt{7}+10\sqrt{2}-7\sqrt{2}$
$=10\sqrt{7}+3\sqrt{2}$ ✓ (2 marks)

b $\sqrt{45}-\sqrt{12}+\sqrt{20}-\sqrt{3}$
$=3\sqrt{5}-2\sqrt{3}+2\sqrt{5}-\sqrt{3}$ ✓
$=5\sqrt{5}-3\sqrt{3}$ ✓ (2 marks)

c $-\sqrt{32}+17-5\sqrt{288}-36$
$=-4\sqrt{2}+17-5\times 12\sqrt{2}-36$ ✓
$=-64\sqrt{2}-19$ ✓ (2 marks)

3 **a** $3\sqrt{6} \times 5\sqrt{2}$
$= 15\sqrt{12}$ ✓
$= 15 \times 2\sqrt{3}$
$= 30\sqrt{3}$ ✓ (2 marks)

b $\sqrt{15} - 2\sqrt{3} \times 3\sqrt{5}$
$= \sqrt{15} - 6\sqrt{15}$ ✓
$= -5\sqrt{15}$ ✓ (2 marks)

4 **a** $h^2 = 16^2 - 9^2$
$= 175$ ✓
$h = \sqrt{175} \quad (h > 0)$
$= 5\sqrt{7}$ ✓ (2 marks)

b

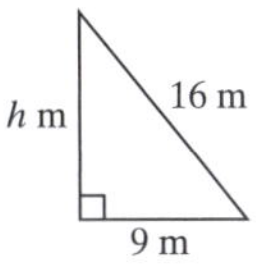

$A = \frac{1}{2}bh$
$= \frac{1}{2} \times 9 \times 5\sqrt{7}$
$= \frac{45\sqrt{7}}{2}\ \text{m}^2$ ✓ (1 mark)

5 $\sqrt{a^2} = -a$ ✓
The square root will always be positive. If a is negative, $-a$ will be positive. ✓ (2 marks)

6 **a** $\sqrt{3456} = \sqrt{36} \times \sqrt{96}$ ✓
$= 6 \times \sqrt{16} \times \sqrt{6}$
$= 24\sqrt{6}$ ✓ (2 marks)
[Any factors could be used to break down the number. In this case it is easy to recognise that 3456 is divisible by 9 and by 4, hence by 36.]

b $\sqrt{2^5 \times 3^3} = \sqrt{2^4 \times 2 \times 3^2 \times 3}$ ✓
$= 2^2 \times \sqrt{2} \times 3 \times \sqrt{3}$
$= 12\sqrt{6}$ ✓ (2 marks)

7 $(2\sqrt{3} - 5\sqrt{2})^2 - (10 - \sqrt{6})^2$
$= 12 - 20\sqrt{6} + 50$
$- (100 - 20\sqrt{6} + 6)$ ✓
$= 62 - 20\sqrt{6} - 106 + 20\sqrt{6}$
$= -44$
$\therefore (2\sqrt{3} - 5\sqrt{2})^2 - (10 - \sqrt{6})^2$ is rational. ✓ (2 marks)

8 **a** $\sqrt{6}(3\sqrt{2} - 2\sqrt{3} - \sqrt{6})$
$= 3\sqrt{12} - 2\sqrt{18} - 6$ ✓
$= 6\sqrt{3} - 6\sqrt{2} - 6$ ✓
(2 marks)

b $(\sqrt{5} - 1)(\sqrt{2} - 3\sqrt{10})$
$= \sqrt{10} - 3\sqrt{50} - \sqrt{2} + 3\sqrt{10}$ ✓
$= 4\sqrt{10} - 15\sqrt{2} - \sqrt{2}$
$= 4\sqrt{10} - 16\sqrt{2}$ ✓
(2 marks)

c $(5\sqrt{3} + \sqrt{7})(5\sqrt{3} - \sqrt{7})$
$-(4 - \sqrt{2})(4 + \sqrt{2})$
$= 75 - 7 - (16 - 2)$ ✓
$= 54$ ✓ (2 marks)

9 **a** $\frac{6\sqrt{2}}{\sqrt{3}} + \frac{\sqrt{12}}{\sqrt{2}}$
$= \frac{6\sqrt{2}}{\sqrt{3}} \times \frac{\sqrt{3}}{\sqrt{3}} + \sqrt{6}$
$= \frac{6\sqrt{6}}{3} + \sqrt{6}$ ✓
$= 3\sqrt{6}$ ✓ (2 marks)

b $\frac{\sqrt{5} - 2}{\sqrt{5} + 2}$
$= \frac{\sqrt{5} - 2}{\sqrt{5} + 2} \times \frac{\sqrt{5} - 2}{\sqrt{5} - 2}$
$= \frac{5 - 4\sqrt{5} + 4}{5 - 4}$ ✓
$= 9 - 4\sqrt{5}$ ✓ (2 marks)

c $\frac{6}{\sqrt{7} + \sqrt{5}}$
$= \frac{6}{\sqrt{7} + \sqrt{5}} \times \frac{\sqrt{7} - \sqrt{5}}{\sqrt{7} - \sqrt{5}}$
$= \frac{6(\sqrt{7} - \sqrt{5})}{7 - 5}$ ✓
$= 3(\sqrt{7} - \sqrt{5})$ ✓ (2 marks)

10 $\frac{5-\sqrt{7}}{3+\sqrt{7}} = a + b\sqrt{7}$
$\frac{5-\sqrt{7}}{3+\sqrt{7}} = \frac{5-\sqrt{7}}{3+\sqrt{7}} \times \frac{3-\sqrt{7}}{3-\sqrt{7}}$
$= \frac{15-5\sqrt{7}-3\sqrt{7}+7}{9-7}$ ✓
$= \frac{22-8\sqrt{7}}{2}$
$= \frac{2(11-4\sqrt{7})}{2}$
$= 11 - 4\sqrt{7}$
$\therefore a = 11$ and $b = -4$ ✓ (2 marks)

11 $\frac{1}{5-2\sqrt{6}} + \frac{1}{5+2\sqrt{6}}$
$= \frac{1}{5-2\sqrt{6}} \times \frac{5+2\sqrt{6}}{5+2\sqrt{6}}$
$+ \frac{1}{5+2\sqrt{6}} \times \frac{5-2\sqrt{6}}{5-2\sqrt{6}}$
$= \frac{5+2\sqrt{6}+5-2\sqrt{6}}{25-24}$ ✓
$= 10$
$\therefore \frac{1}{5-2\sqrt{6}} + \frac{1}{5+2\sqrt{6}}$ is rational ✓
(2 marks)
(Total: 40 marks)

LOGARITHMS
SKILLS CHECK PAGE 10

1 **a** $16 = 4^2$
$\therefore \log_4 16 = 2$
b $2^3 = 8$
$\therefore \log_2 8 = 3$
c $2^4 = 16$
$\therefore \log_2 16 = 4$
d $9^{\frac{1}{2}} = 3$
$\therefore \log_9 3 = \frac{1}{2}$

2 **a** $9 = 3^2$
$\therefore \log_3 9 = 2$
b $4^3 = 64$
$\therefore \log_4 64 = 3$
c $5^{-1} = \frac{1}{5}$
$\therefore \log_5 \frac{1}{5} = -1$
d $36^{\frac{1}{2}} = 6$
$\therefore \log_{36} 6 = \frac{1}{2}$

3 **a** $\log_2 4 = 2$
$\therefore 2^2 = 4$
b $\log_5 125 = 3$
$\therefore 5^3 = 125$
c $\log_2 \frac{1}{2} = -1$
$\therefore 2^{-1} = \frac{1}{2}$
d $\log_{25} 5 = \frac{1}{2}$
$\therefore 25^{\frac{1}{2}} = 5$

4 **a** $\log_2 32 = 5 \quad [32 = 2^5]$

b $\log_3 27 = 3 \quad [27 = 3^3]$

c $\log_5 25 = 2 \quad [25 = 5^2]$

d $\log_7 \frac{1}{7} = -1 \quad [\frac{1}{7} = 7^{-1}]$

e $\log_{16} 4 = \frac{1}{2} \quad [4 = 16^{\frac{1}{2}}]$

f $\log_{10} 10\,000 = 4 \quad [10\,000 = 10^4]$

g $\log_9 \frac{1}{3} = -\frac{1}{2} \quad [\frac{1}{3} = 9^{-\frac{1}{2}}]$

h $\log_{1000} 10 = \frac{1}{3} \quad [10 = 1000^{\frac{1}{3}}]$

i $\log_2 1 = 0 \quad [1 = 2^0]$

j $\log_3 3 = 1 \quad [3 = 3^1]$

k $\log_7 7 = 1 \quad [7 = 7^1]$

l $\log_5 1 = 0 \quad [1 = 5^0]$

5 **a** $\log_a 3 + \log_a 5 = \log_a (3 \times 5)$
$= \log_a 15$

b $\log_b 8 - \log_b 4 = \log_b \frac{8}{4}$
$= \log_b 2$

c $2\log_c 5 = \log_c 5^2$
$= \log_c 25$

6 **a** $\log_6 4 + \log_6 9$
$= \log_6 36 \quad [4 \times 9 = 36]$
$= 2 \quad [36 = 6^2]$

b $\log_2 6 - \log_2 3$
$= \log_2 2 \quad [6 \div 3 = 2]$
$= 1 \quad [2 = 2^1]$

c $2\log_{10} 2 + \log_{10} 25$
$= \log_{10} 4 + \log_{10} 25$
$= \log_{10} 100$
$= 2$

7 **a** $\log_x x^3 = 3$

b $7\log_a x + \log_a x^2$
$= 7\log_a x + 2\log_a x$
$= 9\log_a x \quad [\text{or } \log_a x^9]$

c $3\log_a 4 - 5\log_a 2$
$= 3\log_a 2^2 - 5\log_a 2$
$= 6\log_a 2 - 5\log_a 2$
$= \log_a 2$

8

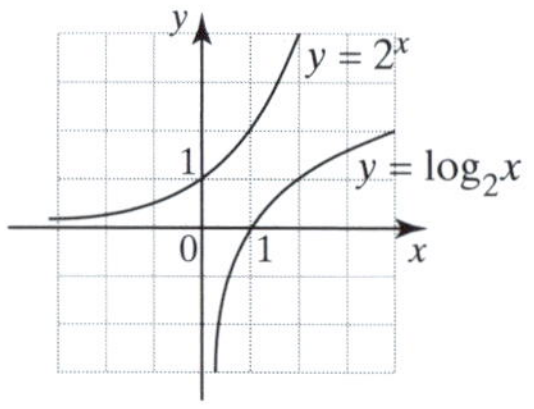

9 **a** $2^{x+1} = 16$
$2^4 = 16$
$\therefore x + 1 = 4$
$x = 3$

b $3^x = \frac{1}{3\sqrt{3}}$
$3^x = 3^{-1.5}$
$x = -1.5$

c $\log_7 x = -2$
$x = 7^{-2}$
$= \frac{1}{49}$

d $\log_8 2 = x$
$8^x = 2$
$x = \frac{1}{3}$

LOGARITHMS
INTERMEDIATE TEST PAGE 11

1 $\frac{\log_m 9}{\log_m 3} = \frac{\log_m 3^2}{\log_m 3}$
$= \frac{2\log_m 3}{\log_m 3}$
$= 2$
$\therefore$ **[A]** ✓ (1 mark)

2 $\log_a 8 + \log_a 4 = \log_a (8 \times 4)$
$= \log_a 32$
$\therefore$ **[C]** ✓ (1 mark)

3 $a^x = y$
$\therefore \log_a y = x$
$\therefore$ **[B]** ✓ (1 mark)

4 Let $\log_8 4 = x$
$\therefore 8^x = 4$
$(2^3)^x = 2^2$
$2^{3x} = 2^2$
$3x = 2$
$x = \frac{2}{3}$
$\therefore$ **[B]** ✓ (1 mark)

5 $\log_a b - \log_a c = \log_a \frac{b}{c}$
$\therefore$ **[A]** ✓ (1 mark)

6 $\log_2 10 + \log_2 6 - \log_2 15$
$= \log_2 \left(\frac{10 \times 6}{15}\right)$
$= \log_2 4$ ✓
$= 2 \quad [\text{since } 2^2 = 4]$ ✓
(2 marks)

7 $\log_3 12 - 2\log_3 2$
$= \log_3 12 - \log_3 2^2$
$= \log_3 12 - \log_3 4$
$= \log_3 \frac{12}{4}$ ✓
$= \log_3 3$
$= 1$ ✓ (2 marks)

8 $\log_a a^4 - \log_a a^3$
$= 4\log_a a - 3\log_a a$ ✓
$= 4 - 3$
$= 1$ ✓ (2 marks)

9 $9^x = \frac{1}{27}$
$(3^2)^x = 3^{-3}$ ✓
$2x = -3$
$x = -\frac{3}{2} \left(\text{or } -1\frac{1}{2}\right)$ ✓ (2 marks)

10

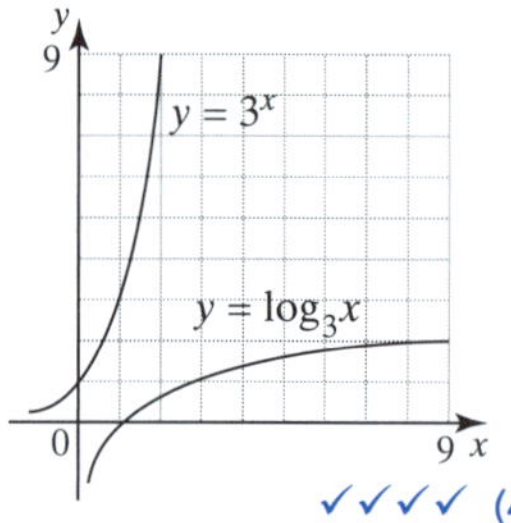

✓✓✓✓ (4 marks)

11 $\log_a \frac{1}{5} = \log_a 5^{-1}$ ✓
$= -\log_a 5$ ✓ (2 marks)

12 $\log_3 x - \log_3 2 = 7$
$\log_3 \frac{x}{2} = 7$ ✓
$\frac{x}{2} = 3^7$
$= 2187$
$x = 4374$ ✓
(2 marks)

13 $\log_5 \frac{x^2}{y} = \log_5 x^2 - \log_5 y$
$= 2\log_5 x - \log_5 y$ ✓
$= 2 \times 0.683 - 1.209$
$= 0.157$ ✓ (2 marks)

14 $\log_2 a = 3.17$
$\log_2 4a = \log_2 (4 \times a)$
$= \log_2 4 + \log_2 a$ ✓
$= 2 + 3.17$
$= 5.17$ ✓ (2 marks)

(Total: 25 marks)

LOGARITHMS
ADVANCED TEST PAGE 12

1 **a** $\log_{32}2 = \frac{1}{5}$ ✓ (1 mark)

$[2^5 = 32 \text{ so } 32^{\frac{1}{5}} = 2]$

b $\log_8\left(\frac{1}{2}\right) = -\frac{1}{3}$ ✓ (1 mark)

c $\log_{100}1000 = \frac{3}{2}$ ✓ (1 mark)

2 If $a^{3.48} = 11$

a $\log_a 11 = 3.48$ ✓ (1 mark)

b $\log_a\left(\frac{1}{11}\right) = \log_a 11^{-1}$

$= -\log_a 11$

$= -3.48$ ✓ (1 mark)

c $\log_a 121 = \log_a 11^2$

$= 2\log_a 11$

$= 2 \times 3.48$

$= 6.96$ ✓ (1 mark)

3 $\log_3 2 = 0.630\,929\,75\ldots$

a $\log_3 16 = \log_3 2^4$

$= 4\log_3 2$

$= 4 \times 0.630\,929\,75\ldots$

$= 2.523\,719\ldots$

$= 2.5237$ (4 d.p.) ✓

(1 mark)

b $\log_3 36$

$= \log_3 6^2$

$= 2\log_3 6$

$= 2\log_3(2 \times 3)$

$= 2(\log_3 2 + \log_3 3)$ ✓

$= 2(0.630\,929\,75\ldots + 1)$

$= 3.261\,859\ldots$

$= 3.2619$ (4 d.p.) ✓

(2 marks)

4 $\log_a 5 = x$, $\log_a 3 = y$ and $\log_a 2 = z$

a $\log_a 50 = \log_a(25 \times 2)$

$= \log_a 5^2 + \log_a 2$ ✓

$= 2\log_a 5 + \log_a 2$

$= 2x + z$ ✓ (2 marks)

b $\log_a\left(\frac{5}{6}\right)$

$= \log_a 5 - \log_a(3 \times 2)$ ✓

$= \log_a 5 - \log_a 3 - \log_a 2$

$= x - y - z$ ✓ (2 marks)

5 **a** $\log_4 8 + \log_9 27 = \frac{3}{2} + \frac{3}{2}$ ✓

$= 3$ ✓

(2 marks)

b $\log_3 48 - 4\log_3 2$

$= \log_3 48 - \log_3 2^4$

$= \log_3 48 - \log_3 16$

$= \log_3\left(\frac{48}{16}\right)$ ✓

$= \log_3 3$

$= 1$ ✓ (2 marks)

c $\log_2 81 \div \log_2(\sqrt[3]{3})$

$= \log_2 3^4 \div \log_2 3^{\frac{1}{3}}$ ✓

$= \dfrac{4\log_2 3}{\frac{1}{3}\log_2 3}$

$= 12$ ✓ (2 marks)

d $\log_6\left(\frac{1}{9}\right) + \log_6\left(\frac{1}{4}\right)$

$= \log_6\left(\frac{1}{36}\right)$ ✓

$= \log_6 6^{-2}$

$= -2$ ✓ (2 marks)

e $-\log_a\left(\frac{1}{a}\right) - \log_a a^6$

$= -\log_a a^{-1} - \log_a a^6$ ✓

$= 1 - 6$

$= -5$ ✓ (2 marks)

f $\dfrac{\log_a 32}{\log_a 8} = \dfrac{\log_a 2^5}{\log_a 2^3}$

$= \dfrac{5\log_a 2}{3\log_a 2}$ ✓

$= \dfrac{5}{3}$

$= 1\frac{2}{3}$ ✓ (2 marks)

6 $\log_a b = c$

$\log_a ab = \log_a a + \log_a b$ ✓

$= 1 + c$ ✓ (2 marks)

7 $\log_4 x = 0.862$

$\therefore x = 4^{0.862}$

$\log_2 x = \log_2 4^{0.862}$ ✓

$= 0.862\log_2 4$

$= 0.862 \times 2$

$= 1.724$ ✓ (2 marks)

8 $y = \log_2 x$

When $x = 8$, $y = \log_2 8$

$= 3$

When $y = 0$,

$\log_2 x = 0$

$x = 2^0$

$= 1$ ✓

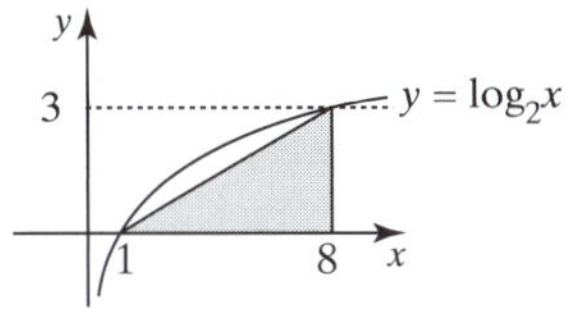

$A = \frac{1}{2}bh$

$= \frac{1}{2} \times 7 \times 3$

$= 10.5 \text{ units}^2$ ✓ (2 marks)

9 $\log_a b = 1.209$, $\log_a c = 0.683$

$\log_a\sqrt{\frac{b}{c}} = \log_a\left(\frac{b}{c}\right)^{\frac{1}{2}}$

$= \frac{1}{2}\log_a\left(\frac{b}{c}\right)$

$= \frac{1}{2}(\log_a b - \log_a c)$ ✓

$= \frac{1}{2}(1.209 - 0.683)$

$= 0.263$ ✓ (2 marks)

10 $\log_a x + \log_a 3 = 3\log_a 6$

$\log_a(3x) = \log_a 6^3$ ✓

$3x = 216$

$x = 72$ ✓ (2 marks)

11 $4^x = \dfrac{1}{2\sqrt{2}}$

$(2^2)^x = 2^{-1.5}$

$2^{2x} = 2^{-1.5}$ ✓

$2x = -1.5$

$x = -0.75$ ✓ (2 marks)

12 $2\log_a x = \log_a(3x + 10)$

$\log_a x^2 = \log_a(3x + 10)$

$x^2 = 3x + 10$ ✓

$x^2 - 3x - 10 = 0$

$(x - 5)(x + 2) = 0$

$x = 5$ or $x = -2$ ✓

But $x > 0$,

$\therefore x = 5$ ✓ (3 marks)

(Total: 40 marks)

FUNCTIONS
SKILLS CHECK PAGE 14

1 A function is a rule where for each input there is only one output.

2 **a** function [A vertical line will only cut the graph in one place, no matter where it is drawn.]
b function
c not a function
d not a function

3 $f(x) = 2x + 3$
a $f(4) = 2 \times 4 + 3$
$= 11$
b $f(0) = 2 \times 0 + 3$
$= 3$
c $f(-1) = 2 \times -1 + 3$
$= 1$

4 **a** $f(x) = 2x^2$
$f(3) = 2 \times 3^2$
$= 2 \times 9$
$= 18$
b $f(x) = 5x - x^2$
$f(3) = 5 \times 3 - 3^2$
$= 15 - 9$
$= 6$
c $f(x) = \dfrac{x^2 - 1}{x^2 + 1}$
$f(3) = \dfrac{3^2 - 1}{3^2 + 1}$
$= 0.8$

5 **a** $f(x) = x$
$f(-2) = -2$
b $f(x) = x^2 - 6x + 5$
$f(-2) = (-2)^2 - 6 \times (-2) + 5$
$= 4 + 12 + 5$
$= 21$
c $f(x) = \dfrac{4}{x}$
$f(-2) = \dfrac{4}{-2}$
$= -2$

6 **a** $y = 2x + 3$
Possible x-values are all real numbers. [Any number can be doubled and have 3 added to it.]
b $y = \dfrac{6}{x}$
All real numbers except $x = 0$.
c $y = x^2$
All real numbers. [Any number can be squared.]
d $y = \sqrt{x}$
$x \geq 0$ [We cannot take the square root of a negative number.]
e $y = 2^x$
All real numbers. [2 can be raised to any power.]

7 **a** $y = 2x + 3$
All real numbers [Any value of y will have a corresponding x value.]
b $y = \dfrac{6}{x}$
All real numbers except $y = 0$. [No number divided into 6 will give 0.]
c $y = x^2$
$y \geq 0$ [No number squared will give a negative.]
d $y = \sqrt{x}$
$y \geq 0$ [The square root of a number is never negative.]
e $y = 2^x$
$y > 0$ [Two raised to any power is positive.]

8 **a** $y = x + 2$
Inverse function is
$x = y + 2$
$y = x - 2$
b $y = \sqrt{x}$ $(x \geq 0)$
Inverse function is
$x = \sqrt{y}$
$y = x^2$ $(x \geq 0)$
c $y = \dfrac{1}{x}$
Inverse function is
$x = \dfrac{1}{y}$
$y = \dfrac{1}{x}$
d $y = x^3$
Inverse function is
$x = y^3$
$y = \sqrt[3]{x}$

FUNCTIONS
INTERMEDIATE TEST PAGE 15

1 $f(x) = 2x^2 - x$
$f(-1) = 2 \times (-1)^2 - (-1)$
$= 2 \times 1 + 1$
$= 3$
$\therefore$ **[A]** ✓ (1 mark)

2 C is not a function.

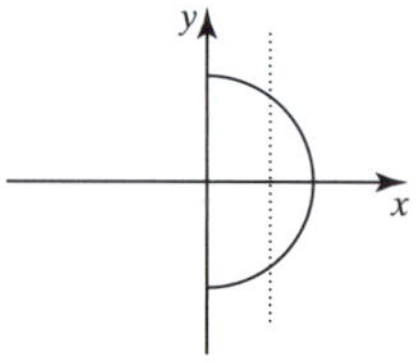

$\therefore$ **[C]** ✓ (1 mark)

3 $y = 2x - 4$
The inverse function is
$x = 2y - 4$
$2y = x + 4$
$y = \dfrac{1}{2}x + 2$
$\therefore$ **[A]** ✓ (1 mark)

4 $y = \dfrac{x}{7}$ is the graph of a straight line. The permissible x-values are all real numbers.
[In A $x \neq 0$. In B $x \geq 0$.
In D $x > 0$. ($x^{-0.5} = \dfrac{1}{\sqrt{x}}$)]
$\therefore$ **[C]** ✓ (1 mark)

5 If $f(x) = 2^x$
$f(0) = 2^0$
$= 1$
$\therefore$ **[D]** ✓ (1 mark)

6 **a** $f(x) = \sqrt{25 - x^2}$
$f(3) = \sqrt{25 - 3^2}$
$= 4$ ✓ (1 mark)
b $f(x) = \dfrac{1}{2x - 1}$
$f(3) = \dfrac{1}{2 \times 3 - 1}$
$= \dfrac{1}{5}$ ✓ (1 mark)
c $f(x) = 2x^3 - 5x^2 - 7x + 4$
$f(3) = 2 \times 3^3 - 5 \times 3^2 - 7 \times 3 + 4$
$= -8$ ✓ (1 mark)

7 $f(x) = \begin{cases} 3x^2 & \text{if } x < 2 \\ x^2 + 3 & \text{if } x \geq 2 \end{cases}$
a $f(-1) = 3 \times (-1)^2$
$= 3$ ✓ (1 mark)
b $f(7) - 1 = 7^2 + 3 - 1$
$= 51$ ✓ (1 mark)
c $f(2) = 2^2 + 3$
$= 7$ ✓ (1 mark)

8 $f(x) = x^2 - 5x$
$f(a) = a^2 - 5a$
Now $f(a) = -4$
$a^2 - 5a = -4$ ✓
$a^2 - 5a + 4 = 0$
$(a - 4)(a - 1) = 0$
$a = 4$ or $a = 1$ ✓
(2 marks)

9 **a**
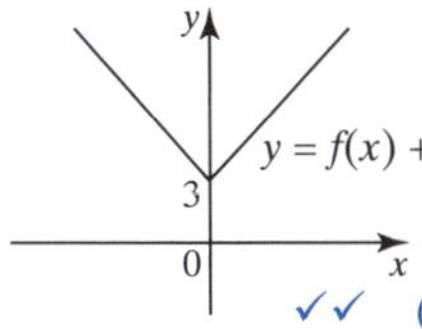

✓✓ (2 marks)

b
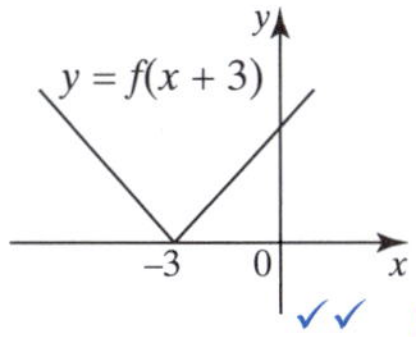

✓✓ (2 marks)

10 $f(x) = 8 - 4x$

a $f(-2) = 8 - 4 \times -2$
$= 16$ ✓ (1 mark)

b $f(x) = 0$
$8 - 4x = 0$
$-4x = -8$
$x = 2$ ✓ (1 mark)

c Permissible x-values are all real numbers. ✓
[Taking four times any number from eight will always give another real number.]
Permissible y-values are all real numbers. ✓ (2 marks)

d $y = 8 - 4x$
Inverse function is
$x = 8 - 4y$
$4y = -x + 8$
$y = -\frac{1}{4}x + 2$ ✓ (1 mark)

e
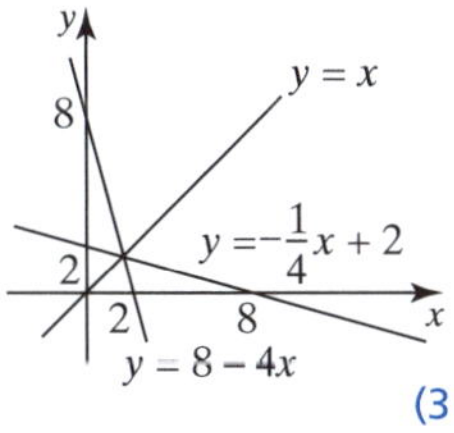

✓✓✓ (3 marks)
(Total: 25 marks)

FUNCTIONS
ADVANCED TEST PAGE 16

1 **a** $y = x^3 - 2$
Function ✓ (1 mark)

b $x^2 + y^2 = 16$
Not a function
[It fails the vertical line test.]
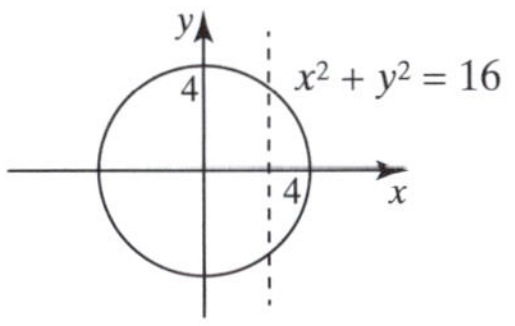

✓ (1 mark)

c $x = y^2$
Not a function
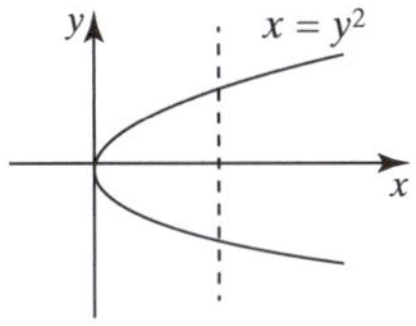

✓ (1 mark)

d $y = 4^{-x}$
Function ✓ (1 mark)

2 No, Jill is not correct. All straight lines are functions except for vertical lines. Any straight line of the form $x = k$, where k is a constant, is not a function. ✓ (1 mark)

3 $f(x) = \begin{cases} 7 - x & \text{for } x \le -1 \\ 5 & \text{for } -1 < x < 0 \\ x^2 & \text{for } x \ge 0 \end{cases}$

a **i** $f(-4) = 7 - (-4)$
$= 11$ ✓ (1 mark)

ii $f\left(-\frac{1}{2}\right) + f\left(\frac{1}{2}\right)$
$= 5 + \left(\frac{1}{2}\right)^2$
$= 5\frac{1}{4}$ ✓ (1 mark)

iii $f(\sqrt{a}) = (\sqrt{a})^2$
$= a$ ✓ (1 mark)
[The square root will not be negative.]

b
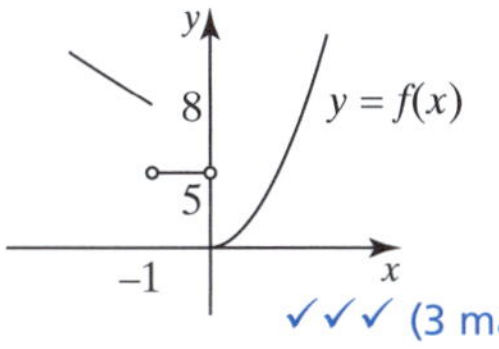

✓✓✓ (3 marks)

4 $f(x) = ax^2 + bx + c$
$f(-x) + f(x)$
$= a(-x)^2 + b(-x) + c$
$+ ax^2 + bx + c$ ✓
$= ax^2 - bx + c + ax^2 + bx + c$
$= 2ax^2 + 2c$ ✓ (2 marks)

5 $f(x) = x^2 + x$

a $f(3) = 3^2 + 3$
$= 12$ ✓ (1 mark)

b $f(c) = 6$
$c^2 + c = 6$
$c^2 + c - 6 = 0$
$(c + 3)(c - 2) = 0$
$c + 3 = 0$ or $c - 2 = 0$
$c = -3$ or $c = 2$
✓✓ (2 marks)

c Permissible x-values are all real numbers. ✓ (1 mark)

d
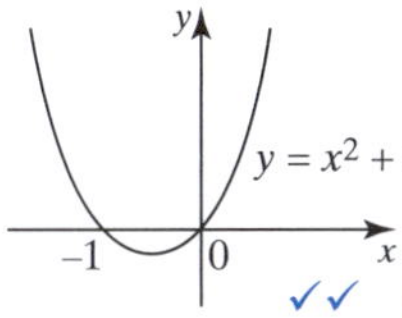

✓✓ (2 marks)

e Axis of symmetry is $x = -0.5$
When $x = -0.5$,
$y = (-0.5)^2 - 0.5$
$= -0.25$
The permissible y-values are $y \ge -0.25$. ✓ (1 mark)

f It would need to be restricted to those values for which $x \ge -0.5$ (or $x \le -0.5$).
[The function will only have an inverse function for half of the parabola. Otherwise there would be two y-values for every x-value and it would not be a function.] ✓ (1 mark)

6 $f(x) = \sqrt{9 - x^2}$

a $9 - x^2 \ge 0$
$x^2 \le 9$
$-3 \le x \le 3$ ✓ (1 mark)

b When $x = 0$, $y = 3$
When $x = \pm 3$, $y = 0$
So $0 \le y \le 3$ ✓ (1 mark)

c No, $f(x)$ will not have an inverse function. There is not a unique x-value for all y-values. ✓ (1 mark)

7 **a** $y = \frac{4}{x}$

Inverse function is $x = \frac{4}{y}$

$xy = 4$ ✓

$y = \frac{x}{4}$ ✓

[It is its own inverse.]

(2 marks)

b $y = 2^x$

Inverse: $x = 2^y$ ✓

$y = \log_2 x$ ✓

(2 marks)

c $y = \sqrt{x}$

Inverse: $x = \sqrt{y}$ $(y \ge 0)$ ✓

$y = x^2$ $(x \ge 0)$ ✓

(2 marks)

8 **a** $f(x)$ does not have an inverse because there is not a unique x-value for every y-value. ✓ (1 mark)

b

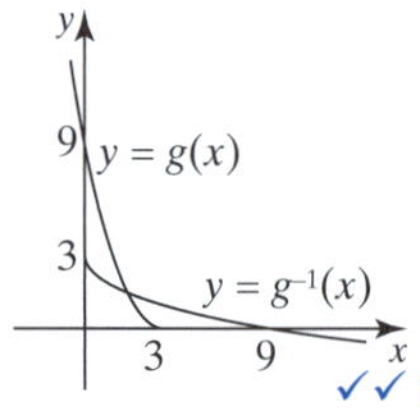

✓✓ (2 marks)

9 **a** [Any function is equal to its inverse when $y = x$.]

$x^3 = x$

$x^3 - x = 0$

$x(x^2 - 1) = 0$ ✓

$x = 0$ or $x^2 = 1$

$x = 0$ or $x = \pm 1$ ✓

(2 marks)

b

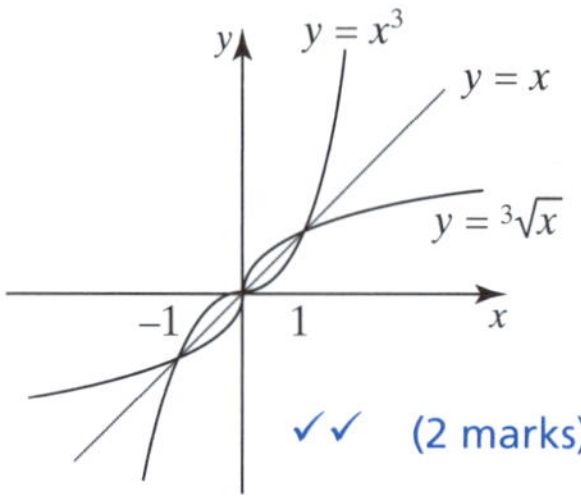

✓✓ (2 marks)

10 $y = x^2 + 2x + 1$

$y = (x + 1)^2$

$(x + 1)^2 = y$

$x + 1 = \pm\sqrt{y}$

$x = -1 \pm \sqrt{y}$ ✓

But $x \ge -1$

$\therefore x = -1 + \sqrt{y}$ ✓

Inverse function is

$y = -1 + \sqrt{x}$ ✓ (3 marks)

(Total: 40 marks)

POLYNOMIAL EXPRESSIONS
SKILLS CHECK PAGE 18

1 **a** $2x^2 + 7x - 5$ True

b $1 - 7x + 4x^2 - x^3$

$= -x^3 + 4x^2 - 7x + 1$ True

c $7 - \sqrt{x}$ False

[$\sqrt{x} = x^{\frac{1}{2}}$. All powers in a polynomial must be whole numbers (non-negative integers)].

d x^6 True

[A polynomial might only have one term.]

e $9x^8 + 7x^3 + \frac{2}{x}$ False

[$\frac{2}{x} = 2x^{-1}$. Powers cannot be negative.]

f $x + 3$ True [binomial]

2 $3x^5 - 2x^4 + 7x^3 + 4x^2 - 8x + 9$

a degree is 5 [highest power]

b leading term is $3x^5$

c leading coefficient is 3

d constant term is 9

3 **a** $4x^7 + 5x^6 - 2x^3 + 6x$

The degree is 7.

b $1 - x^2 = -x^2 + 1$

The degree is 2.

c $x^3 + 2x^2 - 9x - 4$

The degree is 3

d $7x + 2$

The degree is 1. [$7x = 7x^1$]

4 $P(x) = 5x^3 - 7x^2 + 8x + 1$

a $P(0) = 5 \times 0^3 - 7 \times 0^2 + 8 \times 0 + 1$

$= 1$

b $P(1) = 5 \times 1^3 - 7 \times 1^2 + 8 \times 1 + 1$

$= 7$

c $P(-2) = 5 \times (-2)^3 - 7 \times (-2)^2 + 8 \times -2 + 1$

$= -83$

d $P(c) = 5c^3 - 7c^2 + 8c + 1$

5 **a** $7x^3 + x^2 + 9x - 5 + 3x^2 + 17x - 2$

$= 7x^3 + 4x^2 + 26x - 7$

b $x^4 - 2x^3 + 5x^2 - 9 - (7x^3 + 2x^2 - 6x + 3)$

$= x^4 - 2x^3 + 5x^2 - 9 - 7x^3 - 2x^2 + 6x - 3$

$= x^4 - 9x^3 + 3x^2 + 6x - 12$

c $(3x + 2)(x^3 - x^2 + 4x - 1)$

$= 3x(x^3 - x^2 + 4x - 1) + 2(x^3 - x^2 + 4x - 1)$

$= 3x^4 - 3x^3 + 12x^2 - 3x + 2x^3 - 2x^2 + 8x - 2$

$= 3x^4 - x^3 + 10x^2 + 5x - 2$

d $(x - 1)(x^2 - x - 1)$

$= x(x^2 - x - 1) - 1(x^2 - x - 1)$

$= x^3 - x^2 - x - x^2 + x + 1$

$= x^3 - 2x^2 + 1$

6 **a** $P(x) + Q(x)$

$= 6x^2 + 11x + 3 + 2x + 3$

$= 6x^2 + 13x + 6$

b $P(x) - Q(x)$

$= 6x^2 + 11x + 3 - (2x + 3)$

$= 6x^2 + 11x + 3 - 2x - 3$

$= 6x^2 + 9x$

c $P(x) \times Q(x)$

$= (2x + 3)(6x^2 + 11x + 3)$

$= 12x^3 + 22x^2 + 6x + 18x^2 + 33x + 9$

$= 12x^3 + 40x^2 + 39x + 9$

d $6x^2 + 11x + 3$

$= (2x + 3)(3x + 1)$

$\therefore (6x^2 + 11x + 3) \div (2x + 3)$

$= 3x + 1$

[or use the method of long division]

7

```
             x³ − 2x² + 3x − 4
x − 2 ) x⁴ − 4x³ + 7x² − 10x + 8
        x⁴ − 2x³
            −2x³ + 7x²
            −2x³ + 4x²
                   3x² − 10x
                   3x² −  6x
                         −4x + 8
                         −4x + 8
                               0
```

The quotient is $x^3 - 2x^2 + 3x - 4$

8 $P(x) = 2x^3 - 9x^2 + 4$

a **i** $P(1) = 2 \times 1^3 - 9 \times 1^2 + 4$

$= -3$

ii $P(3) = 2 \times 3^3 - 9 \times 3^2 + 4$
$= -23$

iii $P(-2) = 2 \times (-2)^3 - 9 \times (-2)^2 + 4$
$= -48$

b i remainder $= P(1)$
$= -3$

ii remainder $= -23$ $[P(3)]$

iii remainder $= -48$ $[P(-2)]$

9 $P(x) = x^3 - 2x^2 - x + 2$

a i $P(1) = 1^3 - 2 \times 1^2 - 1 + 2$
$= 0$

ii $P(-1) = (-1)^3 - 2 \times (-1)^2 - (-1) + 2$
$= 0$

iii $P(2) = 2^3 - 2 \times 2^2 - 2 + 2$
$= 0$

b $P(x) = (x - 1)(x + 1)(x - 2)$
[Part a gave us the three factors.]

POLYNOMIAL EXPRESSIONS
INTERMEDIATE TEST PAGE 19

1 The degree of the polynomial $P(x) = (x - 1)(x - 2)(x + 3)$ is 3.
[The leading term will be x^3.]
$\therefore$ **[C]** ✓ (1 mark)

2 $P(x) = 2x^3 - 7x^2 + 5x - 4$
$P(-1) = 2(-1)^3 - 7(-1)^2 + 5(-1) - 4$
$= -18$
$\therefore$ **[A]** ✓ (1 mark)

3 $5x^2 - 6x + x^{-1}$ is not a polynomial. [Powers of x must be non-negative integers.]
$\therefore$ **[C]** ✓ (1 mark)

4 $P(x) = x^4 - 5x^3 + 2x^2 - 7x + 1$
$Q(x) = 6x^3 - 3x^2 + 8x - 2$
$P(x) + Q(x)$
$= x^4 - 5x^3 + 2x^2 - 7x + 1 + 6x^3 - 3x^2 + 8x - 2$
$= x^4 + x^3 - x^2 + x - 1$
$\therefore$ **[A]** ✓ (1 mark)

5 $P(x) = x^4 - 7x^3 + 8x^2 + 28x - 48$
$P(0) = -48$
$P(1) = -18$
$P(-1) = -60$
$P(4) = 0$
$(x - 4)$ is a factor of $P(x)$.
$\therefore$ **[D]** ✓ (1 mark)

6 $P(x) = 5x^3 + x^2 - 7x - 3$
$P(-3) = 5(-3)^3 + (-3)^2 - 7(-3) - 3$
$= -108$
$\therefore$ the remainder when $P(x)$ is divided by $x + 3$ is -108
$\therefore$ **[A]** ✓ (1 mark)

7 $(x + 3)(x^2 - 2x + 5)$
$= x(x^2 - 2x + 5) + 3(x^2 - 2x + 5)$
$= x^3 - 2x^2 + 5x + 3x^2 - 6x + 15$
$= x^3 + x^2 - x + 15$
The coefficient of x is -1
[It is not necessary to expand the whole expression. The terms that give x are $x \times 5$ and $3 \times -2x$; $5x - 6x = -1x$]
$\therefore$ **[B]** ✓ (1 mark)

8 Try each option.
A: $P(x) = x^2 - 3x - 10$
$P(2) = 2^2 - 3 \times 2 - 10$
$= -12$
B: $P(x) = x^3 - 8x - 8$
$P(2) = 2^3 - 8 \times 2 - 8$
$= -16$
C: $P(x) = x^4 - 7x^3 + 40$
$P(2) = 2^4 - 7 \times 2^3 + 40$
$= 0$
D: $P(x) = 2x^3 + 5x^2 - 7x - 6$
$P(2) = 2 \times 2^3 + 5 \times 2^2 - 7 \times 2 - 6$
$= 16$
$x^4 - 7x^3 + 40$ has a factor of $x - 2$
$\therefore$ **[C]** ✓ (1 mark)

9 Try each option.
A: $P(x) = x^2 + 3x - 4$
$P(-4) = (-4)^2 + 3 \times (-4) - 4$
$= 0$
B: $P(x) = x^3 + 5x^2 + 4x$
$P(-4) = (-4)^3 + 5 \times (-4)^2 + 4 \times (-4)$
$= 0$
C: $P(x) = x^4 - 11x^2 - 80$
$P(-4) = (-4)^4 - 11 \times (-4)^2 - 80$
$= 0$
D: $P(x) = x^3 + 7x^2 + 12x - 1$
$P(-4) = (-4)^3 + 7 \times (-4)^2 + 12 \times (-4) - 1$
$= -1$
$P(x) = x^3 + 7x^2 + 12x - 1$ does not have a factor of $x + 4$.
$\therefore$ **[D]** ✓ (1 mark)

10 $P(x) = x^3 - 4x^2 - 61x + 280$

a $P(5) = 5^3 - 4 \times 5^2 - 61 \times 5 + 280$
$= 0$ ✓ (1 mark)

b $(x - 5)$ is a factor of $P(x)$ ✓ (1 mark)

c
$$
\begin{array}{r}
x^2 + x - 56 \\
x - 5 \overline{) x^3 - 4x^2 - 61x + 280} \\
\underline{x^3 - 5x^2} \qquad\qquad \\
x^2 - 61x \qquad \\
\underline{x^2 - 5x} \qquad \\
-56x + 280 \\
\underline{-56x + 280} \\
0
\end{array}
$$
✓✓

$P(x) = (x - 5)(x^2 + x - 56)$ ✓
$= (x - 5)(x + 8)(x - 7)$ ✓
(4 marks)

11 $P(x) = 2x^4 - 3x^3 - 7x^2 + x + 9$
$P(-3) = 2 \times (-3)^4 - 3 \times (-3)^3 - 7 \times (-3)^2 - 3 + 9$ ✓
$= 186$
The remainder when $P(x)$ is divided by $x + 3$ is 186 ✓
(2 marks)

12
$$
\begin{array}{r}
6x^2 - 5x + 3 \\
x - 4 \overline{) 6x^3 - 29x^2 + 23x - 10} \\
\underline{6x^3 - 24x^2} \qquad\qquad \\
-5x^2 + 23x \qquad \\
\underline{-5x^2 + 20x} \qquad \\
3x - 10 \\
\underline{3x - 12} \\
2
\end{array}
$$
✓✓

$\therefore Q(x) = 6x^2 - 5x + 3$ ✓
and $c = 2$ ✓ (4 marks)

13 $3x^3 + 8x^2 - 2x - 9 - (2x^3 - 3x^2 - 5x + 4)$
$= 3x^3 + 8x^2 - 2x - 9 - 2x^3 + 3x^2 + 5x - 4$ ✓
$= x^3 + 11x^2 + 3x - 13$ ✓
(2 marks)

14 $P(x) = x^3 + 3x^2 + ax + 7$
The remainder is 1 when divided by $(x + 2)$.
$\therefore P(-2) = 1$ ✓
$P(-2) = (-2)^3 + 3 \times (-2)^2 + a \times -2 + 7$
$= -8 + 12 - 2a + 7$
$= 11 - 2a$ ✓
$\therefore 11 - 2a = 1$
$-2a = -10$
$a = 5$ ✓ (3 marks)

15 **a** $(x-2)(x^3-4x^2+3x)$
$= x(x^3-4x^2+3x)$
$\quad -2(x^3-4x^2+3x)$
$= x^4-4x^3+3x^2-2x^3$
$\quad +8x^2-6x$ ✓
$= x^4-6x^3+11x^2-6x$ ✓
(2 marks)

b $P(x) = x^4-6x^3+11x^2-6x$
$= (x-2)(x^3-4x^2+3x)$ ✓
Now x^3-4x^2+3x
$= x(x^2-4x+3)$
$= x(x-3)(x-1)$
$P(x) = x(x-1)(x-2)(x-3)$
✓ (2 marks)

(Total: 30 marks)

POLYNOMIAL EXPRESSIONS
ADVANCED TEST PAGE 20

1 $P(x) = 6x^4-8x^3+7x^2-9x-2$

a The degree is 4 ✓ (1 mark)

b The leading term is $6x^4$. ✓ (1 mark)

c The constant term is –2. ✓ (1 mark)

d The leading coefficient is 6. ✓ (1 mark)

e No $P(x)$ is not a monic polynomial. The leading coefficient is not 1. ✓ (1 mark)

2 **a** $A(x)+B(x)+C(x)$
$= 3x^3-9x^2+5x-7+8x^4$
$\quad -2x^3-11x^2+x-2$
$\quad +4x^3-6x+8$ ✓
$= 8x^4+5x^3-20x^2-1$ ✓
(2 marks)

b $B(x)-C(x)$
$= 8x^4-2x^3-11x^2+x-2$
$\quad -(4x^3-6x+8)$ ✓
$= 8x^4-2x^3-11x^2+x-2$
$\quad -4x^3+6x-8$
$= 8x^4-6x^3-11x^2+7x-10$
✓ (2 marks)

3 **a** $(x+5)(x^3-2x^2-x+3)$
$= x(x^3-2x^2-x+3)$
$\quad +5(x^3-2x^2-x+3)$
$= x^4-2x^3-x^2+3x+5x^3$
$\quad -10x^2-5x+15$ ✓
$= x^4+3x^3-11x^2-2x+15$
✓ (2 marks)

b $(2x-3)(3x^4-5x^3+4x-6)$
$= 2x(3x^4-5x^3+4x-6)$
$\quad -3(3x^4-5x^3+4x-6)$
$= 6x^5-10x^4+8x^2-12x$
$\quad -9x^4+15x^3-12x+18$ ✓
$= 6x^5-19x^4+15x^3+8x^2$
$\quad -24x+18$ ✓
(2 marks)

4 $P(x) = x^3-3x^2+4x$
$P(4) = 4^3-3\times 4^2+4\times 4$
$= 32$ ✓
$(x-4)$ is not a factor of $P(x)$ because $P(4)$ is not equal to zero. ✓ (2 marks)

5
$$\begin{array}{r|l} & 2x^3-4x^2+3x+1 \\ x+2 & 2x^4+0x^3-5x^2+7x-1 \\ & 2x^4+4x^3 \\ & -4x^3-5x^2 \\ & -4x^3-8x^2 \\ & 3x^2+7x \\ & 3x^2+6x \\ & x-1 \\ & x+2 \\ & -3 \end{array}$$
✓✓

$2x^4-5x^2+7x-1$
$= (x+2)(2x^3-4x^2+3x+1)$
$\quad -3$ ✓ (3 marks)

6 $P(x) = x^3-5x^2-8x+12$

a
$$\begin{array}{r|l} & x^2+x-2 \\ x-6 & x^3-5x^2-8x+12 \\ & x^3-6x^2 \\ & x^2-8x \\ & x^2-6x \\ & -2x+12 \\ & -2x+12 \\ & 0 \end{array}$$
✓

So $Q(x) = x^2+x-2$ ✓
(2 marks)

b $c = 0$
$(x-6)$ is a factor of $P(x)$. ✓
(1 mark)

7 $P(x) = x^4-5x^3+x^2-2x-k$
$P(1) = 1^4-5\times 1^3+1^2-2\times 1-k$
$= -5-k$ ✓
But $(x-1)$ is a factor of $P(x)$.
$P(1) = 0$
$0 = -5-k$
$k = -5$ ✓ (2 marks)

8 $P(x) = (x+2)(x-7)-5$ ✓
$P(-3) = (-3+2)(-3-7)-5$
$= 5$
$\therefore$ the remainder is 5. ✓
(2 marks)

9 $P(x) = x^2+bx+c$
$(x-3)$ is a factor of x^2+bx+c.
$\therefore P(3) = 0$
$3^2+b\times 3+c = 0$
$9+3b+c = 0$ [i] ✓
The remainder when x^2+bx+c is divided by $(x+3)$ is –30.
$\therefore P(-3) = -30$
$(-3)^2+b\times(-3)+c = -30$
$9-3b+c = -30$ [ii]✓
[i] – [ii]: $6b = 30$
$b = 5$
Substitute into [i]:
$9+3\times 5+c = 0$
$c = -24$
$P(x) = x^2+5x-24$ ✓
$= (x-3)(x+8)$
$\therefore$ the other factor is $(x+8)$. ✓
(4 marks)

10 $P(x) = (x^2-x-6)Q(x)+ax+b$
$P(3) = -5$
$-5 = (3^2-3-6)Q(3)+a\times 3+b$
$-5 = 3a+b$ [i] ✓
$P(-2) = 15$
$15 = ((-2)^2-(-2)-6)Q(-2)$
$\quad +a\times(-2)+b$
$15 = -2a+b$ [ii]
[i] – [ii]: $-20 = 5a$
$a = -4$ ✓
Substitute into [i]:
$-5 = 3\times(-4)+b$
$b = 7$ ✓ (3 marks)

11 **a**
$$\begin{array}{r|l} & x^2-10x+24 \\ x+3 & x^3-7x^2-6x+5 \\ & x^3+3x^2 \\ & -10x^2-6x \\ & -10x^2-30x \\ & 24x+5 \\ & 24x+72 \\ & -67 \end{array}$$
✓

$Q(x) = x^2-10x+24$ ✓
$R(x) = -67$ ✓ (3 marks)

b $P(x) = (x+3)Q(x)-67$
$Q(x) = x^2-10x+24$
$= (x-4)(x-6)$
So $Q(4) = 0$ and $Q(6) = 0$
So $P(4)$ and $P(6)$ will both be equal to –67. ✓ (1 mark)

12 **a** $P(x) = x^3 + 4x^2 + x - 6$
$P(1) = 1^3 + 4 \times 1^2 + 1 - 6$
$= 0$
$\therefore (x - 1)$ is a factor of $P(x)$. ✓

$$\begin{array}{r l} & x^2 + 5x + 6 \\ x - 1 \,\big) & x^3 + 4x^2 + x - 6 \\ & \underline{x^3 - x^2} \\ & 5x^2 + x \\ & \underline{5x^2 - 5x} \\ & 6x - 6 \\ & \underline{6x - 6} \\ & 0 \end{array}$$ ✓

$P(x) = (x - 1)(x^2 + 5x + 6)$
$= (x - 1)(x + 2)(x + 3)$ ✓ (3 marks)

b $P(x) = 0$ when $x = 1, x = -2$ or $x = -3$ ✓ (1 mark)

(Total: 40 marks)

FURTHER POLYNOMIALS SKILLS CHECK PAGE 22

1 $P(x) = x^3 - 4x^2 + 7x + 1$
$P(3) = 3^3 - 4 \times 3^2 + 7 \times 3 + 1$
$= 13$
The remainder when $P(x)$ is divided by $x - 3$ is 13.

2 **a** $P(x) = x^4 - 2x^3 - x^2 + 2x$
$P(1) = 1^4 - 2 \times 1^3 - 1^2 + 2 \times 1$
$= 0$
$\therefore (x - 1)$ is a factor of $P(x)$. True

b $P(0) = 0^4 - 2 \times 0^3 - 0^2 + 2 \times 0$
$= 0$
$\therefore x$ is a factor of $P(x)$
True

c $P(-2) = (-2)^4 - 2 \times (-2)^3 - (-2)^2 + 2 \times -2$
$= 24$
$\therefore (x + 2)$ is not a factor of $P(x)$ False

d $P(2) = 2^4 - 2 \times 2^3 - 2^2 + 2 \times 2$
$= 0$
$\therefore (x - 2)$ is a factor of $P(x)$.
True

3

$$\begin{array}{r l} & 5x + 7 \\ x - 2 \,\big) & 5x^2 - 3x + 7 \\ & \underline{5x^2 - 10x} \\ & 7x + 7 \\ & \underline{7x - 14} \\ & 21 \end{array}$$

$\therefore 5x^2 - 3x + 7$
$= (x - 2)(5x + 7) + 21$

4 **a** $(x + 3)(x - 2) = 0$
$x + 3 = 0$ or $x - 2 = 0$
$x = -3$ or $x = 2$

b $x^2(x - 5)(x + 4) = 0$
$x^2 = 0$ or $x - 5 = 0$ or $x + 4 = 0$
$x = 0$ or $x = 5$ or $x = -4$

c $(2x - 1)(3x + 2)(x - 4) = 0$
$2x - 1 = 0$
$2x = 1$
$x = \frac{1}{2}$ or
$3x + 2 = 0$
$3x = -2$
$x = -\frac{2}{3}$ or
$x - 4 = 0$
$x = 4$

d $4(2x + 3)(x - 2) = 0$
$2x + 3 = 0$ or $x - 2 = 0$
$2x = -3$ or $x = 2$
$x = -1\frac{1}{2}$ or $x = 2$

5 **a** 4
b 3
c 5
d 2 [A polynomial of degree n has at most n zeros.]

6 **a** $x^2 + 7x - 18 = (x + 9)(x - 2)$

b $x^3 + 5x^2 + 6x$
$= x(x^2 + 5x + 6)$
$= x(x + 3)(x + 2)$

c $x^4 - 16$
$= (x^2)^2 - 4^2$
$= (x^2 + 4)(x^2 - 4)$
$= (x^2 + 4)(x + 2)(x - 2)$

d $x^3 - x^2 = x^2(x - 1)$

e $x^4 - 25x^2 = x^2(x^2 - 25)$
$= x^2(x + 5)(x - 5)$

f $4x^2 - 4x - 15$
$= 4x^2 - 10x + 6x - 15$
$= 2x(2x - 5) + 3(2x - 5)$
$= (2x - 5)(2x + 3)$

7 **a** $(x - 2)^2 = 0$ has a double root at $x = 2$

b $(x + 8)(x - 2) = 0$ has a single root at $x = 2$

c $x^3(x - 2) = 0$ has a single root at $x = 2$
[It has a triple root at $x = 0$]

d $x(x + 3)(x - 2)(x - 4) = 0$ has a single root at $x = 2$

e $7x(x - 2)^3 = 0$ has a triple root at $x = 2$

8 $y = x^2 + x - 2$

a The leading term is x^2

b The y-intercept is –2. {The constant term}

c $x^2 + x - 2 = 0$
$(x + 2)(x - 1) = 0$
$x + 2 = 0$ or $x - 1 = 0$
$x = -2$ or $x = 1$
The roots are $x = -2$ and $x = 1$.

d The x-intercepts are –2 and 1. [The roots of the equation.]

e

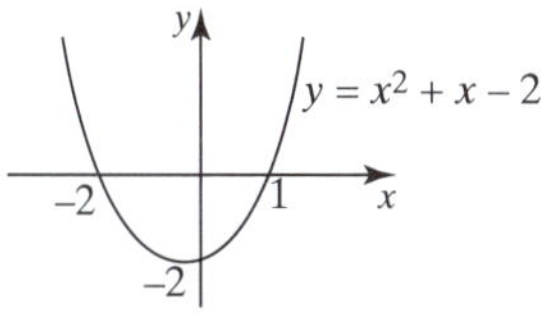

9 $y = x^4 + 3x^3 - 2x + 5$

a as $x \to \infty, y \to \infty$
[Just consider the leading term. If x is a large positive number, so is x^4.]

b as $x \to -\infty, y \to \infty$

10 $y = 3 - 2x + x^2 - x^3$

a as $x \to \infty, y \to -\infty$

b as $x \to -\infty, y \to \infty$

FURTHER POLYNOMIALS INTERMEDIATE TEST PAGE 23

1

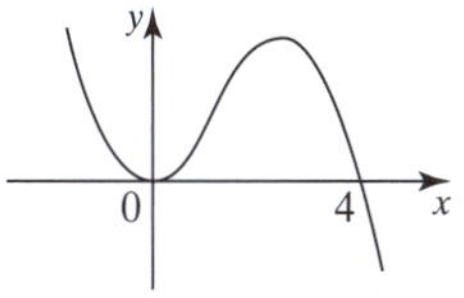

The polynomial has a double root at $x = 0$ and a single root at $x = 4$. [$\therefore$ its equation is either $y = x^2(x - 4)$ or $y = x^2(4 - x)$.]
When x gets large positively, y gets large negatively.
The equation is $y = x^2(4 - x)$
$\therefore$ **[C]** ✓ (1 mark)

2 $y = x^3(x - 1)$ has a triple root at $x = 0$ and a single root at $x = 1$.
When gets large positively, y gets large positively.
When x gets large negatively, y gets large positively.
The only possible graph is B.

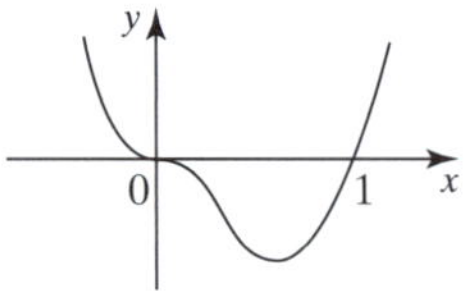

$\therefore$ **[B]** ✓ (1 mark)

3 If both $(x - 3)$ and $(x + 2)$ are factors of $P(x)$ then $P(3) = 0$ and $P(-2) = 0$.
The only polynomial for which both $P(3)$ and $P(-2)$ are zero is $P(x) = x^3 - 2x^2 - 5x + 6$.
$\therefore$ **[B]** ✓ (1 mark)

4 $y = 8 - 2x + 3x^2 - 4x^3 + 7x^4 - 5x^5$
The leading term is $-5x^5$.
If x is a large negative value then $5x^5$ will also have a large negative value so $-5x^5$ will have a large positive value.
y will have a large positive value.
$\therefore$ **[A]** ✓ (1 mark)

5 **a**
$P(x) = x^3 - 5x^2 + 2x + 8$
$P(-1) = (-1)^3 - 5 \times (-1)^2 + 2 \times -1 + 8$
$= -1 - 5 - 2 + 8$
$= 0$ ✓
$\therefore (x + 1)$ is a factor of $P(x)$. (1 mark)

b
$$\begin{array}{r} x^2 - 6x + 8 \\ x + 1 \,\overline{)\, x^3 - 5x^2 + 2x + 8} \\ \underline{x^3 + x^2} \\ -6x^2 + 2x \\ \underline{-6x^2 - 6x} \\ 8x + 8 \\ \underline{8x + 8} \\ 0 \end{array}$$ ✓

$\therefore x^3 - 5x^2 + 2x + 8$
$= (x + 1)(x^2 - 6x + 8)$
$= (x + 1)(x - 2)(x - 4)$ ✓
$x^3 - 5x^2 + 2x + 8 = 0$
$(x + 1)(x - 2)(x - 4) = 0$
$x + 1 = 0$ or $x - 2 = 0$ or $x - 4 = 0$
$\therefore x = -1$ or $x = 2$ or $x = 4$ ✓ (3 marks)

c

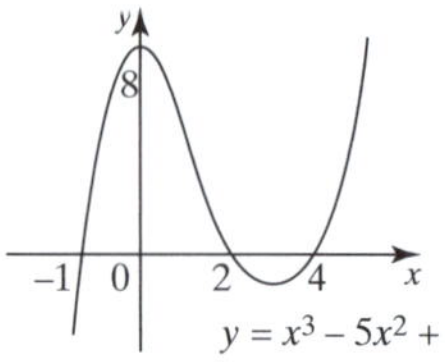

✓✓✓✓ (4 marks)

6 **a** $P(x) = x^3 - 9x^2 + 15x + 25$
$P(-1) = (-1)^3 - 9 \times (-1)^2 + 15 \times -1 + 25$
$= -1 - 9 - 15 + 25$
$= 0$ ✓
$\therefore (x + 1)$ is a factor of $P(x)$.
[It might be necessary to try several different values before finding a value of x such that $P(x) = 0$. You may have found that $P(5) = 0$ instead of $P(-1)$.]

$$\begin{array}{r} x^2 - 10x + 25 \\ x + 1 \,\overline{)\, x^3 - 9x^2 + 15x + 25} \\ \underline{x^3 + x^2} \\ -10x^2 + 15x \\ \underline{-10x^2 - 10x} \\ 25x + 25 \\ \underline{25x + 25} \\ 0 \end{array}$$ ✓

$\therefore x^3 - 9x^2 + 15x + 25$
$= (x + 1)(x^2 - 10x + 25)$ ✓
$= (x + 1)(x - 5)^2$ ✓
(4 marks)

b $y = x^3 - 9x^2 + 15x + 25$
$= (x + 1)(x - 5)^2$
The y-intercept is 25. ✓
The x-intercepts are -1 and 5. ✓
There is a single root at $x = -1$ and a double root at $x = 5$. ✓
When x has a large positive value, y has a large positive value. ✓ (4 marks)

7 $P(x) = 9x^2 - x^4$
a
$P(x) = 0$
$9x^2 - x^4 = 0$
$x^2(9 - x^2) = 0$ ✓
$x^2(3 + x)(3 - x) = 0$
$x^2 = 0$ or $3 + x = 0$ or $3 - x = 0$
$x = 0$ or $x = -3$ or $x = 3$
✓ (2 marks)

b

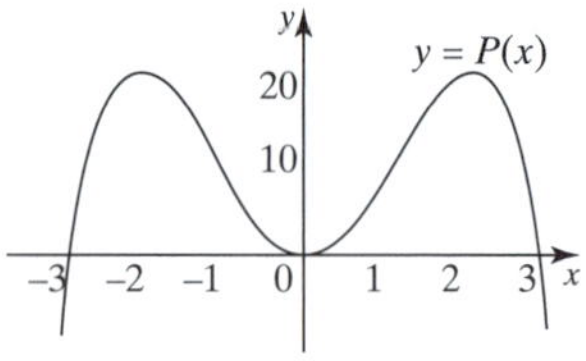

✓✓✓✓ (4 marks)

c **i**

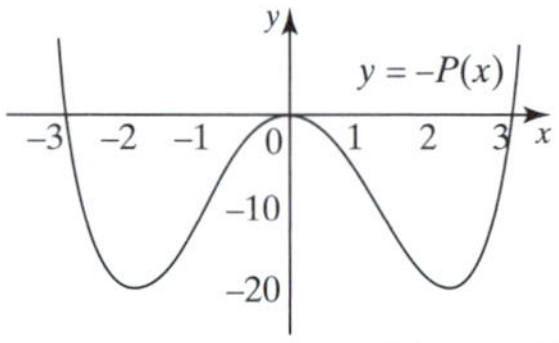

✓ (1 mark)
[Every y-value is the negative of the corresponding y-value in $y = P(x)$.]

ii

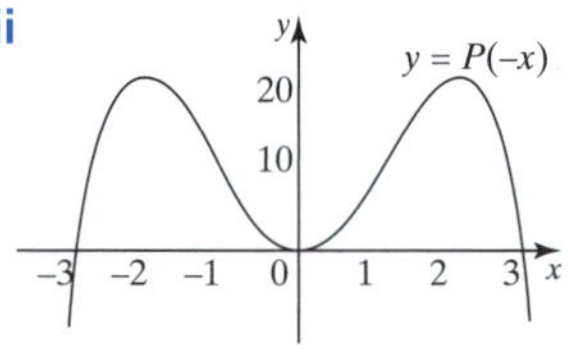

✓ (1 mark)
$[P(-x) = 9(-x)^2 - (-x)^4$
$= 9x^2 - x^4 = P(x)]$

iii

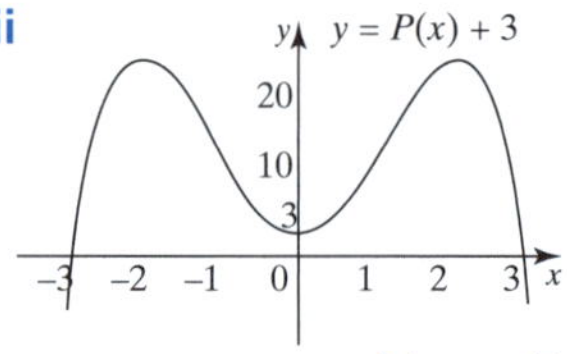

✓ (1 mark)
[The curve moves up 3 units.]

iv

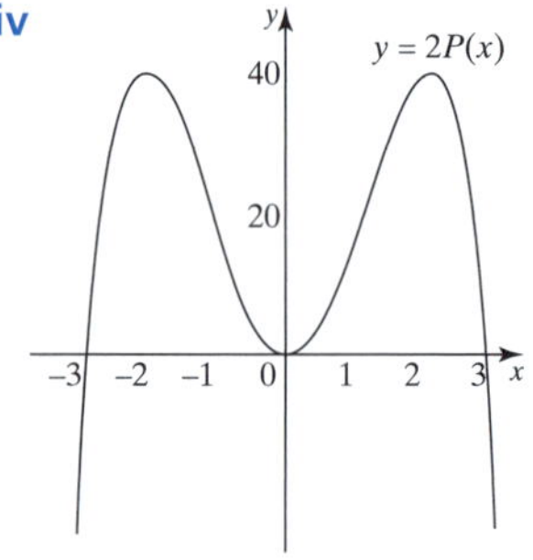

✓ (1 mark)
(Total: 30 marks)

FURTHER POLYNOMIALS
ADVANCED TEST PAGE 24

1 $P(x) = 3x - 4$

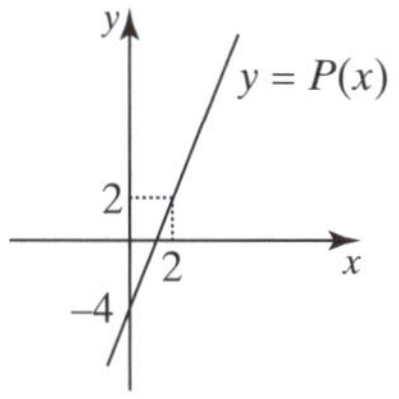

✓✓ (2 marks)

2 **a** $P(x) = 2x^2 + 9x - 5$
$= (2x - 1)(x + 5)$ ✓

$P(x) = 0$ when $x = \frac{1}{2}$ or $x = -5$

The zeroes are $\frac{1}{2}$ and -5 ✓ (2 marks)

b Axis: $x = \dfrac{\frac{1}{2} + (-5)}{2}$
$= -2.25$ ✓
$P(-2.25) = 2(-2.25)^2 + 9(-2.25) - 5$
$= -15.125$
The minimum value is -15.125 (or $-15\frac{1}{8}$). ✓ (2 marks)

c

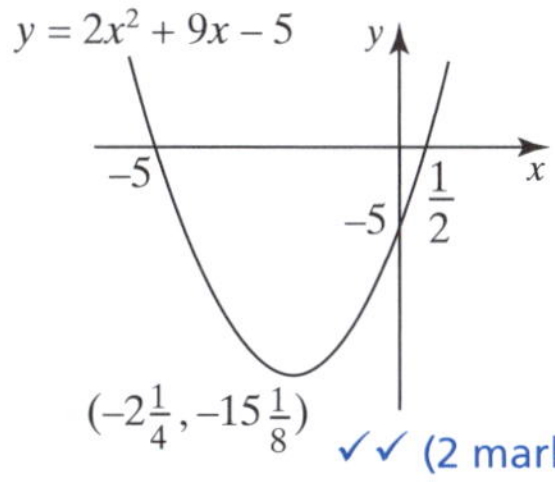

✓✓ (2 marks)

3 **a** $(x + 1)$ is a factor of $P(x) = x^3 - 6x^2 + 3x + 10$

$$\begin{array}{r} x^2 - 7x + 10 \\ x + 1\,\overline{)\,x^3 - 6x^2 + 3x + 10} \\ \underline{x^3 + x^2} \qquad\qquad \\ -7x^2 + 3x \qquad \\ \underline{-7x^2 - 7x} \qquad \\ 10x + 10 \\ \underline{10x + 10} \\ 0 \end{array}$$ ✓

$P(x) = (x + 1)(x^2 - 7x + 10)$
$= (x + 1)(x - 2)(x - 5)$ ✓ (2 marks)

b

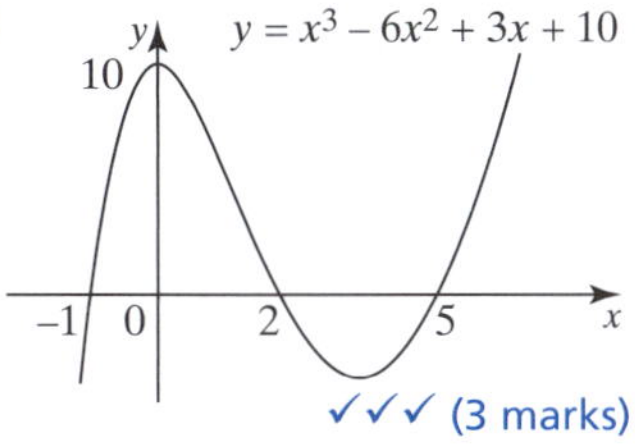

✓✓✓ (3 marks)

4 **a** $P(x) = x^3 + 3x^2 - 9x - 27$
$= x^2(x + 3) - 9(x + 3)$
$= (x + 3)(x^2 - 9)$ ✓
$= (x + 3)(x + 3)(x - 3)$ ✓ (2 marks)

b $y = P(x)$ will meet the x-axis at $x = -3$ and $x = 3$ ✓ (1 mark)

c The curve will cut the x-axis at $x = 3$ but not at $x = -3$. ✓ There is a double root at $x = -3$ so the curve will not cut the axis at that point. ✓ (2 marks)

d

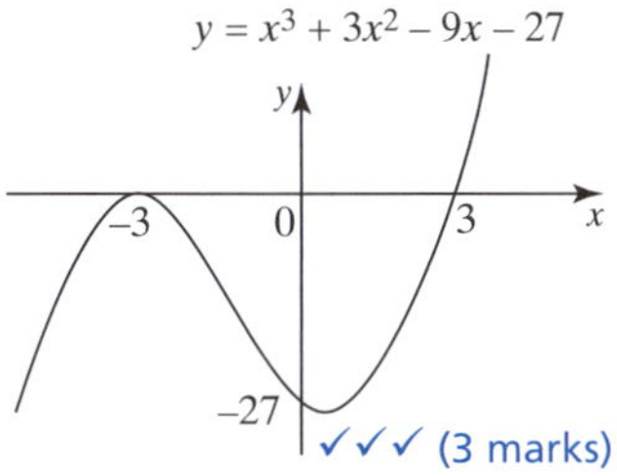

✓✓✓ (3 marks)

5

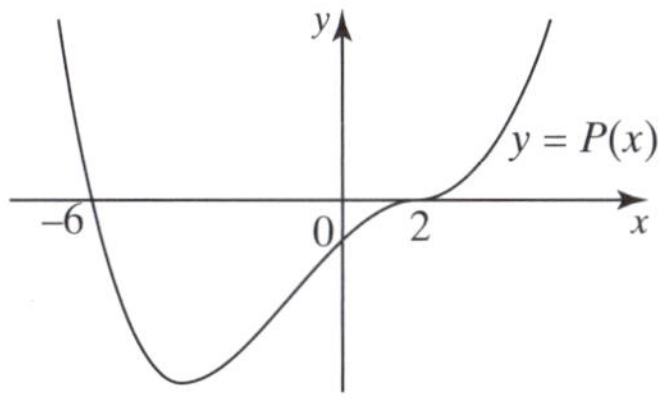

a The root at $x = 2$ is a triple root. ✓ (1 mark)

b $P(x) = (x - 2)^3(x + 6)$ ✓ (1 mark)

c $P(0) = (0 - 2)^3(0 + 6)$
$= -48$ ✓ (1 mark)

6 **a** $A(x) = ax^3 + bx^2 + cx + d$
$A(0) = d$
$-A(0) = -d$
If $A(0) = -A(0)$
then $d = -d$
$2d = 0$
$d = 0$ ✓ (1 mark)

b Assume $A(x)$ is odd.
$\therefore A(x) = ax^3 + bx^2 + cx$
$A(-x) = a(-x)^3 + b(-x)^2 + c(-x)$
$= -ax^3 + bx^2 - cx$
Now $A(x) = -A(x)$
$ax^3 + bx^2 + cx = -(-ax^3 + bx^2 - cx)$ ✓
$ax^3 + bx^2 + cx = ax^3 - bx^2 + cx$
$2bx^2 = 0$
This must be true for all values of x.
$\therefore b = 0$ ✓ (2 marks)

c $B(x) = ax^3 + cx$
$B(x)$ is divisible by $(x - 1)$
$\therefore B(1) = 0$
$a(1)^3 + c(1) = 0$
$a + c = 0$
$a = -c$ ✓
If $a = 1, c = -1$
A possible polynomial is $B(x) = x^3 - x$ ✓ (2 marks)

d No, $B(x)$ is not unique [although every possible polynomial would be a multiple of the above polynomial $B(x)$]. For example, $B(x) = 7x - 7x^3$ could be another possible polynomial. ✓ (1 mark)

7 **a** $P(x) = x^4 - 6x^3 + x^2 + 24x + 16$
$P(-1) = (-1)^4 - 6 \times (-1)^3 + (-1)^2 + 24 \times (-1) + 16$
$= 0$
$\therefore (x + 1)$ is a factor of $P(x)$. ✓ (1 mark)

b

$$\begin{array}{r} x^3 - 7x^2 + 8x + 16 \\ x + 1\,\overline{)\,x^4 - 6x^3 + x^2 + 24x + 16} \\ \underline{x^4 + x^3} \qquad\qquad\qquad\quad \\ -7x^3 + x^2 \qquad\qquad \\ \underline{-7x^3 - 7x^2} \qquad\qquad \\ 8x^2 + 24x \qquad \\ \underline{8x^2 + 8x} \qquad \\ 16x + 16 \\ \underline{16x + 16} \\ 0 \end{array}$$ ✓

$P(x) = (x + 1)(x^3 - 7x^2 + 8x + 16)$
$Q(x) = x^3 - 7x^2 + 8x + 16$
$Q(-1) = (-1)^3 - 7 \times (-1)^2 + 8 \times (-1) + 16$
$= 0$ ✓
$\therefore (x + 1)$ is a factor of $Q(x)$
$\therefore x = -1$ is a double root of $P(x)$ ✓ (3 marks)

c

$$\begin{array}{r} x^2 - 8x + 16 \\ x + 1\,\overline{)\,x^3 - 7x^2 + 8x + 16} \\ \underline{x^3 + x^2\qquad\qquad\quad} \\ -8x^2 + 8x\qquad \\ \underline{-8x^2 - 8x\qquad} \\ 16x + 16 \\ \underline{16x + 16} \\ 0 \end{array}$$

✓

$P(x) = (x + 1)^2(x^2 - 8x + 16)$
$= (x + 1)^2(x - 4)^2$

Roots are $x = -1$ and $x = 4$. ✓ (2 marks)

d $y = x^4 - 6x^3 + x^2 + 24x + 16$

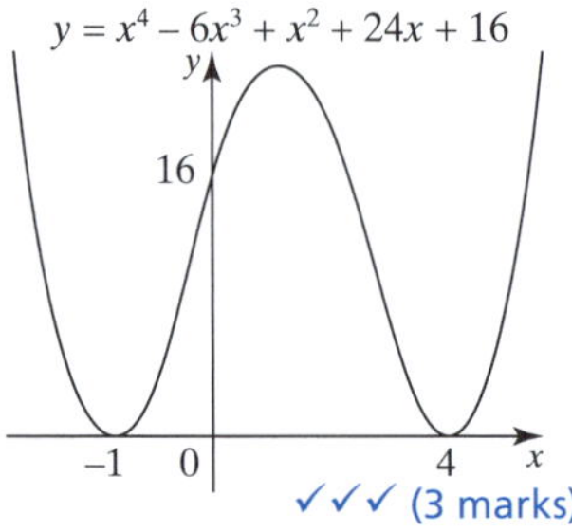

✓✓✓ (3 marks)

e From the graph $P(x) < 0$ for no values of x. ✓ (1 mark)

(Total: 40 marks)

PARABOLAS SKILLS CHECK PAGE 26

1 $y = x^2 - 8x + 12$

a When $y = 0$,
$x^2 - 8x + 12 = 0$
$(x - 2)(x - 6) = 0$
$x - 2 = 0$ or $x - 6 = 0$
$x = 2$ or $x = 6$
The x-intercepts are 2 and 6.

b When $x = 0$,
$y = 0^2 - 8 \times 0 + 12$
$= 12$
The y-intercept is 12.

2 **a** $y = x^2 + 3$
y-intercept $= 3$

b $y = x^2 - 7x - 8$
y-intercept $= -8$

c $y = 15 + 2x - x^2$
y-intercept $= 15$

3 **a** $y = x^2 + 3$
When $y = 0$,
$x^2 + 3 = 0$
$x^2 = -3$
No solution.
$y = x^2 + 3$ does not meet the x-axis.

b $y = x^2 - 7x - 8$
When $y = 0$,
$x^2 - 7x - 8 = 0$
$(x + 1)(x - 8) = 0$
$x + 1 = 0$ or $x - 8 = 0$
$x = -1$ or $x = 8$
∴ the curve meets the x-axis at –1 and 8

c $y = 15 + 2x - x^2$
When $y = 0$
$15 + 2x - x^2 = 0$
$(5 - x)(3 + x) = 0$
$5 - x = 0$ or $3 + x = 0$
$x = 5$ or $x = -3$
∴ the curve meets the x-axis at 5 and –3

4 **a**

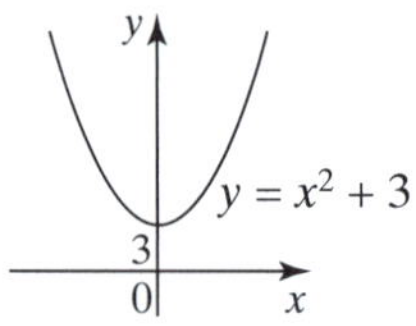

b

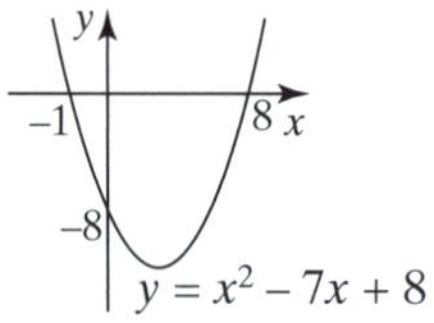

c

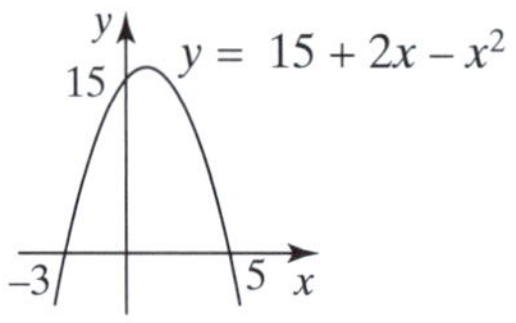

5 **a** y-intercept $= -5$

b x-intercepts at –1 and 5
It cuts the x-axis at (–1, 0) and (5, 0).

c Axis of symmetry: $x = 2$
[Midway between $x = -1$ and $x = 5$.]

d Vertex at (2, –9).
[It lies on the axis of symmetry.]

e concave up

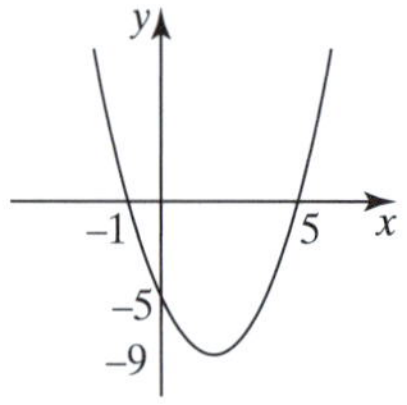

6 **a** x-intercepts at 1 and 5
Axis of symmetry:
$x = \dfrac{1 + 5}{2}$
$x = 3$

b $y = (x + 4)^2 - 9$
Axis of symmetry:
$x + 4 = 0$
$x = -4$

7 $y = 3x^2 - 12x + 7$

a Axis of symmetry:
$x = -\dfrac{b}{2a}$
$x = -\dfrac{-12}{2 \times 3}$
$x = 2$

b When $x = 2$,
$y = 3 \times 2^2 - 12 \times 2 + 7$
$= -5$
Vertex is at (2, –5).

8 The question said that the equations were quadratic so they are of the form $y = ax^2 + bx + c$.

a $y = x^2 + 3$

x	–3	–2	–1	0	1	2	3
y	12	7	4	3	4	7	12

b $y = 3x^2$

x	–3	–2	–1	0	1	2	3
y	27	12	3	0	3	12	27

c $y = (x + 1)^2$
[or $y = x^2 + 2x + 1$]

x	–3	–2	–1	0	1	2	3
y	4	1	0	1	4	9	16

d $y = x^2 + x$

x	–3	–2	–1	0	1	2	3
y	6	2	0	0	2	6	12

9 **a** $y = (x - 3)^2 + 5$
Axis of symmetry is $x = 3$; vertex is at (3, 5).
$y = (x - 3)^2 + 5$
$= x^2 - 6x + 9 + 5$
$= x^2 - 6x + 14$
y-intercept is 14

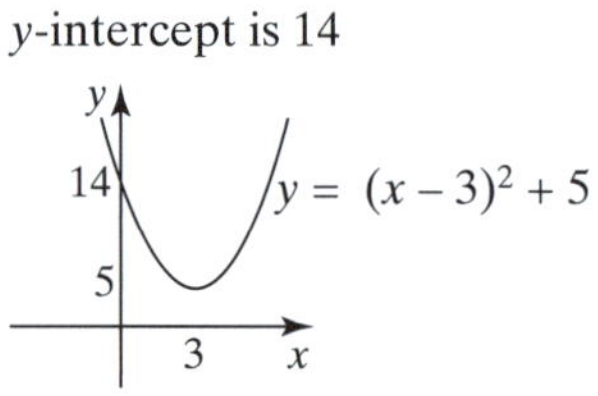

b $y = 2x^2 + 7x - 4$
$= (2x - 1)(x + 4)$
y-intercept is -4
x-intercepts are $\frac{1}{2}$ and -4

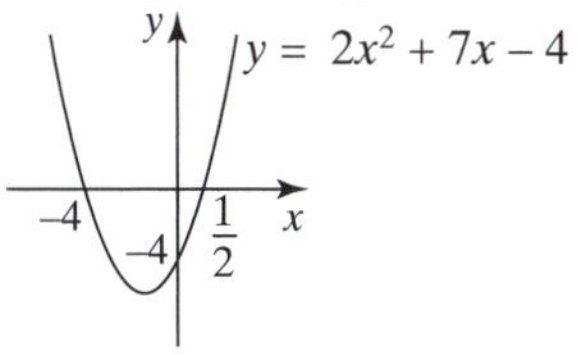

PARABOLAS
INTERMEDIATE TEST PAGE 27

1 $y = (x - 2)^2$
Axis of symmetry at $x = 2$

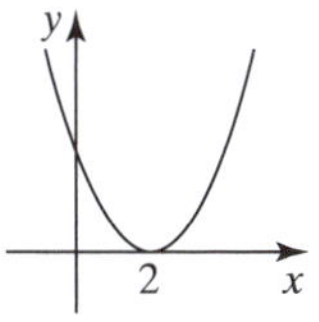

$\therefore$ [D] ✓ (1 mark)

2 $y = (x + 3)^2 - 4$
Axis of symmetry: $x = -3$
Vertex $(-3, -4)$
$\therefore$ [A] ✓ (1 mark)

3 $y = x^2 + x^3$ is not a parabola.
[It is a cubic curve, the highest power of x is 3.]
$\therefore$ [C] ✓ (1 mark)

4 $(-2, 3)$
Try each option.
A: $2(-2)^2 + 3(-2) - 5$
$= -3$ (not 3)
B: $3(-2)^2 + 5(-2) + 2$
$= 4$ (not 3)
C: $2(-2)^2 + 5(-2) + 3$
$= 1$ (not 3)
D: $3(-2)^2 + 2(-2) - 5 = 3$
$y = 3x^2 + 2x - 5$ passes through $(-2, 3)$.
$\therefore$ [D] ✓ (1 mark)

5 $y = 3x^2 - 3x + 2$
$a = 3, b = -3, c = 2$
Axis of symmetry: $x = -\frac{b}{2a}$
$x = -\frac{-3}{2 \times 3}$
$x = \frac{1}{2}$
$\therefore$ [B] ✓ (1 mark)

6 $y = x^2 + 2x - 3$

a When $x = 0$, $y = -3$
y-intercept $= -3$ ✓ (1 mark)

b When $y = 0$,
$x^2 + 2x - 3 = 0$
$(x + 3)(x - 1) = 0$
$x + 3 = 0$ or $x - 1 = 0$
$x = -3$ or $x = 1$
The parabola cuts the x-axis at $(-3, 0)$ and $(1, 0)$. ✓✓ (2marks)

c $a = 1,\ b = 2,\ c = -3$
$x = -\frac{b}{2a}$
$x = -\frac{2}{2 \times 1}$
$x = -1$
The equation of the axis of symmetry is $x = -1$. ✓
[Or use the fact that the axis of symmetry is midway between $x = -3$ and $x = 1$.] (1 mark)

d When $x = -1$,
$y = (-1)^2 + 2 \times -1 - 3$
$= -4$
The vertex is at $(-1, -4)$ ✓ (1 mark)

e

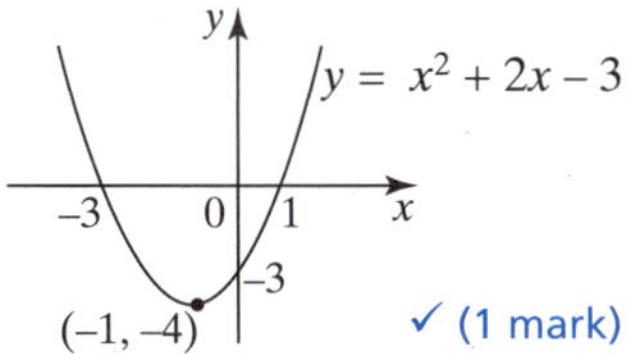

✓ (1 mark)

7 **a** $y = 2x^2 - 8x + 11$
$a = 2,\ b = -8,\ c = 11$
$x = -\frac{b}{2a}$
$x = -\frac{-8}{2 \times 2}$
$x = 2$
The equation of the axis of symmetry is $x = 2$. ✓
When $x = 2$,
$y = 2 \times 2^2 - 8 \times 2 + 11$
$= 3$
Vertex is at $(2, 3)$ ✓ (2 marks)

b No, the parabola will not cut the x-axis. ✓
The curve is concave up ($a > 0$) and the vertex is at $(2, 3)$ so the lowest value of y is when $y = 3$. ✓ (2 marks)

8 $y = 9 + 5x - 2x^2$
$= -2x^2 + 5x + 9$
$a = -2, b = 5, c = 9$
Axis of symmetry: $x = -\frac{b}{2a}$
$x = -\frac{5}{2 \times -2}$
$x = \frac{5}{4}$ ✓
When $x = \frac{5}{4}$,
$y = 9 + 5 \times \frac{5}{4} - 2 \times \left(\frac{5}{4}\right)^2$
$= 12\frac{1}{8}$
The maximum value is $12\frac{1}{8}$. ✓ (2 marks)

9 **a** $4x^2 - 20x = (2x)^2 - 2 \times 2x \times 5$ so to complete the square we need to add 5^2.
$4x^2 - 20x + 11$
$= 4x^2 - 20x + 25 - 14$ ✓
$= (2x - 5)^2 - 14$ ✓ (2 marks)

b $y = 4x^2 - 20x + 11$
[y-intercept $= 11$; vertex $(2.5, -14)$]

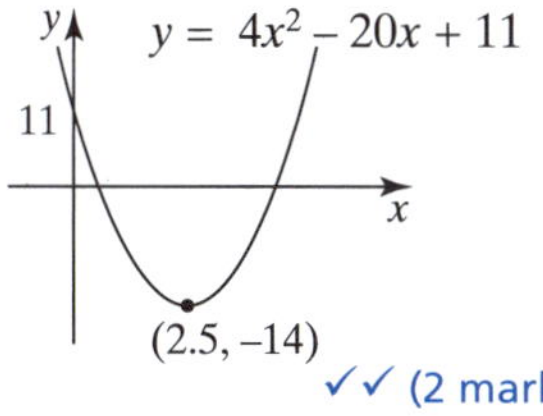

✓✓ (2 marks)

10

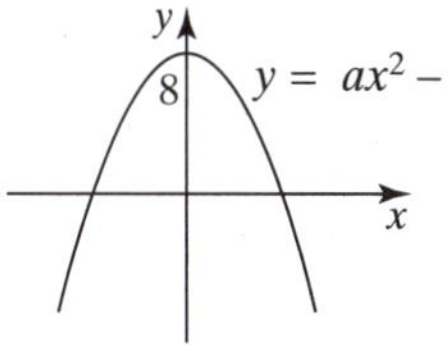

a $c = 8$ ✓ (1 mark)
[It is the y-intercept.]

b $y = ax^2 + 8$
When $x = 1, y = 6$
$6 = a \times 1^2 + 8$
$a = -2$ ✓ (1 mark)

c $y = -2x^2 + 8$
The graph crosses the x-axis when $y = 0$
$0 = -2x^2 + 8$
$2x^2 = 8$
$x^2 = 4$
$x = \pm 2$ ✓✓ (2 marks)
(Total: 25 marks)

PARABOLAS
ADVANCED TEST PAGE 28

1 **a** $y = 27 + 6x - x^2$
$= -x^2 + 6x + 27$
Axis: $x = -\dfrac{b}{2a}$
$x = \dfrac{-6}{2 \times -1}$
$x = 3$ ✓ (1 mark)

b The turning point of the parabola is a maximum. a is negative, so the parabola is concave down. ✓ (1 mark)

c When $x = 3$,
$y = 27 + 6 \times 3 - 3^2$
$= 36$
The vertex is (3, 36). ✓
(1 mark)

2 **a**

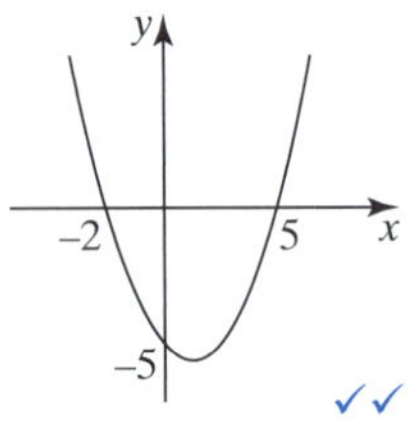

✓✓ (2 marks)

b Equation is of the form:
$y = a(x + 2)(x - 5)$
Now, when $x = 0$, $y = -5$
$-5 = a(0 + 2)(0 - 5)$
$= -10a$
$a = \dfrac{1}{2}$ ✓
So $y = \dfrac{1}{2}(x + 2)(x - 5)$
$= \dfrac{1}{2}(x^2 - 3x - 10)$
$y = \dfrac{1}{2}x^2 - \dfrac{3}{2}x - 5$ ✓
(2 marks)

3

x	0	1	2	3	4	5	6
y	–5	0	9	22	39	60	85

Parabola has equation of the form:
$y = ax^2 + bx + c$
Now, when $x = 0$, $y = -5$
$\therefore c = -5$
The equation is of the form
$y = ax^2 + bx - 5$✓
When $x = 1$, $y = 0$
$a \times 1^2 + b \times 1 - 5 = 0$
$a + b = 5$ [i]
When $x = 2$, $y = 9$
$a \times 2^2 + b \times 2 - 5 = 9$
$4a + 2b = 14$
$2a + b = 7$ [ii]
[ii] – [i]: $a = 2$ ✓
Substitute into [i]:
$2 + b = 5$
$b = 3$
$\therefore$ the equation is
$y = 2x^2 + 3x - 5$ ✓ (3 marks)

4 $y = \dfrac{1}{2}x^2 + 5x - 12$
[Some possible answers (one mark each)]
The y-intercept is –12.
The curve is concave up.
When $\dfrac{1}{2}x^2 + 5x - 12 = 0$
$x^2 + 10x - 24 = 0$
$(x + 12)(x - 2) = 0$
$x = -12$ or $x = 2$
The x-intercepts are –12 and 2.
The axis is $x = -5$
The vertex is $(-5, -24\frac{1}{2})$.
✓✓✓✓ (4 marks)

5 **a** $y = 4x^2 + 4x - 15$
When $x = 0$, $y = -15$
It cuts the y-axis at (0, –15)✓
(1 mark

b When $y = 0$,
$4x^2 + 4x - 15 = 0$
$4x^2 + 10x - 6x - 15 = 0$
$2x(2x + 5) - 3(2x + 5) = 0$
$(2x + 5)(2x - 3) = 0$ ✓
$2x = -5$ or $2x = 3$
$x = -2.5$ or $x = 1.5$
✓ (2 marks)

c

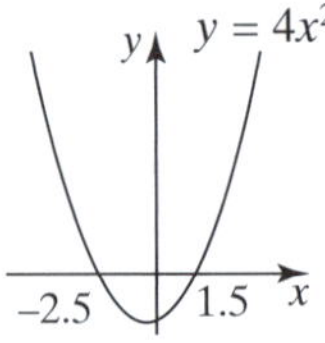

From the graph,
$4x^2 + 4x - 15 > 0$ when
$x < -2.5$ and $x > 1.5$ ✓✓
(2 marks)

6 $y = x^2 - 5x - 6$
When $y = 18$,
$x^2 - 5x - 6 = 18$
$x^2 - 5x - 24 = 0$ ✓
$(x + 3)(x - 8) = 0$
$x = -3$ or $x = 8$ ✓

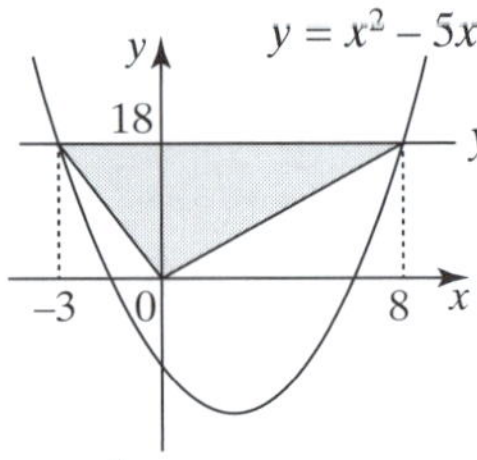

$A = \dfrac{1}{2}bh$
$= \dfrac{1}{2} \times (8 - (-3)) \times 18$
$= 99$ units^2 ✓ (3 marks)

7 **a** $y = x^2 - 8x + 17$ and
$y = x + 9$
At the points of intersection:
$x^2 - 8x + 17 = x + 9$
$x^2 - 9x + 8 = 0$
$(x - 1)(x - 8) = 0$
$x = 1$ or $x = 8$ ✓
When $x = 1$, $y = 1 + 9$
$= 10$
$\therefore P$ is the point (1, 10). ✓
When $x = 8$, $y = 8 + 9$
$= 17$
$\therefore Q$ is the point (8, 17) ✓
(3 marks)

b $y = 2x^2 + bx + c$
At P,
$10 = 2 \times 1^2 + b \times 1 + c$
$b + c = 8$ [i]
At Q,
$17 = 2 \times 8^2 + b \times 8 + c$
$8b + c = -111$ [ii]
[ii] – [i]: $7b = -119$
$b = -17$ ✓
Substitute into [i]:
$-17 + c = 8$
$c = 25$ ✓ (2 marks)

c $y = 3x^2 - 26x + 33$
When $x = 1$,
$y = 3 \times 1^2 - 26 \times 1 + 33$
$= 10$
$\therefore$ the parabola passes through P
When $x = 8$,
$y = 3 \times 8^2 - 26 \times 8 + 33$
$= 17$
$\therefore$ the parabola passes through Q. ✓ (1 mark)

d Passing through the points P and Q are:
$y = x^2 - 8x + 17$
$y = 2x^2 - 17x + 25$
$y = 3x^2 - 26x + 33$
So following the pattern,
$y = 4x^2 - 35x + 41$ will also pass through P and Q.
[i.e. $d = -35$ and $e = 41$] ✓
(1 mark)

8 **a** $ax^2 + bx + c$

$$= a\left[x^2 + \frac{b}{a}x + \frac{c}{a}\right]$$
$$= a\left[x^2 + \frac{b}{a}x + \left(\frac{b}{2a}\right)^2 - \left(\frac{b}{2a}\right)^2 + \frac{c}{a}\right]$$
$$= a\left[\left(x + \frac{b}{2a}\right)^2 - \frac{b^2}{4a^2} + \frac{c}{a}\right]$$
$$= a\left[\left(x + \frac{b}{2a}\right)^2 - \frac{b^2}{4a^2} + \frac{4ac}{4a^2}\right]$$
$$= a\left[\left(x + \frac{b}{2a}\right)^2 - \left(\frac{b^2 - 4ac}{4a^2}\right)\right]$$

✓✓✓ (3 marks)

b Axis of symmetry:
$x = -\frac{b}{2a}$ ✓
When $x + \frac{b}{2a} = 0$ there is only one y-value. ✓ (2 marks)

c $y = ax^2 + bx + c$
$$= a\left[\left(x + \frac{b}{2a}\right)^2 - \left(\frac{b^2 - 4ac}{4a^2}\right)\right]$$
When $x = -\frac{b}{2a}$, ✓
$$y = a\left[-\left(\frac{b^2 - 4ac}{4a^2}\right)\right]$$
$$= -\frac{b^2 - 4ac}{4a}$$
$$= \frac{4ac - b^2}{4a}$$ ✓ (2 marks)

d $y = x^2 - 8x + 21$
$$x = -\frac{b}{2a}$$
$$= -\frac{-8}{2 \times 1}$$
$x = 4$ ✓
$$y = \frac{4ac - b^2}{4a}$$
$$= \frac{4 \times 1 \times 21 - (-8)^2}{4 \times 1}$$
$$= 5$$
Vertex is (4, 5). ✓ (2 marks)

e The parabola $y = x^2 - 8x + 21$ does not cut the x-axis. ✓
The parabola is concave up and has its vertex at (4, 5) meaning that the minimum value is 5. ✓ (2 marks)

(Total: 40 marks)

FURTHER GRAPHS
SKILLS CHECK PAGE 30

1 **a** $y = 6 - x$ straight line
[$y = -x + 6$, gradient -1, y-intercept 6]

b $y = \frac{2}{x}$ hyperbola

c $y = 3^x$ exponential curve

d $y = x^3$ cubic curve

e $y = \frac{x}{2}$ straight line
[$y = \frac{1}{2}x$, gradient $\frac{1}{2}$, y-intercept 0]

f $x^2 + y^2 = 9$ circle
[centre (0, 0), radius 3]

g $y = x^2$ parabola

h $y = 6 - x^2$ parabola
[$y = -x^2 + 6$]

i $y = 2x^2 - 7x + 6$ parabola

j $x + y = 9$ straight line
[$y = -x + 9$]

k $x^2 + 4x + y^2 = 5$ circle
[$(x + 2)^2 + y^2 = 9$
centre (−2, 0), radius 3]

l $y = 5^{-x}$ exponential curve

2 circle centre (0, 0)
radius 5 units
$x^2 + y^2 = 5^2$
$x^2 + y^2 = 25$

3 **a** $y = 2^{-x}$

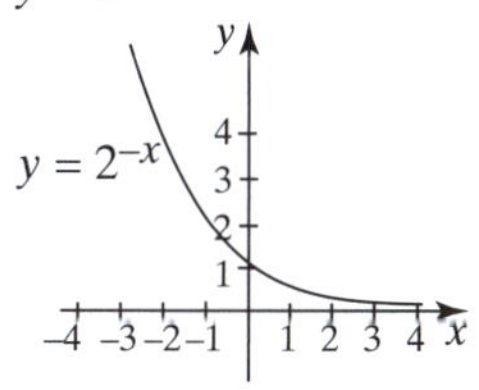

b $y = x^3 - 1$

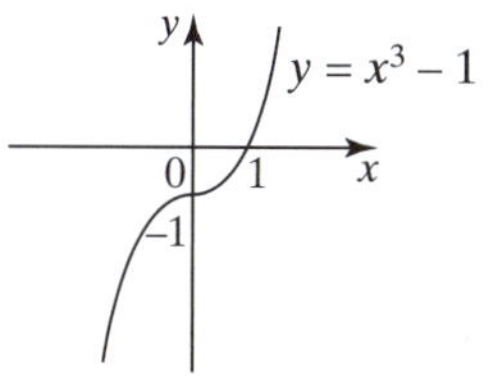

c $y = (x - 3)^2$

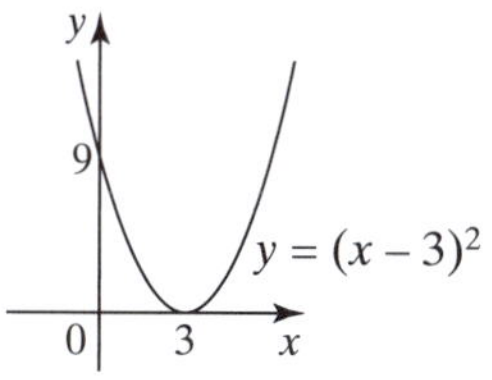

d $x^2 + y^2 = 64$

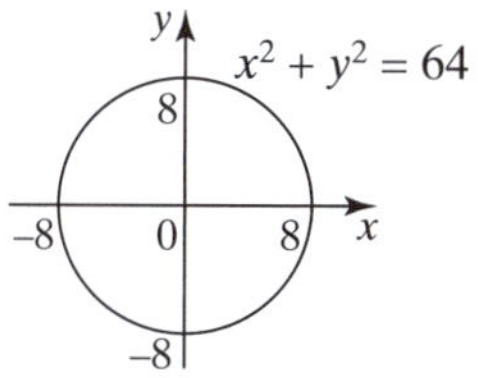

4 $x^2 + y^2 = 196$
a centre at origin (0, 0)
b $r = \sqrt{196}$
radius = 14 units

5 y-values are: −1, −1.2, −1.5, −2, −3, −6, 6, 3, 2, 1.5, 1.2, 1

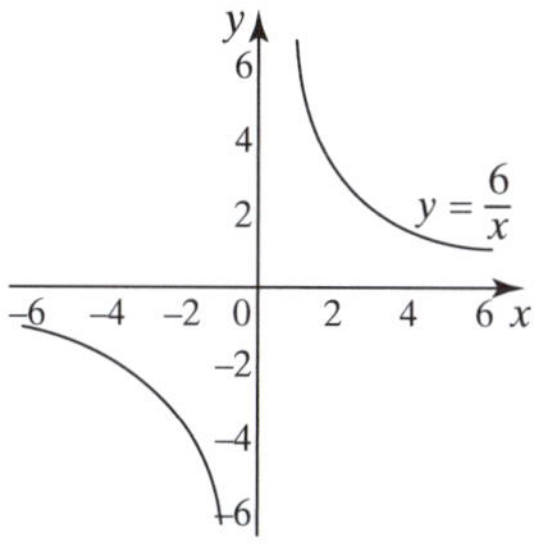

6 **a** parabola
b hyperbola
c exponential curve
d parabola
e cubic curve

7 **a** $y = \frac{x}{4}$
[Straight line, gradient $\frac{1}{4}$, y-intercept 0]

b $y = \frac{-4}{x}$
[hyperbola, passes through (2, −2)]

c $y = x^2 - 4$
[parabola, vertex is at (0, –4)]

d $y = 4x$
[straight line, gradient 4, y-intercept 0]

e $x^2 + y^2 = 4$
[circle, centre (0, 0), radius 2]

f $y = 4 - x^2$
[of form $y = -ax^2 + c$; $c = 4$]

g $y = 4^x$
[exponential curve, pass through (1, 4)]

h $y = 4x^2$
[parabola, vertex (0, 0), passes through (1, 4)]

i $y = \frac{4}{x}$
[hyperbola, passes through (2, 2)]

j $y = 4 - x$
[straight line, gradient –1, y-intercept 4]

FURTHER GRAPHS
INTERMEDIATE TEST PAGE 31

1 A: $x^2 + y^2 = 25$
When $x = 3$ and $y = 4$,
$x^2 + y^2 = 3^2 + 4^2$
$= 9 + 16$
$= 25$

B: $y = \frac{x^3}{3} - 5$
When $x = 3$,
$y = \frac{3^3}{3} - 5$
$= 9 - 5$
$= 4$

C: $y = (x - 1)^2$
When $x = 3$,
$y = (3 - 1)^2$
$= 4$

D: $y = 2^x$
When $x = 3$,
$y = 2^3$
$= 8 \quad (\neq 4)$

∴ [D] ✓ (1 mark)

2 $y = -2^x$
When $x = 0$, $y = -2^0$
$= -1$
The curve goes through the point (0, –1).
When $x = 2$, $y = -2^2$
$= -4$
The curve goes through the point (2, –4).
The only possible curve is C.

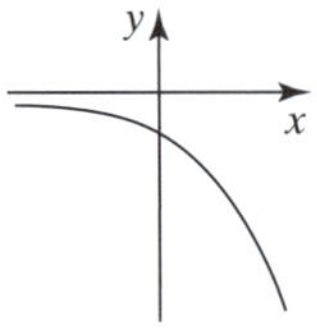

∴ [C] ✓ (1 mark)

3 $y = 2x^3 + 3$
When $x = -1$,
$y = 2 \times (-1)^3 + 3$
$= 2 \times -1 + 3$
$= 1 \quad$ (not –5)
[The other points should also be checked to make sure they do satisfy the equation.]
∴ [A] ✓ (1 mark)

4 a

x	–8	–4	–2	–1	$-\frac{1}{2}$	$\frac{1}{2}$	1	2	4	8
y	$-\frac{1}{2}$	–1	–2	–4	–8	8	4	2	1	$\frac{1}{2}$

✓ (1 mark)

b

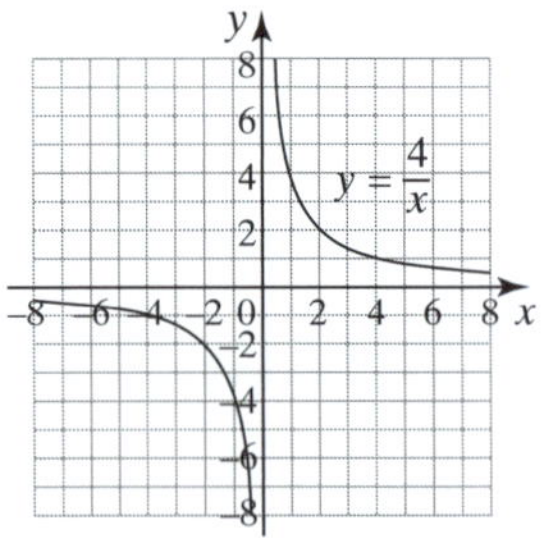

✓✓ (2 marks)

c As the x-values get very large the y-values get closer to 0. ✓ (1 mark)

d As the x-values get close to zero, the y-values get very large. ✓
Large positive values when x is positive and large negative values when x is negative. ✓ (2 marks)

e

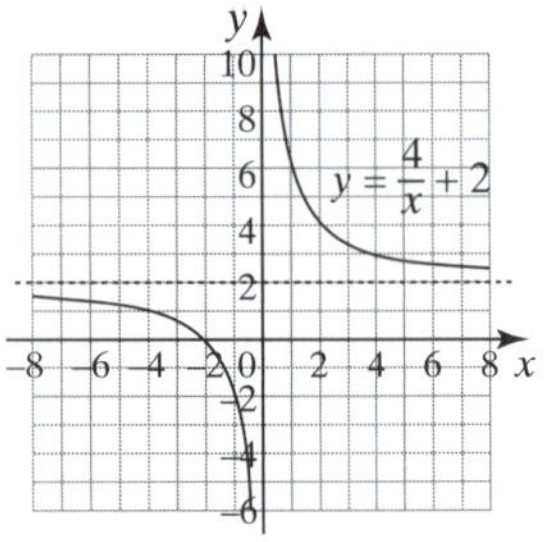

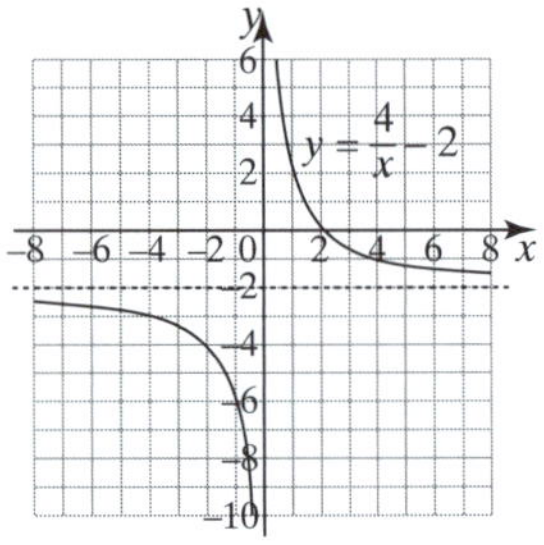

✓✓ (2 marks)

f The addition of a constant, c, moves the curve up (or down if c is negative) the y-axis. The horizontal asymptote also moves up c units. ✓ (1 mark)

5 a $P(2, -1)$, $Q(5, 4)$
$d = \sqrt{(x_2 - x_1)^2 + (y_2 - y_1)^2}$
$= \sqrt{(5-2)^2 + (4-(-1))^2}$
$= \sqrt{34}$ ✓ (1 mark)

b Centre (2, –1), radius $\sqrt{34}$
$(x - 2)^2 + (y + 1)^2 = 34$ ✓ (1 mark)

6 a

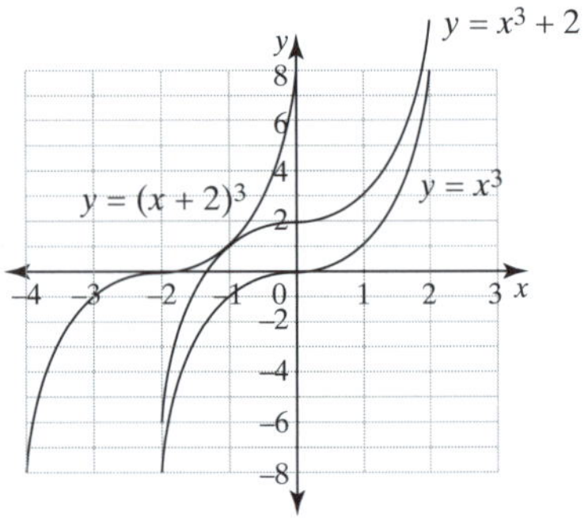

✓✓✓ (3 marks)

b Adding a constant, c, to the end of the equation ($y = x^3 + c$) moves the curve up the y-axis c units. ✓
Adding a constant, c, to the expression that is cubed

($y = (x + c)^3$) moves the curve to the left along the x-axis. The curve cuts the x-axis at $(-c, 0)$. ✓
[The curve $y = (x - c)^3$ has moved to the right c units and cuts the x-axis at $(c, 0)$.]
(2 marks)

7 $x^2 - 6x + y^2 + 4y = 36$
$x^2 - 6x + 9 + y^2 + 4y + 4 = 36 + 9 + 4$
$(x - 3)^2 + (y + 2)^2 = 49$ ✓
Centre $(3, -2)$, radius 7 ✓
(2 marks)

8 **a** $y = \dfrac{-6}{x - 3}$
When $x = 0$, $y = \dfrac{-6}{0 - 3}$
$y = 2$ ✓
(1 mark)

b $x - 3 \neq 0$
$x \neq 3$
Vertical asymptote is $x = 3$ ✓
Horizontal asymptote is $y = 0$. ✓ (2 marks)

c

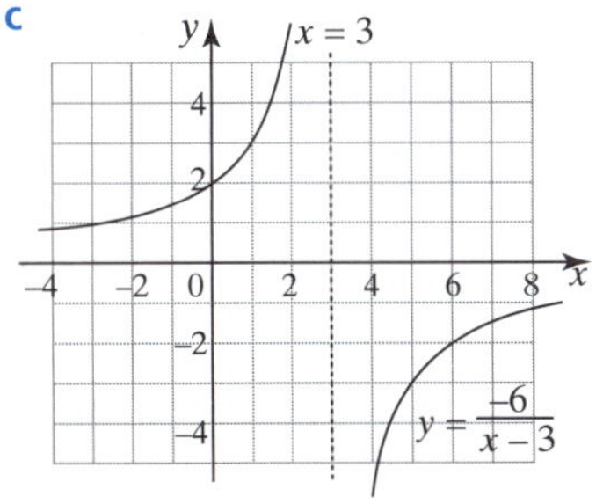

✓ (1 mark)

(Total: 25 marks)

FURTHER GRAPHS
ADVANCED TEST PAGE 32

1 **a** $(x - 3)^2 + (y + 1)^2 = 25$
Circle, centre $(3, -1)$, radius 5 units ✓✓ (2 marks)

b When $y = -1$,
$(x - 3)^2 + (-1 + 1)^2 = 25$
$(x - 3)^2 = 25$
$x - 3 = -5$ or $x - 3 = 5$
$x = -2$ or $x = 8$ ✓
The points of intersection are $(-2, -1)$ and $(8, -1)$. ✓
(2 marks)

c

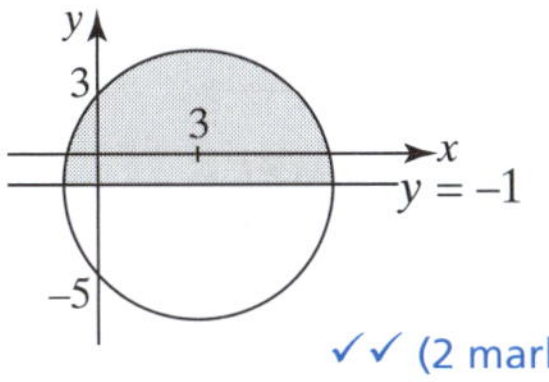

✓✓ (2 marks)

d Shaded part is a semicircle.
$A = \dfrac{1}{2}\pi r^2$
$= \dfrac{1}{2} \times \pi \times 5^2$
$= 39.269\,908\ldots$
$= 39.3$ units^2 (1 d.p.) ✓
(1 mark)

2 **a**

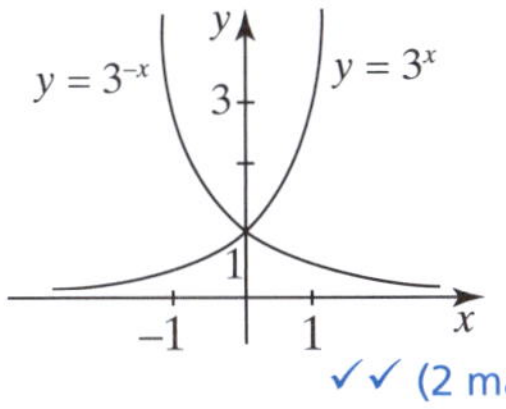

✓✓ (2 marks)

b From the diagram $3^x = 3^{-x}$ when $x = 0$. ✓ (1 mark)

3 **a** $y = x^3$ and $y = 6x - x^2$
At the point of intersection:
$x^3 = 6x - x^2$ ✓
$x^3 + x^2 - 6x = 0$
$x(x^2 + x - 6) = 0$
$x(x + 3)(x - 2) = 0$
$x = 0$ or $x = -3$ or $x = 2$ ✓
When $x = 0$, $y = 0^3 = 0$
When $x = -3$, $y = (-3)^3 = -27$
When $x = 2$, $y = 2^3 = 8$
The points of intersection are $(0, 0)$, $(-3, -27)$ and $(2, 8)$. ✓ (3 marks)

b

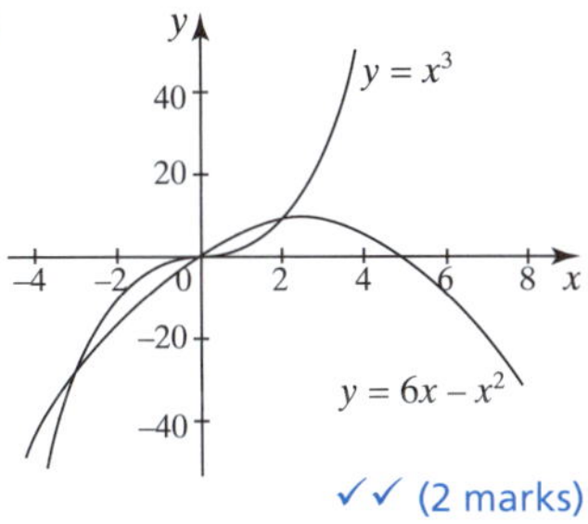

✓✓ (2 marks)

4 $y = a^x + k$
When $x = 0$, $y = 3$
$-3 = a^0 + k$
$= 1 + k$
$k = -4$ ✓
$\therefore y = a^x - 4$

When $x = 2$, $y = 0$
$0 = a^2 - 4$
$a^2 = 4$
$a = \pm 2$ ✓
But $a > 0$,
$\therefore a = 2$ ✓ (3 marks)

5 **a** $y = \dfrac{2}{x} + 3$
As x gets very large, $\dfrac{2}{x}$ gets closer to zero and y gets closer to 3. ✓ (1 mark)

b When $x = 0$, y is undefined. There will be a gap in the curve when $x = 0$. ✓ (1 mark)

c For every permissible x-value there is a y-value. ✓
(1 mark)

d If $y = 7$,
$7 = \dfrac{2}{x} + 3$
$4 = \dfrac{2}{x}$
$x = \dfrac{1}{2}$ ✓ (1 mark)

e The equations of the asymptotes are $x = 0$ and $y = 3$. ✓✓ (2 marks)

f

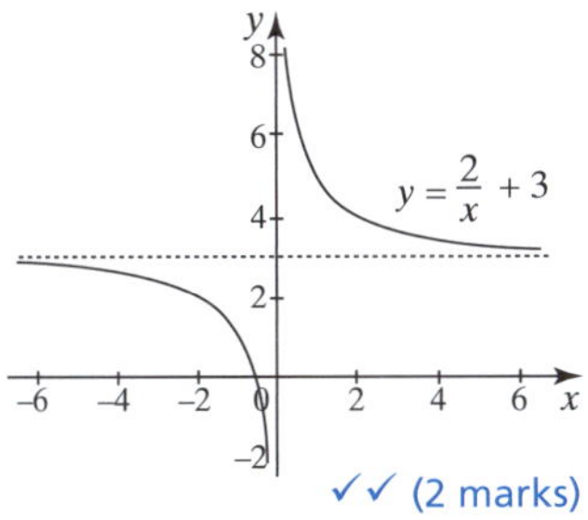

✓✓ (2 marks)

g The constant 3 has the effect of moving the curve $y = \dfrac{2}{x}$ up 3 units. ✓ (1 mark)

6 **a** $y = \dfrac{2}{x + 3}$
When the x-values get very large the y-values get close to zero. ✓ (1 mark)

b When $x = 0$, $y = \dfrac{2}{3}$ ✓
(1 mark)

c Yes, there is a y-value for every permissible x-value. ✓ (1 mark)

d When $y = -1$,

$$-1 = \frac{2}{x+3}$$

$x + 3 = -2$

$x = -5$ ✓ (1 mark)

e The equations of the asymptotes are $x = -3$ and $y = 0$. ✓✓ (2 marks)

f

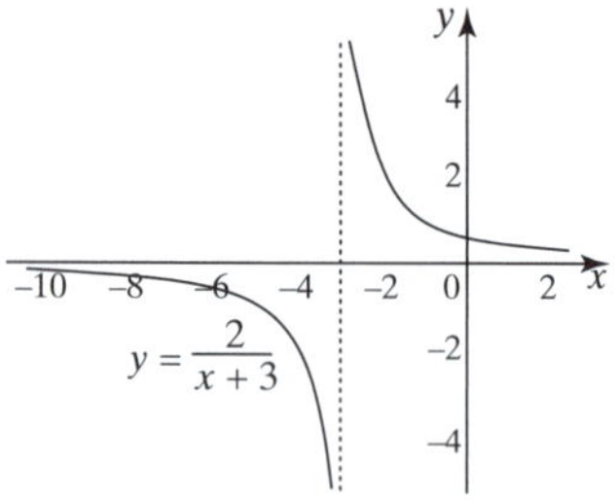

✓✓ (2 marks)

g The curve $y = \frac{2}{x}$ moves 3 units to the left. ✓ (1 mark)

7 $y = -x^4 + 18x^2 - 32$

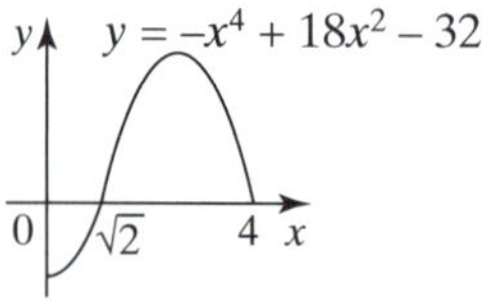

a The y-intercept is –32. ✓ (1 mark)

b The equation only involves x^2 and x^4 and both of those are positive whether x is negative or positive. So the value of the equation when x is a negative number is the same as the value when x is the corresponding positive number. ✓ (1 mark)

c When $x = 3$,

$y = -3^4 + 18 \times 3^2 - 32$

$= 49$

So $q = 49$ ✓ (1 mark)

d Yes, 49 is the maximum value of $-x^4 + 18x^2 - 32$. Because the curve is symmetrical y will also be 49 when $x = -3$. Because it is a polynomial of degree 4 there are only 4 roots so y will have a negative value when $x > 4$ and when $x < -4$. ✓ (1 mark)

(Total: 40 marks)

CURVE-SKETCHING TECHNIQUES
SKILLS CHECK PAGE 34

1 **a**

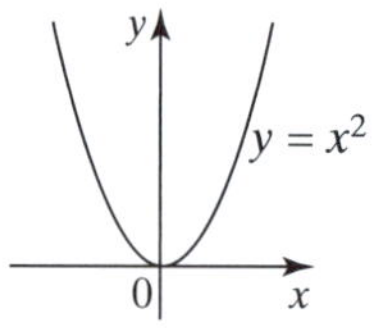

b

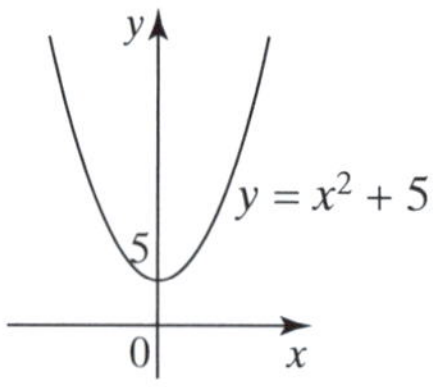

c

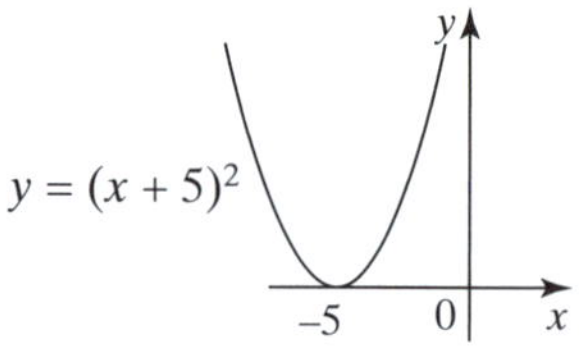

2 **a** $y = x + 5$

The y-intercept is 5.

b $y = x^2 - 2x - 3$

The y-intercept is –3. [The constant term.]

c $y = x^3 + 5x^2 + 6x$

The y-intercept is 0. [When $x = 0$, $y = 0$.]

3 **a** $y = x + 5$

When $y = 0$, $x + 5 = 0$

$x = -5$

The x-intercept is –5.

b $y = x^2 - 2x - 3$

When $y = 0$,

$x^2 - 2x - 3 = 0$

$(x - 3)(x + 1) = 0$

$x - 3 = 0$ or $x + 1 = 0$

$x = 3$ or $x = -1$

The x-intercepts are –1 and 3.

c $y = x^3 + 5x^2 + 6x$

When $y = 0$,

$x^3 + 5x^2 + 6x = 0$

$x(x^2 + 5x + 6) = 0$

$x(x + 3)(x + 2) = 0$

$x = 0$ or

$x + 3 = 0 \rightarrow x = -3$ or

$x + 2 = 0 \rightarrow x = -2$

The x-intercepts are 0, –2 and –3.

4 **a** $(x - 2)^2 + (y - 1)^2 = 49$

$(x - 2)^2 + (y - 1)^2 = 7^2$

Centre (2, 1), radius = 7 units

b $(x + 3)^2 + y^2 = 36$

$(x + 3)^2 + (y - 0)^2 = 6^2$

Centre (–3, 0), radius 6 units

c $(x - 1)^2 + (y + 4)^2 = 1$

$(x - 1)^2 + (y + 4)^2 = 1^2$

Centre (1, –4), radius 1 unit

5 **a** Centre (3, 2), radius 5 units

Equation is

$(x - 3)^2 + (y - 2)^2 = 5^2$

b Centre (–1, 4), radius 2 units

Equation is

$(x + 1)^2 + (y - 4)^2 = 2^2$

6 **a** $x^2 + 8x$

$8x = 2 \times x \times 4$

We need to add $4^2 = 16$

$[x^2 + 8x + 16 = (x + 4)^2]$

b $x^2 - 10x$

$10x = 2 \times x \times 5$

We need to add $5^2 = 25$

$[x^2 - 10x + 25 = (x - 5)^2]$

c $y^2 - 12y$

$12y = 2 \times y \times 6$

We need to add $6^2 = 36$

$[y^2 - 12y + 36 = (y - 6)^2]$

7 **a** $x^2 - 6x + y^2 + 8y = 39$

$x^2 - 6x + 9 + y^2 + 8y + 16 = 39 + 9 + 16$

$(x - 3)^2 + (y + 4)^2 = 64$

$(x - 3)^2 + (y + 4)^2 = 8^2$

b $x^2 + 2x + y^2 - 2y + 1 = 0$

$x^2 + 2x + 1 + y^2 - 2y + 1 = 1$

$(x + 1)^2 + (y - 1)^2 = 1^2$

8 **a** $y = \frac{6}{x}$ [i]

$y = x + 1$ [ii]

$\frac{6}{x} = x + 1$

$6 = x^2 + x$

$x^2 + x - 6 = 0$

$(x + 3)(x - 2) = 0$

$x + 3 = 0$ or $x - 2 = 0$

$x = -3$ or $x = 2$

Substitute into [ii]:

When $x = -3$, $y = -3 + 1$

$= -2$

When $x = 2$, $y = 2 + 1$

$= 3$

b $x^2 + y^2 = 50$ [i]
$x = 2y + 5$ [ii]
Substitute [ii] into [i];
$(2y + 5)^2 + y^2 = 50$
$4y^2 + 20y + 25 + y^2 = 50$
$5y^2 + 20y - 25 = 0$
$y^2 + 4y - 5 = 0$
$(y + 5)(y - 1) = 0$
$y + 5 = 0$ or $y - 1 = 0$
$y = -5$ or $y = 1$
Substitute into [ii]:
When $y = -5$, $x = -5$
When $y - 1$, $x = 7$

9 $y = 3^x$

a y gets very large as x gets large

b y gets closer to 0 as 3^x gets smaller

c Every x-value has a corresponding y-value

d When $x = 0$, $y = 1$

CURVE-SKETCHING TECHNIQUES
INTERMEDIATE TEST PAGE 35

1 $y = 2(x - 1)(x + 3)(x - 4)$
When $x = 0$,
$y = 2(0 - 1)(0 + 3)(0 - 4)$
$= 24$
∴ **[D]** ✓ (1 mark)

2 The circle $(x - 2)^2 + y^2 = 9$ has centre at (2, 0) and radius 3 units.

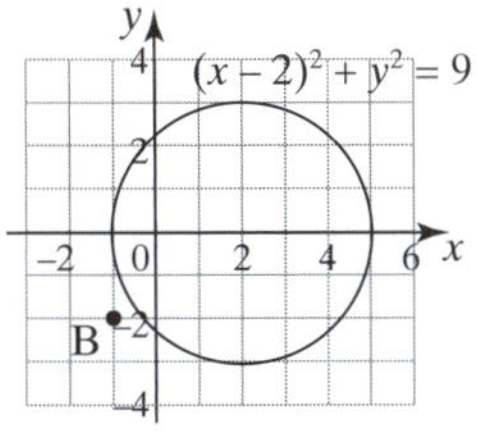

(–1, –2) lies outside the circle.
∴ **[B]** ✓ (1 mark)

3 The hyperbola $y = \frac{3}{x - 6}$ has asymptotes at $x = 6$ and $y = 0$.
∴ **[A]** ✓ (1 mark)

4 $y = 4x^3 - 12x^2 - 16x + 48$

a $4(x - 2)(x + 2)(x - 3)$
$= 4(x^2 - 4)(x - 3)$
$= 4(x^3 - 3x^2 - 4x + 12)$
$= 4x^3 - 12x^2 - 16x + 48$ ✓
(1 mark)

b When $x = 0$, $y = 48$
The y-intercept is 48. ✓
(1 mark)

c $y = 4(x - 2)(x + 2)(x - 3)$
When $y = 0$
$x - 2 = 0 \rightarrow x = 2$ or
$x + 2 = 0 \rightarrow x = -2$ or
$x - 3 = 0 \rightarrow x = 3$
The x-intercepts are –2, 2 and 3. ✓✓ (2 marks)

d When x gets large, y gets large [If x is a large number, $4x^3$ is a much larger number.] ✓ (1 mark)

e

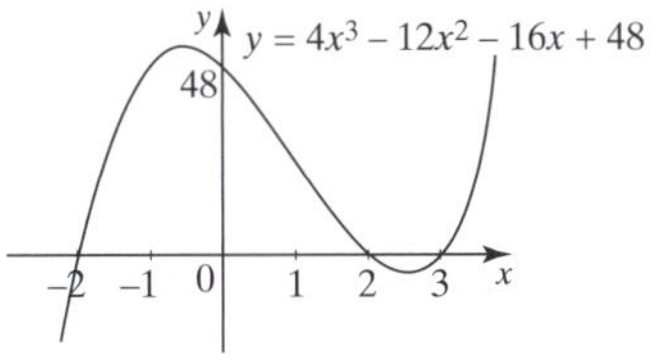

✓✓✓ (3 marks)

5 **a**

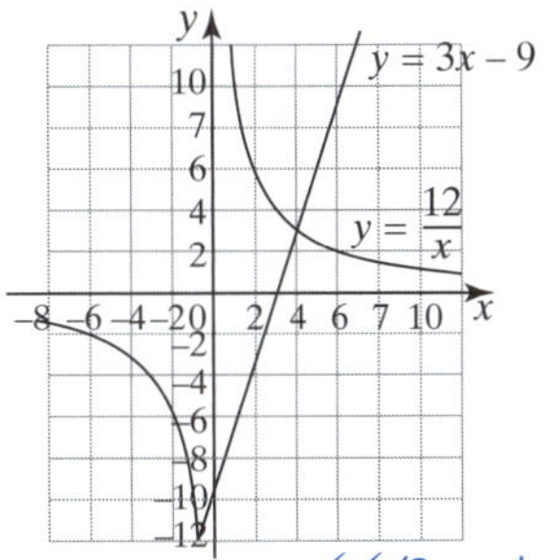

✓✓ (2 marks)

b $\frac{12}{x} = 3x - 9$ when $x = -1$ and $x = 4$. ✓✓ (2 marks)

6 **a** $x^2 - 6x + y^2 + 8y = 0$
$x^2 - 6x + 9$
$+ y^2 + 8y + 16 = 9 + 16$ ✓
$(x - 3)^2 + (y + 4)^2 = 25$
$(x - 3)^2 + (y + 4)^2 = 5^2$ ✓
(2 marks)

b Centre (3, –4) ✓
Length of radius
$= 5$ units ✓ (2 marks)

c $(x - 3)^2 + (y + 4)^2 = 25$ (i)
$x - 2y - 6 = 0$ (ii)
From (ii) $x = 2y + 6$
Substitute into (i)
$(2y + 6 - 3)^2 + (y + 4)^2 = 25$
$(2y + 3)^2 + (y + 4)^2 = 25$
$4y^2 + 12y + 9$
$+ y^2 + 8y + 16 = 25$
$5y^2 + 20y = 0$
$5y(y + 4) = 0$
$y = 0$ or $y + 4 = 0$
$y = 0$ or $y = -4$ ✓
When $y = 0$,
$x = 2 \times 0 + 6$
$= 6$
When $y = -4$,
$x = 2 \times -4 + 6$
$= -2$
∴ The points of intersection are (6, 0) and (–2, –4). ✓✓
(3 marks)

7 All of the graphs are cubic curves.
$y = x^3$ passes through the origin. When x has a large positive value y has a large positive value and when x has a large negative value, y has a large negative value.
$y = 4x^3$ also passes through the origin and y has a large positive value when x has a large positive value and y has a large negative value when x has a large negative value.
$y = 4x^3$ is narrower than $y = x^3$. ✓
$y = x^3 + 4$ has y-intercept 4. It is the same graph as $y = x^3$ but moved up 4 units. Every y-value in $y = x^3 + 4$ is 4 more than the corresponding y-value in $y = x^3$. ✓
$y = (x - 4)^3$ crosses the x-axis at $x = 4$. It is the same graph as $y = x^3$ but moved across 4 units to the right. ✓ (3 marks)

8 **a**

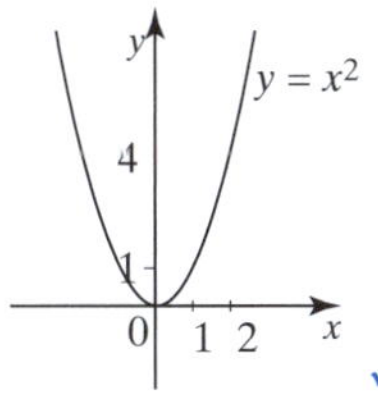

✓

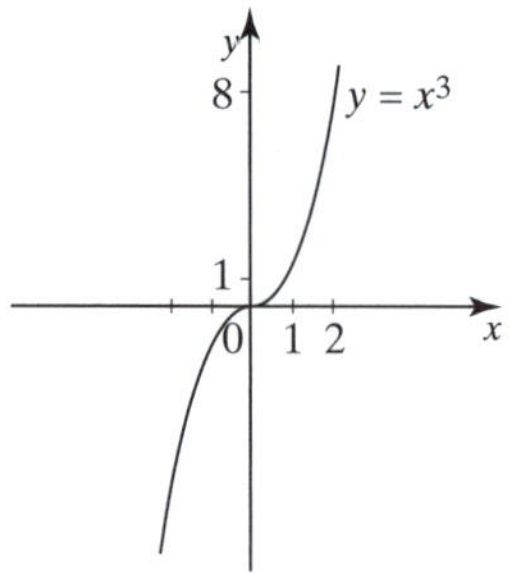

✓

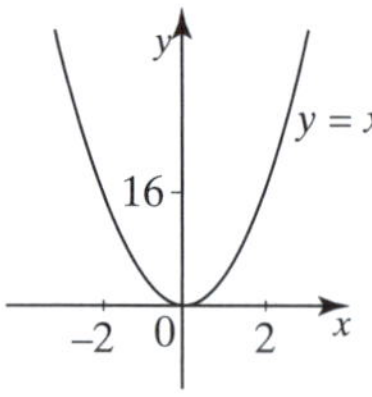

✓

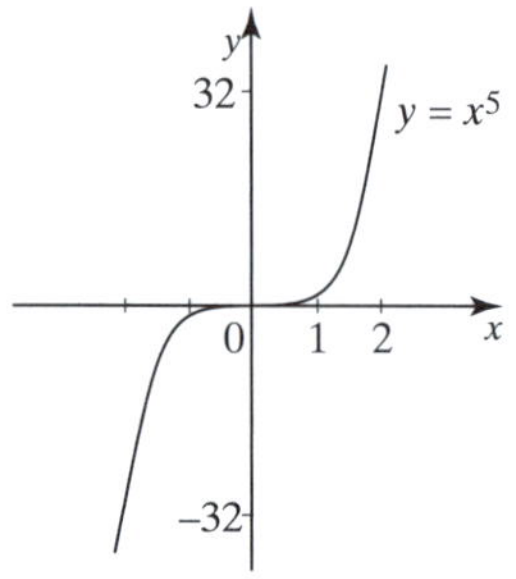

✓

(4 marks)

b $y = x^2$ and $y = x^4$ are symmetrical about the y-axis. ✓ (1 mark)

(Total: 30 marks)

CURVE-SKETCHING TECHNIQUES
ADVANCED TEST PAGE 36

1 **a** $y = \frac{12}{x}$ and $y = 2x + 5$

At the points of intersection:

$$2x + 5 = \frac{12}{x}$$
$$2x^2 + 5x = 12$$
$$2x^2 + 5x - 12 = 0 \checkmark$$
$$(2x - 3)(x + 4) = 0$$
$$x = 1.5 \text{ or } x = -4$$

When $x = 1.5$, $y = \frac{12}{1.5} = 8$ ✓

When $x = -4$, $y = \frac{12}{-4} = -3$

∴ the points of intersection are (1.5, 8) and (−4, −3) ✓

(3 marks)

b

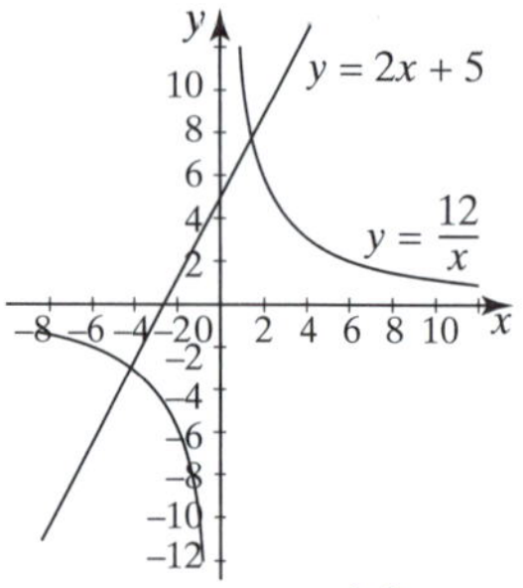

✓✓ (2 marks)

c Any line with negative gradient that passes through the origin will only pass through the second and fourth quadrants while the hyperbola is only in the first and third quadrants. ✓

(1 mark)

d Let $y = mx$ be the equation of the line where $m < 0$.

At any point of intersection:

$$mx = \frac{12}{x}$$
$$x^2 = \frac{12}{m}$$

Now, if $m < 0$ then $\frac{12}{m} < 0$

But $x^2 \geq 0$ for all values of x

So $x^2 \neq \frac{12}{m}$

So there is no point of intersection. ✓ (1 mark)

2

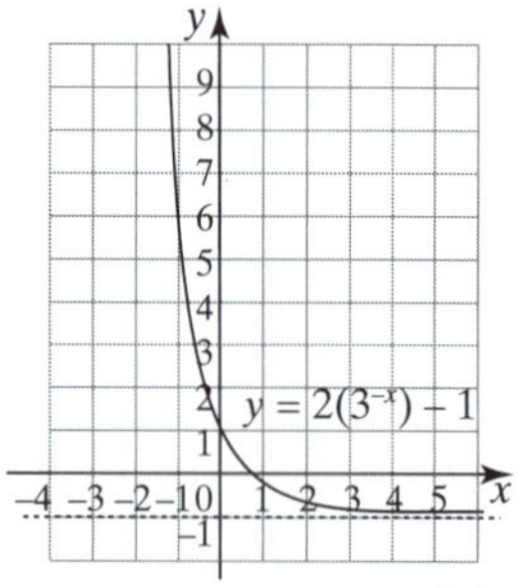

✓✓ (2 marks)

3 **a**

$$x^2 + 6x + y^2 - 10y - 15 = 0$$
$$x^2 + 6x + 9 + y^2 - 10y + 25 = 15 + 9 + 25$$
$$(x + 3)^2 + (y - 5)^2 = 49 \checkmark$$

Centre (−3, 5) ✓

Radius = 7 units ✓

(3 marks)

b (3, 5), (2, 10)

$$d = \sqrt{(x_2 - x_1)^2 + (y_2 - y_1)^2}$$
$$= \sqrt{(2 - (-3))^2 + (10 - 5)^2}$$
$$= \sqrt{50} \checkmark$$

Now $\sqrt{50} > 7$

So the point (2, 10) lies outside the circle. ✓

(2 marks)

4

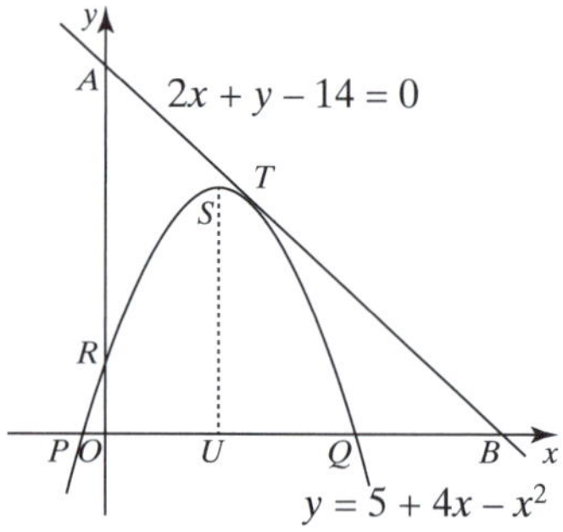

a $2x + y - 14 = 0$

$$y = -2x + 14$$

At the point of intersection:

$$-2x + 14 = 5 + 4x - x^2$$
$$x^2 - 6x + 9 = 0$$
$$(x - 3)^2 = 0$$
$$x = 3 \checkmark$$

When $x = 3$, $y = -2 \times 3 + 14$
$= 8$

T is the point (3, 8). ✓

(2 marks)

b The y-intercept of the parabola is at R.

So R is the point (0, 5). ✓

A circle, centre (0, 0) and radius 5 units will pass through R.

$x^2 + y^2 = 25$ ✓ (2 marks)

c $y = 5 + 4x - x^2$

When $y = 0$,

$0 = (5 - x)(1 + x)$

$x = 5$ or $x = -1$

So $P(-1, 0)$ and $Q(5, 0)$.

As Q is also 5 units from the origin, the circle will pass through Q. ✓ (1 mark)

d The y-intercept of the line $y = -2x + 14$ is at A, so A is the point (0, 14).

$AR = (14 - 5)$ units
$= 9$ units ✓

Axis of parabola:

$$x = \frac{-1 + 5}{2}$$
$$x = 2$$

When $x = 2$,
$y = 5 + 4 \times 2 - 2^2$
$= 9$
So $SU = 9$ units
$\therefore AR = SU$ ✓ (2 marks)

5 **a**

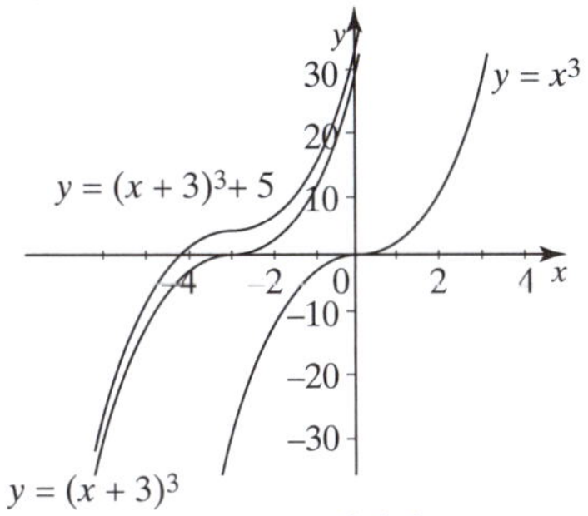

✓✓✓ (3 marks)

b

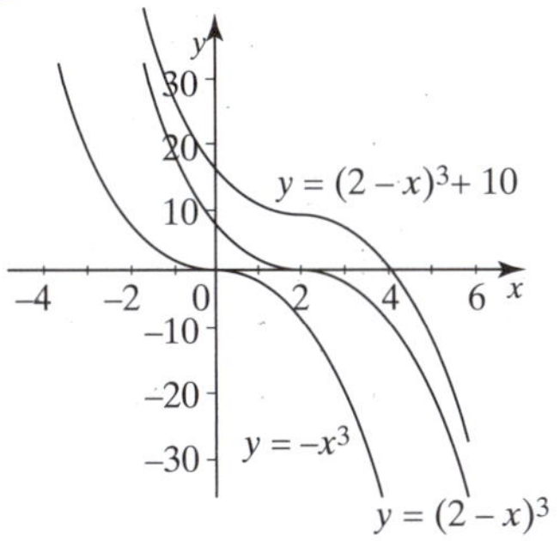

✓✓✓ (3 marks)

c The curve $y = (x - 4)^3 - 2$ is the same shape and size as $y = x^3$, but is moved 4 units to the right and 2 units down. ✓ (1 mark)

6 **a** $y = \dfrac{x^6 + 8}{4}$

$= \dfrac{x^6}{4} + \dfrac{8}{4}$

$= \dfrac{x^6}{4} + 2$ ✓

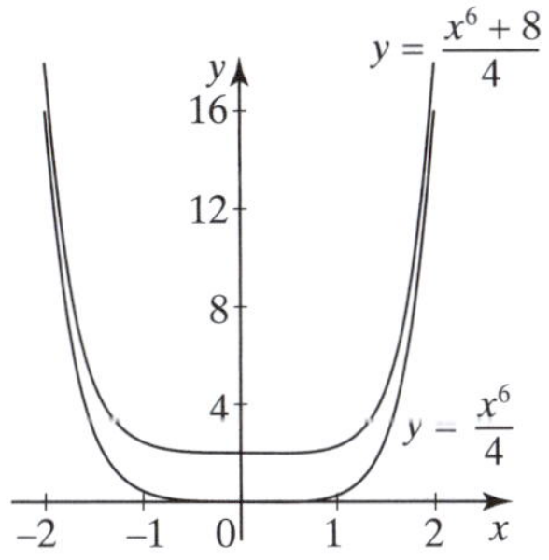

✓✓ (3 marks)

b

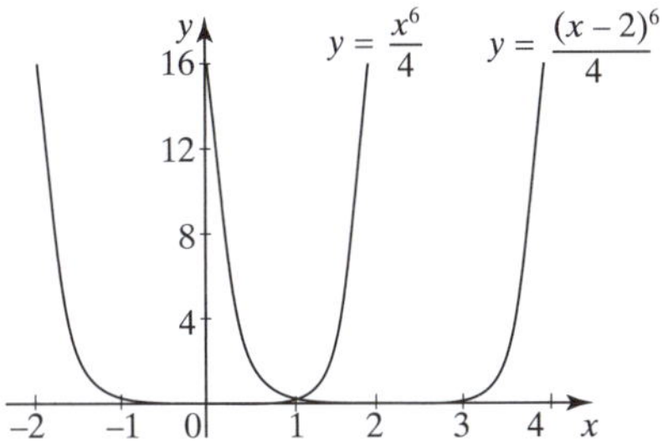

✓✓✓ (3 marks)

c Thc graphs $y = \frac{1}{4}x^6 - 3$ and $y = \frac{(x+1)^6}{4}$ will be the same shape and size as $y = \frac{x^6}{4}$.
$y = \frac{1}{4}x^6 - 3$ will have moved down 3 units and
$y = \frac{(x+1)^6}{4}$ will have moved 1 unit to the left. ✓✓ (2 marks)

7 $y = ax^3 + bx^2 + cx + d$
When $x = 0$, $y = -10$
$\therefore d = -10$ ✓
The curve has x-intercepts at -2, 2 and 5.
$y = a(x + 2)(x - 2)(x - 5)$
$= a(x^2 - 4)(x - 5)$
$= a(x^3 - 5x^2 - 4x + 20)$
$= ax^3 - 5ax^2 - 4ax + 20a$ ✓
But $d = -10$
$20a = -10$
$a = -\dfrac{1}{2}$ ✓
So $y = -\dfrac{1}{2}x^3 + \dfrac{5}{2}x^2 + 2x - 10$
$b = \dfrac{5}{2}$ and $c = 2$ ✓ (4 marks)
(Total: 40 marks)

THE EQUATION OF A STRAIGHT LINE
SKILLS CHECK

PAGE 38

1 **a** $m = 4, b = -2$
$y = 4x - 2$

b $m = \dfrac{1}{2}, b = 5$
$y = \dfrac{1}{2}x + 5$

c $m = -2, b = 0$
$y = -2x$

2 **a** $m = 2, (3, 1)$
$y - y_1 = m(x - x_1)$
$y - 1 = 2(x - 3)$
$y - 1 = 2x - 6$
$y = 2x - 5$

b $m = -3, (1, -4)$
$y - y_1 = m(x - x_1)$
$y - (-4) = -3(x - 1)$
$y + 4 = -3x + 3$
$y = -3x - 1$

c $m = \dfrac{2}{3}, (-3, -5)$
$y - y_1 = m(x - x_1)$
$y - (-5) = \dfrac{2}{3}(x - (-3))$
$y + 5 = \dfrac{2}{3}(x + 3)$
$y + 5 = \dfrac{2}{3}x + 2$
$y = \dfrac{2}{3}x - 3$

3 (3, 2) and (−1, 6)

a $m = \dfrac{y_2 - y_1}{x_2 - x_1}$
$= \dfrac{6 - 2}{-1 - 3}$
$= \dfrac{4}{-4}$
$= -1$

b $m = -1$, (3, 2)
$y - y_1 = m(x - x_1)$
$y - 2 = -1(x - 3)$
$y - 2 = -x + 3$
$y = -x + 5$

4 (1, 3), (2, 6)
$m = \dfrac{y_2 - y_1}{x_2 - x_1}$
$= \dfrac{6 - 3}{2 - 1}$
$= 3$
$y - y_1 = m(x - x_1)$
$y - 3 = 3(x - 1)$
$y - 3 = 3x - 3$
$y = 3x$

5 **a** $2x + y - 5 = 0$
$y = -2x + 5$

b $3x + 2y - 6 = 0$
$2y = -3x + 6$
$y = -\dfrac{3}{2}x + 3$

6 **a** $y = 3x - 1$
$3x - y - 1 = 0$

b $y = -\frac{2}{3}x + 4$
$3y = -2x + 12$
$2x + 3y - 12 = 0$

7 **a** $y = 2x + 8$
y-intercept = 8
When $y = 0$, $2x + 8 = 0$
$2x = -8$
$x = -4$
The x-intercept is –4

b $x + 3y - 6 = 0$
When $x = 0$, $3y - 6 = 0$
$3y = 6$
$y = 2$
When $y = 0$, $x - 6 = 0$
$x = 6$
The x-intercept is 6 and the y-intercept is 2.

8 **a** If $m_1 = m_2$ the lines are parallel.

b If $m_1 m_2 = -1$, the lines are perpendicular.

9 **a** $-\frac{1}{2}$ [Parallel lines have the same gradient.]

b 2 [Perpendicular lines have gradients which are negative reciprocals.]

THE EQUATION OF A STRAIGHT LINE
INTERMEDIATE TEST PAGE 39

1 $y = 2x - 5$
Gradient = 2 [coefficient of x]
Any perpendicular line has gradient $-\frac{1}{2}$.
[The negative reciprocal of 2.]
∴ [D] ✓ (1 mark)

2 $y = -3x + 6$
When $y = 0$, $-3x + 6 = 0$
$-3x = -6$
$x = 2$
The x-intercept is 2.
∴ [C] ✓ (1 mark)

3 $y = 2 - x$
$y = -x + 2$
Gradient = –1
Any parallel line has $m = -1$.
Any line that passes through the origin has y-intercept 0.
Equation is $y = -x$
∴ [B] ✓ (1 mark)

4 **a** $4x + 2y - 1 = 0$
$2y = -4x + 1$
$y = -2x + \frac{1}{2}$ ✓
(1 mark)

b Gradient = –2 ✓
y-intercept is $\frac{1}{2}$ ✓
(2 marks)

c
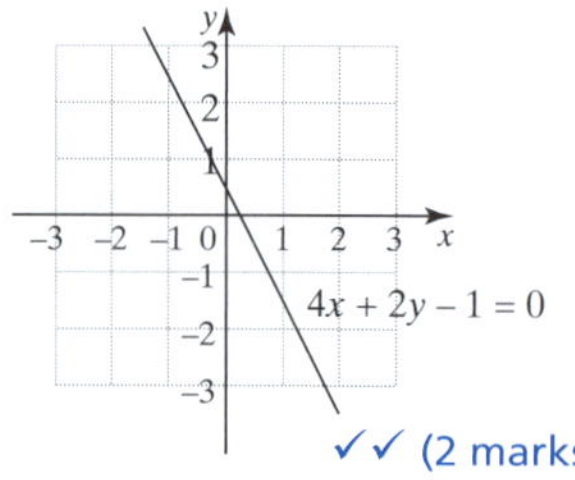

✓✓ (2 marks)

5
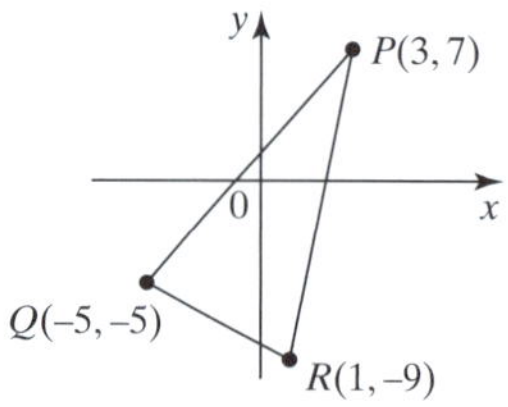

$P(3, 7)$, $Q(-5, -5)$

$m_1 = \frac{y_2 - y_1}{x_2 - x_1}$
$= \frac{-5 - 7}{-5 - 3}$
$= \frac{-12}{-8}$
$= \frac{3}{2}$ ✓

$Q(-5, -5)$, $R(1, -9)$

$m_2 = \frac{y_2 - y_1}{x_2 - x_1}$
$= \frac{-9 - (-5)}{1 - (-5)}$
$= \frac{-4}{6}$
$= -\frac{2}{3}$ ✓

$m_1 m_2 = \frac{3}{2} \times -\frac{2}{3}$
$= -1$ ✓

PQ is perpendicular to QR.
∴ $\angle PQR = 90°$. ✓
[Or use the distance formula to find the lengths of the three sides and show by Pythagoras' theorem that the triangle is right-angled.] (4 marks)

6 **a** $3x + 4y - 12 = 0$
When $x = 0$,
$3 \times 0 + 4y - 12 = 0$
$4y = 12$
$y = 3$
The y-intercept is 3. ✓
(1 mark)

b $3x + 4y - 12 = 0$
When $y = 0$,
$3x + 4 \times 0 - 12 = 0$
$3x = 12$
$x = 4$
The x-intercept is 4. ✓
(1 mark)

c
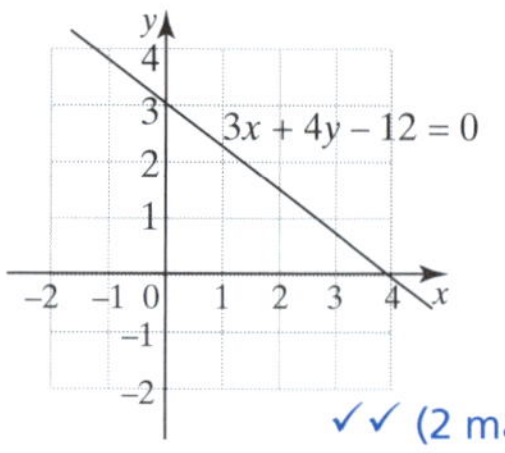

✓✓ (2 marks)

7
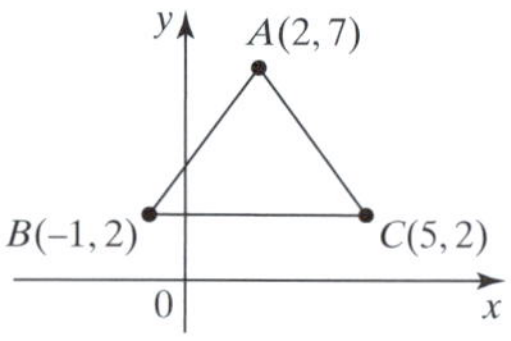

a $B(-1, 2)$, $C(5, 2)$
Line BC: $y = 2$ ✓
(1 mark)

b A line perpendicular to $y = 2$ is of the form $x = k$.
The line passes through (2, 7). Equation is $x = 2$. ✓
(1 mark)

c $A(2, 7)$, $C(5, 2)$
gradient $= \frac{y_2 - y_1}{x_2 - x_1}$
$= \frac{2 - 7}{5 - 2}$
$= -\frac{5}{3}$
Perpendicular line has gradient $m = \frac{3}{5}$. ✓

$(-1, 2),\ m = \frac{3}{5}$

$y - y_1 = m(x - x_1)$

$y - 2 = \frac{3}{5}(x - (-1))$

$y - 2 = \frac{3}{5}(x + 1)$

$y - 2 = \frac{3}{5}x + \frac{3}{5}$

$y = \frac{3}{5}x + 2\frac{3}{5}$ ✓

[Or $3x - 5y + 13 = 0$]

(2 marks)

d $x = 2$ [i]

$y = \frac{3}{5}x + 2\frac{3}{5}$ [ii]

Substitute [i] into [ii]:

When $x = 2,\ y = \frac{3}{5} \times 2 + 2\frac{3}{5}$

$= 3\frac{4}{5}$

The lines meet at $(2, 3\frac{4}{5})$

✓✓ (2 marks)

e $A(2, 7), B(-1, 2)$

gradient $= \frac{y_2 - y_1}{x_2 - x_1}$

$= \frac{2 - 7}{-1 - 2}$

$= \frac{5}{3}$

Perpendicular line has gradient $m = -\frac{3}{5}$. ✓

$(5, 2),\ m = -\frac{3}{5}$

$y - y_1 = m(x - x_1)$

$y - 2 = -\frac{3}{5}(x - 5)$

$y - 2 = -\frac{3}{5}x + 3$

$y = -\frac{3}{5}x + 5$

[Or $3x + 5y - 25 = 0$] ✓

(2 marks)

f $y = -\frac{3}{5}x + 5$

When $x = 2$,

$y = -\frac{3}{5} \times 2 + 5$

$= 3\frac{4}{5}$

The line in part **e** passes through $(2, 3\frac{4}{5})$. ✓ (1 mark)

(Total: 25 marks)

THE EQUATION OF A STRAIGHT LINE
ADVANCED TEST PAGE 40

1

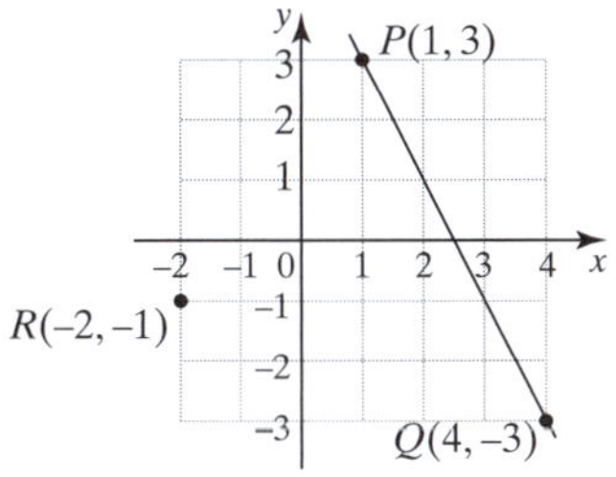

a $P(1, 3), Q(4, -3)$

$m = \frac{y_2 - y_1}{x_2 - x_1}$

$m_{PQ} = \frac{-3 - 3}{4 - 1}$

$= -2$

$y - y_1 = m(x - x_1)$

$y - 3 = -2(x - 1)$

$y - 3 = -2x + 2$

$y = -2x + 5$ ✓

$SR \perp PQ$

$\therefore m_{SR} = \frac{1}{2}$ ✓

$R(-2, -1)$

$y - y_1 = m(x - x_1)$

$y + 1 = \frac{1}{2}(x + 2)$

$y + 1 = \frac{1}{2}x + 1$

$y = \frac{1}{2}x$ ✓

At S,

$\frac{1}{2}x = -2x + 5$

$2\frac{1}{2}x = 5$

$x = 2$

$y = \frac{1}{2} \times 2$

$= 1$

So S is the point $(2, 1)$. ✓

(4 marks)

b $m_{QR} = \frac{-1 - (-3)}{-2 - 4}$

$= -\frac{1}{3}$

Perpendicular line has $m = 3$ ✓

$P(1, 3)$

$y - y_1 = m(x - x_1)$

$y - 3 = 3(x - 1)$

$y - 3 = 3x - 3$

$y = 3x$

So both lines have y-intercept 0, so they meet at the origin. ✓ (2 marks)

c $O(0, 0), Q(4, -3)$

$m_{QO} = \frac{-3 - 0}{4 - 0}$

$= -\frac{3}{4}$

y-intercept $= 0$

Equation is $y = -\frac{3}{4}x$ ✓

(1 mark)

d $P(1, 3), R(-2, -1)$

$m_{PR} = \frac{-1 - 3}{-2 - 1}$

$= \frac{4}{3}$

Now $-\frac{3}{4} \times \frac{4}{3} = -1$

$\therefore$ the lines are perpendicular. ✓ (1 mark)

2 **a** $2x - ky - 9 = 0$

$m = \frac{-2}{-k}$

$= \frac{2}{k}$ ✓

$x + 3y + 1 = 0$

$m = -\frac{1}{3}$

If the lines are parallel

$\frac{2}{k} = -\frac{1}{3}$

$6 = -k$

$k = -6$ ✓ (2 marks)

b If the lines are perpendicular

$\frac{2}{k} \times -\frac{1}{3} = -1$ ✓

$\frac{-2}{3k} = -1$

$3k = 2$

$k = \frac{2}{3}$ ✓ (2 marks)

3 **a** $2x + y - 6 = 0$

$y = -2x + 6$

Gradient $= -2$ ✓

When $y = 0$,

$-2x + 6 = 0$

$2x = 6$

$x = 3$ ✓

Perpendicular line has $m = \frac{1}{2}$. It passes through the point $(3, 0)$. cont.

$y - y_1 - m(x - x_1)$

$y - 0 = \frac{1}{2}(x - 3)$

$y = \frac{1}{2}x - \frac{3}{2}$ ✓ (3 marks)

b

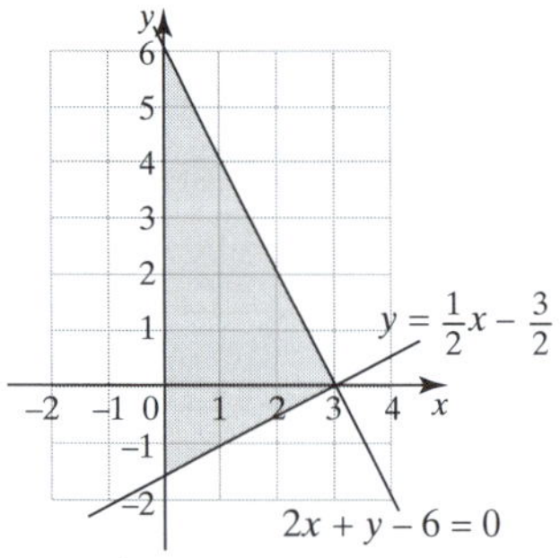

$A = \frac{1}{2}bh$

$= \frac{1}{2} \times (6 + \frac{3}{2}) \times 3$ ✓

$= 11\frac{1}{4}$ units² ✓ (2 marks)

4 **a** AB: $4x + 3y - 25 = 0$

AC: $3x - 4y = 0$

BC: $2x - 11y + 25 = 0$

$m_{AB} = -\frac{4}{3}$

$m_{AC} = \frac{3}{4}$ ✓

Now $-\frac{4}{3} \times \frac{3}{4} = -1$

$\therefore AB \perp AC$

∴ the triangle is right-angled at A. ✓ (2 marks)

b $4x + 3y - 25 = 0$ [i]

$2x - 11y + 25 = 0$ [ii]

[ii] × 2: $4x - 22y + 50 = 0$ [iii]

[i] – [iii]: $25y - 75 = 0$

$25y = 75$

$y = 3$ ✓

Substitute into [i]:

$4x + 3 \times 3 - 25 = 0$

$4x = 16$

$x = 4$

∴ B is the point (4, 3) ✓ (2 marks)

c $3x - 4y = 0$ [i]

$2x - 11y + 25 = 0$ [ii]

[i] × 2: $6x - 8y = 0$ [iii]

[ii] × 3: $6x - 33y + 75 = 0$ [iv]

[iii] – [iv]: $25y - 75 = 0$

$y = 3$ ✓

Substitute into [i]:

$3x - 4 \times 3 = 0$

$3x = 12$

$x = 4$

∴ C is the point (4, 3). ✓ (2 marks)

d The triangle cannot exist. ✓ B and C are the same point. But the triangle is right-angled at A. This can only happen if A is also the same point. So, A, B and C are the same point and the triangle does not exist. ✓ (2 marks)

5 **a** $P(8, 7)$, $R(-6, 5)$

$M\left(\frac{x_1 + x_2}{2}, \frac{y_1 + y_2}{2}\right)$

$= \left(\frac{8-6}{2}, \frac{7+5}{2}\right)$

$= (1, 6)$ ✓ (1 mark)

b $Q(-2, -1)$, $R(-6, 5)$

$m = \frac{y_2 - y_1}{x_2 - x_1}$

$m_{QR} = \frac{5 - (-1)}{-6 - (-2)}$

$= -\frac{3}{2}$ ✓

So k has $m = -\frac{3}{2}$ and passes through (8, 7).

$y - y_1 = m(x - x_1)$

$y - 7 = -\frac{3}{2}(x - 8)$

$2y - 14 = -3x + 24$

$3x + 2y - 38 = 0$ ✓ (2 marks)

c $Q(-2, -1)$, $M(1, 6)$

$m = \frac{y_2 - y_1}{x_2 - x_1}$

$m_{QM} = \frac{6 - (-1)}{1 - (-2)}$

$= \frac{7}{3}$ ✓

$y - y_1 = m(x - x_1)$

$y - (-1) = \frac{7}{3}(x - (-2))$

$3(y + 1) = 7(x + 2)$

$3y + 3 = 7x + 14$

$7x - 3y + 11 = 0$ ✓ (2 marks)

d $3x + 2y - 38 - 0$ [i]

$7x - 3y + 11 = 0$ [ii]

[i] × 3: $9x + 6y - 114 = 0$ [iii]

[ii] × 2: $14x - 6y + 22 = 0$ [iv]

[iii] + [iv]: $23x - 92 = 0$

$23x = 92$

$x = 4$ ✓

Substitute into [i]:

$3 \times 4 + 2y - 38 = 0$

$2y = 26$

$y = 13$

S is the point (4, 13). ✓ (2 marks)

e $S(4, 13)$, $R(-6, 5)$

$m = \frac{y_2 - y_1}{x_2 - x_1}$

$m_{SR} = \frac{5 - 13}{-6 - 4}$

$= \frac{4}{5}$ ✓

$P(8, 7)$, $Q(-2, -1)$

$m = \frac{y_2 - y_1}{x_2 - x_1}$

$m_{PQ} = \frac{-1 - 7}{-2 - 8}$

$= \frac{4}{5}$

$m_{SR} = m_{PQ}$

So $SR \parallel PQ$ ✓ (2 marks)

6 **a** $A(-3, 1)$, $B(5, -1)$

$M\left(\frac{x_1 + x_2}{2}, \frac{y_1 + y_2}{2}\right)$

$= \left(\frac{-3+5}{2}, \frac{1+(-1)}{2}\right)$

$= (1, 0)$ ✓ (1 mark)

b $m = \frac{y_2 - y_1}{x_2 - x_1}$

$m_{AB} = \frac{-1 - 1}{5 - (-3)}$

$= -\frac{1}{4}$ ✓ (1 mark)

c Gradient of perpendicular line = 4.

The perpendicular bisector will pass through the midpoint (1, 0).

$y - y_1 = m(x - x_1)$

$y - 0 = 4(x - 1)$

$y = 4x - 4$ ✓ (1 mark)

d $C(-5, -24)$
$y = 4x - 4$
When $x = -5$,
$y = 4 \times (-5) - 4$
$= -24$
$\therefore$ C lies on the perpendicular bisector of AB. ✓ (1 mark)

e If D is equidistant from A and B it must lie on the perpendicular bisector of AB.
$2x + y - 14 = 0$ [i]
$y = 4x - 4$ [ii]
Substitute [ii] into [i]:
$2x + 4x - 4 - 14 = 0$
$6x = 18$
$x = 3$ ✓
Substitute into [ii]:
$y = 4 \times 3 - 4$
$= 8$
$\therefore$ D is the point $(3, 8)$. ✓
(2 marks)

(Total: 40 marks)

EQUATIONS SKILLS CHECK PAGE 42

1 **a** $x + 38 = 102$ [-38]
$x = 64$

b $x - 23 = -5$ [$+23$]
$x = 18$

c $13x = 533$ [$\div 13$]
$x = 41$

d $\frac{x}{18} = 21$ [$\times 18$]
$x = 378$

e $2x + 17 = 56$
$2x = 39$
$x = 19.5$

f $\frac{x}{5} - 19 = 81$
$\frac{x}{5} = 100$
$x = 500$

2 **a** $x + a = b$
$x = b - a$

b $x - y = z$
$x = z + y$

c $px = q$
$x = \frac{q}{p}$

d $\frac{x}{n} = k$
$x = kn$

e $rx + s = t$
$rx = t - s$
$x = \frac{t - s}{r}$

f $\frac{x}{m} - u = h$
$\frac{x}{m} = h + u$
$x = m(h + u)$
$x = hm + mu$

3 **a** $6m - (2m + 3)$
$= 6m - 2m - 3$
$= 4m - 3$

b $3(x + 2) + 2(x - 7)$
$= 3x + 6 + 2x - 14$
$= 5x - 8$

c $5(2a - 1) - 4(3a + 2)$
$= 10a - 5 - 12a - 8$
$= -2a - 13$

d $6(3p - q) - 3(5p - 2q)$
$= 18p - 6q - 15p + 6q$
$= 3p$

4 If $x = -1$,
$5(x + 1) - 2(3 - 2x)$
$= 5(-1 + 1) - 2(3 - 2 \times -1)$
$= -10$ ($\neq 10$) false

5 **a** If $y = \sqrt{x}$ then $x \geq 0$
Values of x that cannot be used are $x < 0$.

b If $y = \frac{1}{x - 2}$
then $x - 2 \neq 0$
$\therefore x \neq 2$
The value of x that cannot be used is $x = 2$.

6 If $x = 2p$,
a $3x + 2 = 3(2p) + 2$
$= 6p + 2$

b $x^2 - x = (2p)^2 - 2p$
$= 4p^2 - 2p$

c $px - x + 2p$
$= p(2p) - 2p + 2p$
$= 2p^2$

7 **a** $\frac{x}{5} + \frac{3x}{5} = \frac{4x}{5}$

b $\frac{a}{2} - \frac{a}{3} = \frac{3a}{6} - \frac{2a}{6}$
$= \frac{a}{6}$

c $\frac{x + 2}{3} + \frac{x + 1}{4}$
$= \frac{4(x + 2) + 3(x + 1)}{12}$
$= \frac{4x + 8 + 3x + 3}{12}$
$= \frac{7x + 11}{12}$

d $\frac{2n - 3}{2} - \frac{n - 2}{3}$
$= \frac{3(2n - 3) - 2(n - 2)}{6}$
$= \frac{6n - 9 - 2n + 4}{6}$
$= \frac{4n - 5}{6}$

8 **a** $64 = 2^6$

b $\frac{1}{8} = \frac{1}{2^3}$
$= 2^{-3}$

c $\frac{1}{\sqrt{2}} = \frac{1}{2^{\frac{1}{2}}}$
$= 2^{-\frac{1}{2}}$

d $4\sqrt{2} = 2^2 \times 2^{\frac{1}{2}}$
$= 2^{2.5}$

9 **a** $\frac{2x - 1}{3} = 7$
$2x - 1 = 21$
$2x = 22$
$x = 11$

b $4(3x - 2) = 9 - (x + 4)$
$12x - 8 = 9 - x - 4$
$12x - 8 = 5 - x$
$13x = 13$
$x = 1$

c $\frac{x + 1}{5} = \frac{x - 1}{4}$
$4(x + 1) = 5(x - 1)$
$4x + 4 = 5x - 5$
$9 = x$

10 **a** $27^{\frac{1}{3}} = \sqrt[3]{27} = 3$

b $25^{-\frac{1}{2}} = \frac{1}{\sqrt{25}} = \frac{1}{5}$

c $\left(\frac{1}{2}\right)^{-4} = (2^{-1})^{-4} = 2^4 = 16$

d $\log_4\left(\frac{1}{8}\right) = \log_4\left(\frac{1}{2^3}\right)$
$= \log_4 2^{-3}$
$= \log_4 (2^2)^{-\frac{3}{2}}$
$= \log_4\left(4^{-\frac{3}{2}}\right)$
$= -\frac{3}{2}$
$= -1.5$

EQUATIONS
INTERMEDIATE TEST PAGE 43

1 $x^2 = 4ay$
When $x = 2ap$, $(2ap)^2 = 4ay$
$4a^2p^2 = 4ay$
$ap^2 = y$
$\therefore$ [C] ✓ (1 mark)

2 $T = \sqrt{\frac{4g}{l}}$
$T^2 = \frac{4g}{l}$
$lT^2 = 4g$
$\frac{lT^2}{4} = g$
$\therefore$ [C] ✓ (1 mark)

3 $\log_8 2 = x$
$\therefore 8^x = 2$
$(2^3)^x = 2$
$2^{3x} = 2^1$
$3x = 1$
$x = \frac{1}{3}$
$\therefore$ [D] ✓ (1 mark)

4 $1.2^n = 539$
$1.2^{25} = 95.3962166\ldots$
So $n > 25$
$1.2^{30} = 237.376313\ldots$
So $n > 30$
$1.2^{35} = 590.668229\ldots$
So $n < 35$
n lies between 30 and 35.
$\therefore$ [C] ✓ (1 mark)

5 **a** $S = V - Dn$
If $V = 12000$, $D = 500$, $n = 7$,
$S = 12000 - 500 \times 7$
$= 12000 - 3500$
$= 8500$ ✓ (1 mark)

b $S = V - Dn$
$S + Dn = V$
$Dn = V - S$
$D = \frac{V - S}{n}$ ✓ (1 mark)

c $D = \frac{V - S}{n}$
When $S = 1175$, $V = 1650$ and $n = 5$,
$D = \frac{1650 - 1175}{5}$
$= \frac{475}{5}$
$= 95$ ✓ (1 mark)

6 $\frac{5n-1}{3} - \frac{2n-3}{4} = 1$
$12 \times \left(\frac{5n-1}{3}\right) - 12 \times \left(\frac{2n-3}{4}\right) = 12 \times 1$
$4(5n - 1) - 3(2n - 3) = 12$ ✓
$20n - 4 - 6n + 9 = 12$
$14n + 5 = 12$ ✓
$14n = 7$
$n = 0.5$ ✓
(3 marks)

7 **a** $d = \sqrt{b^2 - 4ac}$
$b^2 - 4ac \geq 0$
$b^2 \geq 4ac$ ✓ (1 mark)

b $d = \sqrt{b^2 - 4ac}$
$d^2 = b^2 - 4ac$
$d^2 + 4ac = b^2$ ✓
$4ac = b^2 - d^2$
$c = \frac{b^2 - d^2}{4a}$ ✓
(2 marks)

c If $c < 0$ and $a > 0$,
then $b^2 - d^2 < 0$
$b^2 < d^2$ ✓ (1 mark)

8 Let x be the number.
$5(x - 15) = 3x + 7$ ✓
$5x - 75 = 3x + 7$
$2x - 75 = 7$
$2x = 82$
$x = 41$
The number is 41. ✓ (2 marks)

9 $9^{x-1} = \frac{1}{27}$
$(3^2)^{x-1} = 3^{-3}$ ✓
$3^{2x-2} = 3^{-3}$
$2x - 2 = -3$ ✓
$2x = -1$
$x = -\frac{1}{2}$ ✓ (3 marks)

10 $\frac{x-2}{2} - \frac{2x+1}{5} = \frac{3}{4}$
$20 \times \left(\frac{x-2}{2}\right) - 20 \times \left(\frac{2x+1}{5}\right) = 20 \times \frac{3}{4}$
$10(x - 2) - 4(2x + 1) = 15$ ✓
$10x - 20 - 8x - 4 = 15$
$2x - 24 = 15$ ✓
$2x = 39$
$x = 19.5$ ✓
(3 marks)

11 $2^x = 3.4$
$\log_{10} 2^x = \log_{10} 3.4$ ✓
$x \log_{10} 2 = \log_{10} 3.4$
$x = \log_{10} 3.4 \div \log_{10} 2$ ✓
$= 1.7655347\ldots$
[by calculator]
$= 1.77$ (2 d.p.) ✓
[Or use a trial and error method.] (3 marks)

(Total: 25 marks)

EQUATIONS
ADVANCED TEST PAGE 44

1 **a** $7(x - 4) = 8 - 3(x - 2)$
$7x - 28 = 8 - 3x + 6$ ✓
$10x = 42$
$x = 4.2$ ✓ (2 marks)

b $5 - (6 - a) = 3a$
$5 - 6 + a = 3a$ ✓
$-1 = 2a$
$a = -\frac{1}{2}$ ✓ (2 marks)

c $\frac{3t}{4} - \frac{t}{9} = 2$
$27t - 4t = 72$ ✓
$23t = 72$
$t = 3\frac{3}{23}$ ✓ (2 marks)

d $\frac{5h}{2} - \frac{h-3}{4} = h + 2$
$10h - (h - 3) = 4(h + 2)$
$10h - h + 3 = 4h + 8$ ✓
$5h = 5$
$h = 1$ ✓ (2 marks)

e $\frac{3m}{m+2} = \frac{5}{2}$
$2(3m) = 5(m + 2)$ ✓
$6m = 5m + 10$
$m = 10$ ✓ (2 marks)

f $\frac{1}{2}(p-5)-\frac{1}{3}(p+8)=7$
$3(p-5)-2(p+8)=42$
$3p-15-2p-16=42$ ✓
$p-31=42$
$p=73$ ✓
(2 marks)

2 **a** $\frac{2x}{5}-3\leq 1$
$\frac{2x}{5}\leq 4$
$2x\leq 20$ ✓
$x\leq 10$ ✓ (2 marks)

b $6(3-2a)-5(a-4)>4$
$18-12a-5a+20>4$
$38-17a>4$ ✓
$-17a>-34$
$a<2$ ✓
(2 marks)

c $\frac{3(n+2)}{5}-\frac{4(n-1)}{3}<4$
$9(n+2)-20(n-1)<60$
$9n+18-20n+20<60$ ✓
$-11n<22$
$n>-2$ ✓
(2 marks)

3 $x=\frac{-b\pm\sqrt{b^2-4ac}}{2a}$
$a=3$, $b=-6$ and $c=1$
$x=\frac{-(-6)\pm\sqrt{(-6)^2-4\times3\times1}}{2\times3}$
$=\frac{6\pm\sqrt{24}}{6}$
$=\frac{6\pm2\sqrt{6}}{6}$
$=\frac{3\pm\sqrt{6}}{3}$ ✓ (1 mark)

4 **a** $2\pi r=37$
$r=\frac{37}{2\pi}$
$=5.888\,732\,89\ldots$
$=5.889$ (3 d.p.) ✓
(1 mark)

b $d^2=57$
$d=\pm\sqrt{57}$
$=\pm7.549\,834\,43\ldots$
$=\pm7.550$ (3 d.p.) ✓
(1 mark)

c $7x^3=18$
$x^3=\frac{18}{7}$
$x=\sqrt[3]{\frac{18}{7}}$
$=1.370\,013\,42\ldots$
$=1.370$ (3 d.p.) ✓
(1 mark)

5 **a** $x^2=64$
$x=\pm8$ ✓ (1 mark)

b $x^3=64$
$x=4$ ✓ (1 mark)

c If either a negative or positive number is squared the result is positive. So, when finding a number that when squared gives a particular result the number could be either positive or negative. If a positive number is cubed the result is positive and if a negative number is cubed the result is negative. So, when finding the number that when cubed gives a particular result there is only one possible value. ✓
(1 mark)

6 **a** $5x+2y-1=0$ [i]
$4x-3y+36=0$ [ii]
[i] × 3: $15x+6y-3=0$ [iii]
[ii] × 2: $8x-6y+72=0$ [iv]
[iii] + [iv]: $23x+69=0$ ✓
$23x=-69$
$x=-3$ ✓
Substitute into [i]:
$5\times(-3)+2y-1=0$
$2y=16$
$y=8$
$\therefore x=-3$ and $y=8$ ✓
(3 marks)

b $y=x^2+11x+5$
$y=3x-2$
$x^2+11x+5=3x-2$
$x^2+8x+7=0$ ✓
$(x+7)(x+1)=0$
$x=-7$ or $x=-1$ ✓
If $x=-7$, $y=3\times(-7)-2$
$=-23$
If $x=-1$, $y=3\times(-1)-2$
$=-5$ ✓
$\therefore x=-7$ and $y=-23$ or
$x=-1$ and $y=-5$ (3 marks)

c $y=\frac{8}{x}$
$y=x+2$
$x+2=\frac{8}{x}$
$x^2+2x=8$
$x^2+2x-8=0$ ✓
$(x+4)(x-2)=0$
$x=-4$ or $x=2$ ✓
If $x=-4$, $y=-4+2$
$=-2$
If $x=2$, $y=2+2$
$=4$ ✓
$\therefore x=-4$ and $y=-2$ or
$x=2$ and $y=4$ (3 marks)

7 **a** $V=\frac{4}{3}\pi r^3$
$r^3=\frac{3V}{4\pi}$ ✓
$r=\sqrt[3]{\frac{3V}{4\pi}}$ ✓ (2 marks)

b $A=P(1+r)^n$
$(1+r)^n=\frac{A}{P}$ ✓
$1+r=\sqrt[n]{\frac{A}{P}}$
$r=\sqrt[n]{\frac{A}{P}}-1$ ✓
(2 marks)

c $S=\frac{a}{1-r}$
$S(1-r)=a$
$S-Sr=a$ ✓
$S-a=Sr$
$r=\frac{S-a}{S}$ $\left[\text{or } 1-\frac{a}{S}\right]$
✓ (2 marks)

(Total: 40 marks)

FURTHER FACTORISATION
SKILLS CHECK PAGE 46

1 **a** $9x-3y=3(3x-y)$
b $x^2-6x=x(x-6)$
c $10xy-5y=5y(2x-1)$
[Both 5 and y are common factors. To factorise fully we need to take out both factors.]
d $12x^2-8x=4x(3x-2)$
e $x^2y+xy^2=xy(x+y)$
f $8ab^2-4abc=4ab(2b-c)$

g $15x^7 - 3x^2 = 3x^2(5x^5 - 1)$

h $16p^3q^2 - 2p^4q + 4pq^2$
$= 2pq(8p^2q - p^3 + 2q)$

2 **a** $x^2 - 49 = x^2 - 7^2$
$= (x + 7)(x - 7)$

b $a^2 - 1 = (a + 1)(a - 1)$

c $4p^2 - 25 = (2p)^2 - 5^2$
$= (2p + 5)(2p - 5)$

d $x^2 - y^2 = (x + y)(x - y)$

e $9m^2 - 16 = (3m)^2 - 4^2$
$= (3m + 4)(3m - 4)$

f $36 - t^2 = (6 + t)(6 - t)$

g $64p^2 - 81q^2$
$= (8p)^2 - (9q)^2$
$= (8p + 9q)(8p - 9q)$

h $a^4 - 25b^2$
$= (a^2)^2 - (5b)^2$
$= (a^2 + 5b)(a^2 - 5b)$

3 **a** $x^2 + 8x + 16 = (x + 4)^2$

b $a^2 - 10a + 25 = (a - 5)^2$

c $4n^2 + 4n + 1$
$= (2n)^2 + 2 \times 2n \times 1 + 1^2$
$= (2n + 1)^2$

d $9x^2 - 42x + 49$
$= (3x)^2 - 2 \times 3x \times 7 + 7^2$
$= (3x - 7)^2$

4 **a** $x^2 + 18x + 81 = (x + 9)^2$

b $y^2 + 12y + 36 = (y + 6)^2$

c $b^2 - 14b + 49 = (b - 7)^2$

d $a^2 - 5a + 6.25 = (a - 2.5)^2$

e $100 + 220m + 121m^2$
$= (10 + 11m)^2$

f $16x^2 - 40xy + 25y^2$
$= (4x)^2 - 2 \times 4x \times 5y + (5y)^2$
$= (4x - 5y)^2$

5 **a** $x^2 + 8x + 15 = (x + 3)(x + 5)$
[Look for two numbers that add to 8 and multiply to 15. The numbers are 3 and 5.]

b $a^2 - 14a + 48 = (a - 6)(a - 8)$
[Look for two numbers that add to –14 and multiply to 48. They are –6 and –8.]

c $p^2 + 2p - 35 = (p + 7)(p - 5)$
[The two numbers that add to 2 and multiply to –35 are 7 and –5.]

d $m^2 - 3m - 28 = (m - 7)(m + 4)$

e $n^2 + n - 6 = (n + 3)(n - 2)$

f $k^2 - 5k - 6 = (k - 6)(k + 1)$

6 **a** $x^2 + xy + 2x + 2y$
$= x(x + y) + 2(x + y)$
$= (x + y)(x + 2)$

b $ab + 3a - 5b - 15$
$= a(b + 3) - 5(b + 3)$
$= (b + 3)(a - 5)$

c $a^2 - ab - ac + bc$
$= a(a - b) - c(a - b)$
$= (a - b)(a - c)$

d $2mn - 2m + 3n - 3$
$= 2m(n - 1) + 3(n - 1)$
$= (n - 1)(2m + 3)$

e $xz - yz - x + y$
$= z(x - y) - 1(x - y)$
$= (x - y)(z - 1)$
[It is not necessary to write the 1 in the 2nd line but it can't be left out in the last line.]

f $6kn - 3k - 2n + 1$
$= 3k(2n - 1) - 1(2n - 1)$
$= (2n - 1)(3k - 1)$

FURTHER FACTORISATION
INTERMEDIATE TEST PAGE 47

1 $a^2 - 2a - 63 = (a - 9)(a + 7)$
[Look for two numbers that add to –2 and multiply to –63. The numbers are –9 and 7.]
∴ [C] ✓ (1 mark)

2 $x^2 - 5x + 6 = (x - 3)(x - 2)$
[What two numbers multiply to 6 and add to –5?]
∴ [A] ✓ (1 mark)

3 $x^2 - 18x + 81 = (x - 9)^2$
∴ [B] ✓ (1 mark)

4 $xy - 4x - 6y + 24$
$= x(y - 4) - 6(y - 4)$
$= (y - 4)(x - 6)$
∴ [A] ✓ (1 mark)

5 $m^2 - 1 = m^2 - 1^2$
$= (m + 1)(m - 1)$
∴ [C] ✓ (1 mark)

6 $6k(k + 2) - (k + 2)$
$= 6k(k + 2) - 1(k + 2)$
$= (k + 2)(6k - 1)$
∴ [D] ✓ (1 mark)

7 $4n^2 + 4n - 3$
$= 4n^2 + 6n - 2n - 3$
$= 2n(2n + 3) - 1(2n + 3)$
$= (2n + 3)(2n - 1)$
∴ [B] ✓ (1 mark)

8 $\frac{x^2 - 64}{8x - 64} = \frac{(x + 8)(x - 8)}{8(x - 8)}$
$= \frac{x + 8}{8}$
[After cancelling the common factor $(x - 8)$. The fraction $\frac{x + 8}{8}$ cannot be simplified. 8 cannot be cancelled because it is not a common factor.]
∴ [C] ✓ (1 mark)

9 $5m^2 + 8m - 4$
$= 5m^2 + 10m - 2m - 4$
[Ask: What two terms add to $8m$ and multiply to $-20m^2$ ($5m^2 \times -4$)? All the expressions are equal to $5m^2 + 8m - 4$ but only C can then be factorised by grouping: $5m(m + 2) - 2(m + 2) = (m + 2)(5m - 2)$]
∴ [C] ✓ (1 mark)

10 $\frac{g^2 + 7g + 6}{g + 6} = \frac{(g + 6)(g + 1)}{g + 6}$
$= g + 1$
∴ [D] ✓ (1 mark)

11 $2x^2 - 4x - 30$
$= 2(x^2 - 2x - 15)$ ✓
$= 2(x - 5)(x + 3)$ ✓ (2 marks)

12 $6m^2 + 17m + 12$
$= 6m^2 + 9m + 8m + 12$
$= 3m(2m + 3) + 4(2m + 3)$ ✓
$= (2m + 3)(3m + 4)$ ✓
(2 marks)

13 $3x^4 - 3$
$= 3(x^4 - 1)$
$= 3((x^2)^2 - 1^2)$
$= 3\,(x^2 + 1)(x^2 - 1)$ ✓
$= 3(x^2 + 1)(x + 1)(x - 1)$ ✓
(2 marks)

14 $a^2 + ab - a - b$
$= a(a + b) - 1(a + b)$ ✓
$= (a + b)(a - 1)$ ✓ (2 marks)

15 $9p^2 - 30pq + 25q^2$
$= (3p)^2 - 2 \times 3p \times 5q + (5q)^2$
$= (3p - 5q)^2$ ✓✓ (2 marks)

16 $\frac{2n + 6}{6} \times \frac{3n}{n^2 + 3n}$
$= \frac{2(n + 3)}{6} \times \frac{3n}{n(n + 3)}$ ✓
$= \frac{6}{6}$
$= 1$ ✓ (2 marks)

17 $\frac{5}{x^2 - 4} + \frac{3}{x^2 + 2x}$
$= \frac{5}{(x + 2)(x - 2)} + \frac{3}{x(x + 2)}$ ✓
$= \frac{5x}{x(x + 2)(x - 2)}$
$+ \frac{3(x - 2)}{x(x + 2)(x - 2)}$ ✓
$= \frac{5x + 3x - 6}{x(x + 2)(x - 2)}$
$= \frac{8x - 6}{x(x + 2)(x - 2)}$ ✓(3 marks)

(Total: 25 marks)

FURTHER FACTORISATION ADVANCED TEST PAGE 48

1 $6a^2b^3c^5 - 4ab^5c^3 + 12a^4b^2c^6$
$= 2ab^2c^3(3abc^2 - 2b^3 + 6a^3c^3)$ ✓
(1 mark)

2 **a** $ab + ac - 9b - 9c$
$= a(b + c) - 9(b + c)$
$= (b + c)(a - 9)$ ✓
(1 mark)

b $1 - 16x^2$
$= (1 + 4x)(1 - 4x)$ ✓
(1 mark)

c $x^2 - 20x + 100$
$= (x - 10)^2$ ✓ (1 mark)

d $x^2 + 11xy - 60y^2$
$= (x + 15y)(x - 4y)$ ✓
(1 mark)

e $5a^2 - 13a + 6$
$= (5a - 3)(a - 2)$ ✓
(1 mark)

f $4m^2 - 12m - 27$
$= (2m - 9)(2m + 3)$ ✓
(1 mark)

3 **a** $x^3 - 9x$
$= x(x^2 - 9)$ ✓
$= x(x + 3)(x - 3)$ ✓
(2 marks)

b $2x^2 + 22x + 48$
$= 2(x^2 + 11x + 24)$ ✓
$= 2(x + 3)(x + 8)$ ✓
(2 marks)

c $m^4 - 1$
$= (m^2 + 1)(m^2 - 1)$ ✓
$= (m^2 + 1)(m + 1)(m - 1)$ ✓
(2 marks)

d $\frac{4}{3}\pi r^3 + \frac{1}{3}\pi r^2 h$
$= \frac{1}{3}\pi r^2(4r + h)$ ✓✓
(2 marks)

e $u^4 - 13u^2 + 36$
$= (u^2 - 4)(u^2 - 9)$ ✓
$= (u + 2)(u - 2)(u + 3)(u - 3)$
✓ (2 marks)

f $12x^2 - 60x + 75$
$= 3(4x^2 - 20x + 25)$ ✓
$= 3(2x - 5)^2$ ✓ (2 marks)

4 **a** $\frac{x + 7}{x^2 + 8x + 7}$
$= \frac{x + 7}{(x + 7)(x + 1)}$ ✓
$= \frac{1}{x + 1}$ ✓ (2 marks)

b $\frac{x^2 - 81}{x^2 - x - 72}$
$= \frac{(x + 9)(x - 9)}{(x + 8)(x - 9)}$ ✓
$= \frac{x + 9}{x + 8}$ ✓ (2 marks)

c $\frac{x^2 + 2x - 15}{x^2 + 8x + 15}$
$= \frac{(x + 5)(x - 3)}{(x + 5)(x + 3)}$ ✓
$= \frac{x - 3}{x + 3}$ ✓ (2 marks)

d $\frac{2a^2 + 3a - 2}{2a^2 + 5a - 3}$
$= \frac{(2a - 1)(a + 2)}{(2a - 1)(a + 3)}$ ✓
$= \frac{a + 2}{a + 3}$ ✓ (2 marks)

e $\frac{8n^3 - 24n^2 - 144n}{2n^2 + 6n}$
$= \frac{8n(n^2 - 3n - 18)}{2n(n + 3)}$ ✓
$= \frac{4(n + 3)(n - 6)}{n + 3}$
$= 4n - 24$ ✓ (2 marks)

f $\frac{8p^2 - 6pq - 28p + 21q}{16p^2 - 24pq + 9q^2}$
$= \frac{2p(4p - 3q) - 7(4p - 3q)}{(4p - 3q)^2}$ ✓
$= \frac{(4p - 3q)(2p - 7)}{(4p - 3q)^2}$
$= \frac{2p - 7}{4p - 3q}$ ✓ (2 marks)

5 **a** $\frac{4x^2 - 9}{2x^2 - 5x - 12} \times \frac{3x^2 - 13x + 4}{6x^2 - 11x + 3}$
$= \frac{(2x + 3)(2x - 3)}{(2x + 3)(x - 4)}$
$\times \frac{(3x - 1)(x - 4)}{(3x - 1)(2x - 3)}$ ✓✓
$= 1$ ✓ (3 marks)

b $\frac{x^2 + xy + xz}{xy - xz - yz + z^2}$
$\div \frac{x^3 + x^2y + x^2z + xyz}{xy + y^2 - xz - yz}$
$= \frac{x(x + y + z)}{x(y - z) - z(y - z)}$
$\div \frac{x(x^2 + xy + xz + yz)}{y(x + y) - z(x + y)}$ ✓
$= \frac{x(x + y + z)}{(y - z)(x - z)}$
$\times \frac{(x + y)(y - z)}{x(x(x + y) + z(x + y))}$
$= \frac{x + y + z}{x - z}$
$\times \frac{x + y}{(x + y)(x + z)}$ ✓
$= \frac{x + y + z}{(x - z)(x + z)}$ ✓
(3 marks)

6 $\dfrac{x+8}{x^2+x-6} - \dfrac{x+4}{x^2+5x+6}$

$= \dfrac{x+8}{(x+3)(x-2)} - \dfrac{x+4}{(x+3)(x+2)}$

$= \dfrac{(x+8)(x+2)-(x+4)(x-2)}{(x+3)(x-2)(x+2)}$ ✓

$= \dfrac{x^2+10x+16-(x^2+2x-8)}{(x+3)(x-2)(x+2)}$

$= \dfrac{x^2+10x+16-x^2-2x+8}{(x+3)(x-2)(x+2)}$ ✓

$= \dfrac{8x+24}{(x+3)(x-2)(x+2)}$

$= \dfrac{8(x+3)}{(x+3)(x-2)(x+2)}$

$= \dfrac{8}{(x-2)(x+2)}$

$= \dfrac{8}{x^2-4}$ ✓ (3 marks)

(Total: 40 marks)

QUADRATIC EQUATIONS
SKILLS CHECK PAGE 50

1 a $x^2 = 16$
$x = 4$ or $x = -4$

b $e^2 = 9$
$e = 3$ or $e = -3$

c $4k^2 = 1$
$k^2 = \dfrac{1}{4}$
$k = \dfrac{1}{2}$ or $k = -\dfrac{1}{2}$

2 a $x(x+4) = 0$
$x = 0$ or $x + 4 = 0$
$x = 0$ or $x = -4$

b $(x-7)(x+1) = 0$
$x - 7 = 0$ or $x + 1 = 0$
$x = 7$ or $x = -1$

c $(2a+5)(3a-2) = 0$
$2a + 5 = 0$ or $3a - 2 = 0$
$2a = -5$ or $3a = 2$
$a = -2\dfrac{1}{2}$ or $a = \dfrac{2}{3}$

3 a $x^2 + 8x + 15 = (x+5)(x+3)$
[What 2 numbers add to 8 and multiply to 15?]

b $t^2 + 3t - 10 = (t+5)(t-2)$
[What 2 numbers add to 3 and multiply to −10?]

c $2m^2 - 7m + 6$
$= 2m^2 - 4m - 3m + 6$
$= 2m(m-2) - 3(m-2)$
$= (m-2)(2m-3)$

4 $a = 8, b = 10, c = -3$

$x = \dfrac{-b \pm \sqrt{b^2-4ac}}{2a}$

$= \dfrac{-10 \pm \sqrt{10^2 - 4\times8\times-3}}{2\times8}$

$= \dfrac{-10 \pm \sqrt{196}}{16}$

$x = \dfrac{-10+14}{16}$ or $x = \dfrac{-10-14}{16}$

$x = \dfrac{1}{4}$ or $x = -1\dfrac{1}{2}$

5 a $(x+5)^2 = x^2 + 10x + 25$
$[x^2 + 2 \times x \times 5 + 5^2]$

b $(x-3)^2 = x^2 - 6x + 9$

c $(a+b)^2 = a^2 + 2ab + b^2$

6 a $x^2 + 14x + 49 = (x+7)^2$

b $n^2 - 12n + 36 = (n-6)^2$

7 a $8x = 2 \times x \times 4$
Need to add $4^2 = 16$
$[x^2 + 8x + 16 = (x+4)^2]$

b $-6a = 2 \times a \times -3$
Need to add $(-3)^2 = 9$
$[a^2 - 6a + 9 = (a-3)^2]$

c $4p^2 + 20p = (2p)^2 + 20p$
$20p = 2 \times 2p \times 5$
Need to add $5^2 = 25$
$[4p^2 + 20p + 25 = (2p+5)^2]$

8 a When $x = 3$,
$x^2 + 7x + 12$
$= 3^2 + 7 \times 3 + 12$
$= 42 \quad (\neq 0)$ False

b When $y = -3$,
$y^2 - 4y - 21$
$= (-3)^2 - 4 \times -3 - 21$
$= 0$ True

9 a $(x+2)^2 = 9$ has 2 solutions
$[x + 2 = 3$ or $x + 2 = -3]$

b $(x+2)^2 = -4$ has no solution
[No real number is negative when squared.]

c $(x+2)^2 = 0$ has one solution
[when $x + 2 = 0$]

10 a $n^2 - 3n = 0$
$n(n-3) = 0$
$n = 0$ or $n - 3 = 0$
$n = 0$ or $n = 3$

b $x^2 + 6x + 8 = 0$
$(x+2)(x+4) = 0$
$x + 2 = 0$ or $x + 4 = 0$
$x = -2$ or $x = -4$

c $a^2 - 4a - 45 = 0$
$(a-9)(a+5) = 0$
$a - 9 = 0$ or $a + 5 = 0$
$a = 9$ or $a = -5$

QUADRATIC EQUATIONS
INTERMEDIATE TEST PAGE 51

1 $4x^2 - 4x + 1 = 0$
$(2x-1)^2 = 0$
There is only one solution.
$[2x - 1 = 0$ and hence $x = 0.5]$
∴ [B] ✓ (1 mark)

2 $p^2 + 8p - 9 = 0$
$(p+9)(p-1) = 0$
$p + 9 = 0$ or $p - 1 = 0$
$p = -9$ or $p = 1$
$p = -9$ is a solution
[Or substitute each value into the equation to see which solution satisfies it.]
∴ [D] ✓ (1 mark)

3 $x = 3$ or $x = -2$
$(x-3) = 0$ or $(x+2) = 0$
$(x-3)(x+2) = 0$
$x^2 + 2x - 3x - 6 = 0$
$x^2 - x - 6 = 0$
∴ [D] ✓ (1 mark)

4 $(x-5)(x-2) = 28$
$x^2 - 2x - 5x + 10 = 28$
$x^2 - 7x - 18 = 0$
$(x+2)(x-9) = 0$
$x + 2 = 0$ or $x - 9 = 0$
$x = -2$ or $x = 9$
[Or substitute to see which solutions work.]
∴ [C] ✓ (1 mark)

5 $4x^2 = 1$
$x^2 = \dfrac{1}{4}$
$x = \pm\sqrt{\dfrac{1}{4}}$ ✓
$x = \pm\dfrac{1}{2}$ ✓ (2 marks)

6 a $x^2 - 18x + 81 = 0$
$(x-9)^2 = 0$ ✓
$x - 9 = 0$
$x = 9$ ✓
(2 marks)

b $x(x+5) = 14$
$x^2 + 5x = 14$
$x^2 + 5x - 14 = 0$ ✓
$(x+7)(x-2) = 0$
$x + 7 = 0$ or $x - 2 = 0$
$x = -7$ or $x = 2$
✓✓ (3 marks)

c $2m^2 + 9m + 4 = 0$
$2m^2 + 8m + m + 4 = 0$
$2m(m+4) + 1(m+4) = 0$
$(m+4)(2m+1) = 0$ ✓
$m + 4 = 0$ or $2m + 1 = 0$
$m = -4$ or $2m = -1$
$m = -4$ or $m = -\frac{1}{2}$
✓✓ (3 marks)

7 **a** $x^2 + 6x = 16$
$x^2 + 6x + 9 = 16 + 9$
$(x+3)^2 = 25$ ✓
$x + 3 = -5$ or $x + 3 = 5$
$x = -8$ or $x = 2$
✓✓ (3 marks)

b $a^2 + 12a = 7$
$a^2 + 12a + 36 = 7 + 36$
$(a+6)^2 = 43$ ✓
$a + 6 = \pm\sqrt{43}$ ✓
$a = -6 \pm \sqrt{43}$ ✓
(3 marks)

c $9n^2 - 6n - 5 = 0$
$9n^2 - 6n = 5$
$9n^2 - 6n + 1 = 5 + 1$
$(3n-1)^2 = 6$ ✓
$3n - 1 = \pm\sqrt{6}$ ✓
$3n = 1 \pm \sqrt{6}$
$n = \frac{1 \pm \sqrt{6}}{3}$ ✓
(3 marks)

8 **a** $2x^2 - 6x + 3 = 0$
$a = 2, b = -6, c = 3$
$x = \frac{-b \pm \sqrt{b^2 - 4ac}}{2a}$
$= \frac{-(-6) \pm \sqrt{(-6)^2 - 4 \times 2 \times 3}}{2 \times 2}$ ✓
$= \frac{6 \pm \sqrt{12}}{4}$ ✓
$= \frac{6 \pm \sqrt{4} \times \sqrt{3}}{4}$
$= \frac{6 \pm 2\sqrt{3}}{4}$ ✓
$= \frac{2(3 \pm \sqrt{3})}{4}$
$= \frac{3 \pm \sqrt{3}}{2}$ ✓ (4 marks)

b $3x^2 + 7x - 1 = 0$
$a = 3, b = 7, c = -1$
$x = \frac{-b \pm \sqrt{b^2 - 4ac}}{2a}$
$= \frac{-7 \pm \sqrt{7^2 - 4 \times 3 \times -1}}{2 \times 3}$ ✓
$= \frac{-7 \pm \sqrt{61}}{6}$ ✓

Either $x = \frac{-7 + \sqrt{61}}{6}$
$= 0.135\,041\,612\ldots$
$= 0.14$ (2 d.p.) ✓
Or $x = \frac{-7 - \sqrt{61}}{6}$
$= -2.468\,374\,946\ldots$
$= -2.47$ (2 d.p.)
✓ (4 marks)

9 $(x+3)(x+5) = 35$
$x^2 + 5x + 3x + 15 = 35$
$x^2 + 8x - 20 = 0$ ✓
$(x+10)(x-2) = 0$
$x + 10 = 0$ or $x - 2 = 0$
$x = -10$ or $x = 2$ ✓✓
(3 marks)

10 **a** Let the width, b, be x m.
The length: $l = (x + 12)$ m ✓
$A = lb$
$= x(x + 12)$ ✓
But the area is 45 m^2
$\therefore x(x + 12) = 45$
$x^2 + 12x - 45 = 0$ ✓
The width, x m, can be found by solving that equation. (3 marks)

b $x^2 + 12x - 45 = 0$
$(x-3)(x+15) = 0$
$x - 3 = 0$ or $x + 15 = 0$
$x = 3$ or $x = -15$ ✓
But $x > 0$ [because it is the width of the rectangle]
$\therefore x = 3$ ✓
$l = x + 12$
$= 3 + 12$
$= 15$
The length of the rectangle is 15 m and the width is 3 m. ✓
(3 marks)

(Total: 40 marks)

QUADRATIC EQUATIONS
ADVANCED TEST PAGE 52

1 **a** $3x^2 + 17x + 10 = 0$
$(3x+2)(x+5) = 0$ ✓
$3x + 2 = 0$ or $x + 5 = 0$
$x = -\frac{2}{3}$ or $x = -5$
✓ (2 marks)

b $5a^2 + 50a = 120$
$a^2 + 10a = 24$
$a^2 + 10a - 24 = 0$
$(a+12)(a-2) = 0$ ✓
$a = -12$ or $a = 2$ ✓
(2 marks)

2 **a** $x^2 + 8x = 20$
$x^2 + 8x + 16 = 20 + 16$
$(x+4)^2 = 36$ ✓
$x + 4 = -6$ or $x + 4 = 6$
$x = -10$ or $x = 2$ ✓
(2 marks)

b $9x^2 = 12x + 1$
$9x^2 - 12x = 1$
$9x^2 - 12x + 4 = 1 + 4$
$(3x-2)^2 = 5$ ✓
$3x - 2 = \pm\sqrt{5}$
$3x = 2 \pm \sqrt{5}$
$x = \frac{2 \pm \sqrt{5}}{3}$ ✓
(2 marks)

3 **a** $x^2 + 6x + 7 = 0$
$a = 1, b = 6, c = 7$
$x = \frac{-b \pm \sqrt{b^2 - 4ac}}{2a}$
$= \frac{-6 \pm \sqrt{6^2 - 4 \times 1 \times 7}}{2 \times 1}$
$= \frac{-6 \pm \sqrt{8}}{2}$ ✓
$= \frac{-6 \pm 2\sqrt{2}}{2}$ ✓
$= \frac{2(-3 \pm \sqrt{2})}{2}$
$= -3 \pm \sqrt{2}$ ✓ (3 marks)

b $4x^2 - 12x + 3 = 0$

$a = 4, b = -12, c = 3$

$$x = \frac{-b \pm \sqrt{b^2 - 4ac}}{2a}$$
$$= \frac{-(-12) \pm \sqrt{(-12)^2 - 4 \times 4 \times 3}}{2 \times 4}$$
$$= \frac{12 \pm \sqrt{96}}{8} \checkmark$$
$$= \frac{12 \pm 4\sqrt{6}}{8} \checkmark$$
$$= \frac{4(3 \pm \sqrt{6})}{8}$$
$$= \frac{3 \pm \sqrt{6}}{2} \checkmark$$

(3 marks)

4 **a** $3x^2 + 5x - 7 = 0$

$a = 3, b = 5, c = -7$

$$x = \frac{-b \pm \sqrt{b^2 - 4ac}}{2a}$$
$$= \frac{-5 \pm \sqrt{5^2 - 4 \times 3 \times -7}}{2 \times 3} \checkmark$$

$x = 0.90671775\ldots$

$= 0.907$ (3 d.p.) or

$x = -2.5733844\ldots$

$= -2.573$ (3 d.p.) ✓✓

(3 marks)

b $x^2 + x = 1$

$x^2 + x - 1 = 0$

$a = 1, b = 1, c = -1$

$$x = \frac{-b \pm \sqrt{b^2 - 4ac}}{2a}$$
$$= \frac{-1 \pm \sqrt{1^2 - 4 \times 1 \times -1}}{2 \times 1} \checkmark$$

$x = 0.6180339\ldots$

$x = 0.618$ (3 d.p.) or

$x = -1.6180339\ldots$

$x = -1.618$ (3 d.p.) ✓✓

(3 marks)

5 **a** $(x - 4)(x - 10) = -8$

$x^2 - 10x - 4x + 40 = -8$

$x^2 - 14x + 48 = 0$ ✓

$(x - 6)(x - 8) = 0$

$x = 6$ or $x = 8$ ✓

(2 marks)

b $2^{2x} - 12(2^x) + 32 = 0$

$(2^x)^2 - 12(2^x) + 32 = 0$

Let $u = 2^x$

$u^2 - 12u + 32 = 0$

$(u - 4)(u - 8) = 0$

$u = 4$ or $u = 8$ ✓

$2^x = 4$ or $2^x = 8$

$x = 2$ or $x = 3$ ✓✓

(3 marks)

c $(n^2 + 5n)^2 - 8(n^2 + 5n) - 84 = 0$

Let $u = n^2 + 5n$

$u^2 - 8u - 84 = 0$

$(u + 6)(u - 14) = 0$ ✓

Either $u + 6 = 0$

$n^2 + 5n + 6 = 0$

$(n + 2)(n + 3) = 0$

$n = -2$ or $n = -3$ ✓

Or $u - 14 = 0$

$n^2 + 5n - 14 = 0$

$(n + 7)(n - 2) = 0$

$n = -7$ or $n = 2$ ✓

So $n = -2$ or -3 or -7 or 2

(3 marks)

6 **a** $x^2 - 4x + y^2 + 6y = 12$ [i]

$y = x - 4$ [ii]

Substitute [ii] into [i]:

$x^2 - 4x + (x - 4)^2 + 6(x - 4) = 12$

$x^2 - 4x + x^2 - 8x + 16 + 6x - 24 = 12$

$2x^2 - 6x - 20 = 0$ ✓

$x^2 - 3x - 10 = 0$

$(x - 5)(x + 2) = 0$

$x = 5$ or $x = -2$ ✓

Substitute into [ii]:

When $x = 5$, $y = 5 - 4$

$= 1$

When $x = -2$, $y = -2 - 4$

$= -6$ ✓

$\therefore$ $x = 5$ and $y = 1$ or

$x = -2$ and $y = -6$ (3 marks)

b $x^2 - 4x + y^2 + 6y = 12$

$x^2 - 4x + 4 + y^2 + 6y + 9 = 12 + 4 + 9$

$(x - 2)^2 + (y + 3)^2 = 25$

This is a circle,

centre (2, –3) and radius 5.

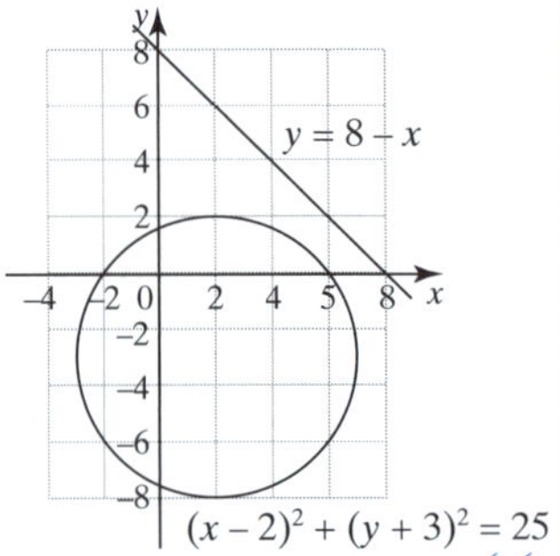

✓✓

The circle and the line do not intersect so there is no simultaneous solution. ✓

(3 marks)

7 **a** $ax^2 + bx + c = 0$

$$x^2 + \frac{b}{a}x = -\frac{c}{a}$$
$$x^2 + \frac{b}{a}x + \left(\frac{b}{2a}\right)^2 = \left(\frac{b}{2a}\right)^2 - \frac{c}{a} \checkmark$$
$$\left(x + \frac{b}{2a}\right)^2 = \frac{b^2}{4a^2} - \frac{c}{a}$$
$$= \frac{b^2 - 4ac}{4a^2} \checkmark$$
$$x + \frac{b}{2a} = \pm\sqrt{\frac{b^2 - 4ac}{4a^2}}$$
$$= \pm\frac{\sqrt{b^2 - 4ac}}{2a}$$
$$x = -\frac{b}{2a} \pm \frac{\sqrt{b^2 - 4ac}}{2a}$$
$$= \frac{-b \pm \sqrt{b^2 - 4ac}}{2a}$$

✓ (3 marks)

b If $b^2 - 4ac < 0$, $ax^2 + bx + c = 0$ has no real solutions. ✓ (1 mark)

c If $b^2 - 4ac = 0$, $ax^2 + bx + c = 0$ has only one solution, $x = -\dfrac{b}{2a}$. ✓ (1 mark)

d If $b^2 - 4ac$ is a perfect square, the equation $ax^2 + bx + c = 0$ has rational roots. ✓ (1 mark)

(Total: 40 marks)

VOLUME OF PYRAMIDS, CONES AND SPHERES
SKILLS CHECK PAGE 54

1 **a**

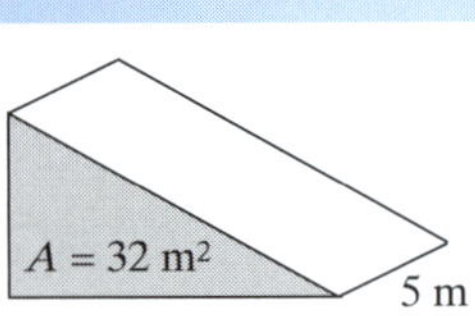

$V = Ah$

$= 32 \times 5$

$\therefore$ volume $= 160\text{ m}^3$

b

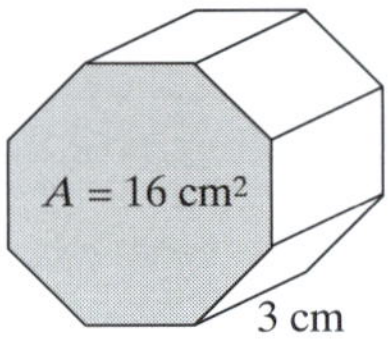

$V = Ah$

$= 16 \times 3$

$\therefore$ volume $= 48\text{ cm}^3$

c

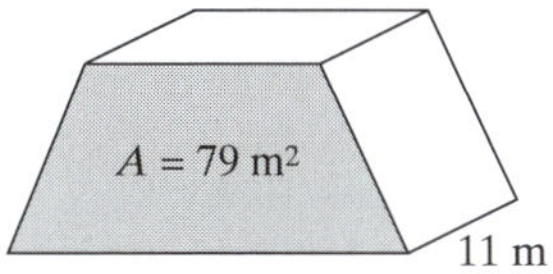

$V = Ah$
$= 79 \times 11$
$= 869$
$\therefore$ volume $= 869$ m^3

2 **a**

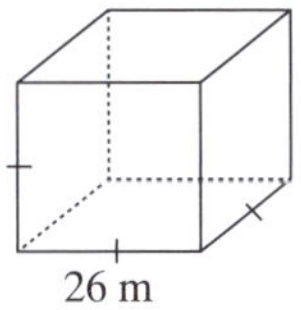

$V = s^3$
$= 26^3$
$= 17\,576$
$\therefore$ volume $= 17\,576$ m^3

b

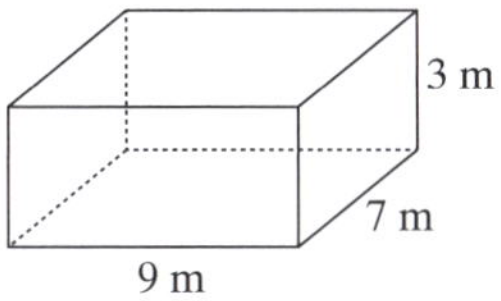

$V = lbh$
$= 9 \times 7 \times 3$
$= 189$
$\therefore$ volume $= 189$ m^3

c

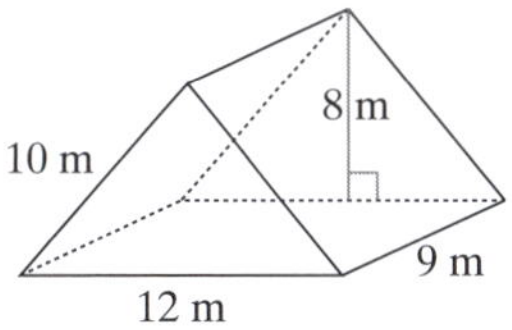

$V = \frac{1}{2}bhH$
$= \frac{1}{2} \times 12 \times 8 \times 9$
$= 432$
$\therefore$ volume $= 432$ m^3

3 **a**

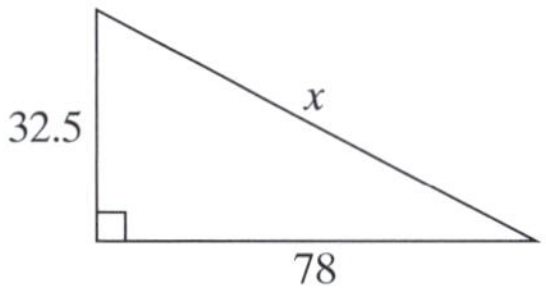

$x^2 = 32.5^2 + 78^2$
$= 7140.25$
$x = \sqrt{7140.25} \quad (x > 0)$
$= 84.5$

b

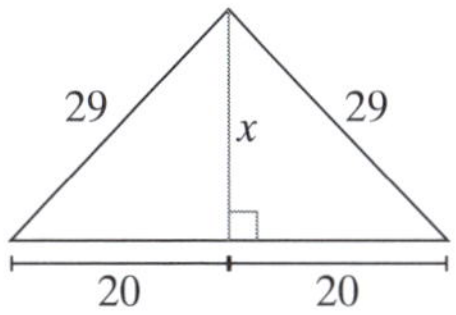

$x^2 = 29^2 - 20^2$
$= 441$
$x = \sqrt{441} \quad (x > 0)$
$= 21$

4 **a** $V = \frac{1}{3}Ah$
$= \frac{1}{3} \times 18 \times 10$
$= 60$
$\therefore$ volume $= 60$ cm^3

b $V = \frac{1}{3}Ah$
$= \frac{1}{3} \times 6 \times 2$
$= 4$
$\therefore$ volume $= 4$ m^3

c $V = \frac{1}{3}Ah$
$= \frac{1}{3} \times 78 \times 39$
$= 1014$
$\therefore$ volume $= 1014$ mm^3

5 **a** $V = \pi r^2 h$
$= \pi \times 9^2 \times 2$
$= 508.938\,0099\ldots$
$\therefore$ volume $= 508.9$ m^3 (1 d.p.)

b $V = \frac{1}{3}\pi r^2 h$
$= \frac{1}{3} \times \pi \times 5^2 \times 6$
$= 157.079\,6327\ldots.$
$\therefore$ volume $= 157.1$ m^3 (1 d.p.)

6 **a**

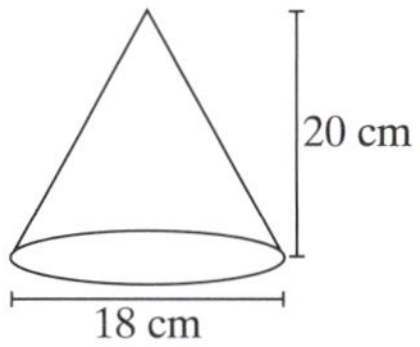

$r = 18 \div 2 = 9$

$V = \frac{1}{3}\pi r^2 h$
$= \frac{1}{3} \times \pi \times 9^2 \times 20$
$= 1696.460\,033\ldots$
$\therefore$ volume $= 1696.5$ cm^3 (1 d.p.)

b

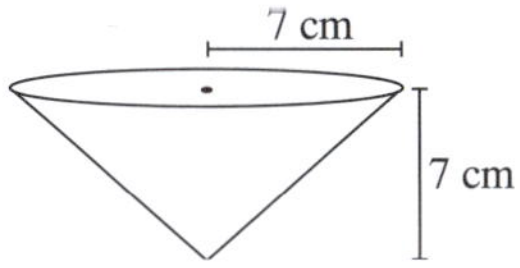

$V = \frac{1}{3}\pi r^2 h$
$= \frac{1}{3} \times \pi \times 7^2 \times 7$
$= 359.188\,7601\ldots$
$\therefore$ volume $= 359.2$ cm^3 (1 d.p.)

7 **a** $V = \frac{4}{3}\pi r^3$
$= \frac{4}{3} \times \pi \times 9^3$
$= 3053.628\,059\ldots$
$\therefore$ volume $= 3053.6$ m^3 (1 d.p.)

b diameter = 35 cm
radius = 17.5 cm
$V = \frac{4}{3}\pi r^3$
$= \frac{4}{3} \times \pi \times 17.5^3$
$= 22\,449.2975\ldots$
$\therefore$ volume $= 22\,449.3$ cm^3 (1 d.p.)

8 1 m^3 = 1000 L
8 m^3 = 8000 L

VOLUME OF PYRAMIDS, CONES AND SPHERES
INTERMEDIATE TEST PAGE 55

1 $V = Ah$ only in C.

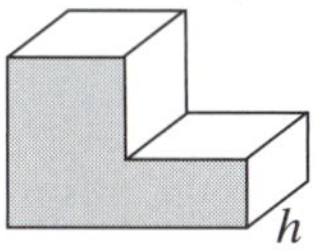

[In A, the shaded face is not the cross-section or 'base' of the prism, (slices of the prism parallel to the shaded face get smaller). In B, the solid is a cone, $V = \frac{1}{3}Ah$. In D, h is not the perpendicular height; it is not perpendicular to the shaded face.]
$\therefore$ **[C]** ✓ (1 mark)

2 The volume of a cone is $\frac{1}{3}$ the volume of a cylinder with the same radius and height.
The cone could be filled 3 times from a cylinder of the same height. It could be filled 3 × 3 times from a cylinder 3 times the height. 9 times.
$\therefore$ **[C]** ✓ (1 mark)

3 $V = \frac{4}{3}\pi r^3$
When $r = 2.7$,
$V = \frac{4}{3} \times \pi \times 2.7^3$
$= 82.447\,957\ldots$
The volume is 80 m³ to the nearest ten cubic metres.
∴ **[C]** ✓ (1 mark)

4 **a** Let h cm be the height of the triangle.

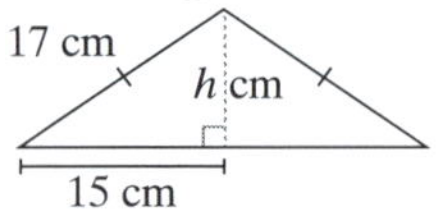

$17^2 = h^2 + 15^2$
$h^2 = 17^2 - 15^2$ ✓
$= 289 - 225$
$= 64$
$h = 8 \quad (h > 0)$
The perpendicular height of the shaded face is 8 cm. ✓
(2 marks)

b

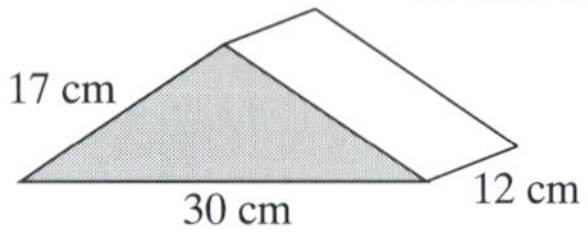

Area of the shaded face:
$A = \frac{1}{2}bh$
$= \frac{1}{2} \times 30 \times 8$
$= 120$ ✓
Volume of the prism:
$V = AH$
$= 120 \times 12$
$= 1440$
The volume of the prism is 1440 cm³. ✓ (2 marks)

c $V = \frac{1}{3}AH$
$= \frac{1}{3} \times 1440$
$= 480$
The volume of the pyramid is 480 cm³. ✓ (1 mark)

5 $V = 90, \; h = 15$
$V = \frac{1}{3}Ah$
$90 = \frac{1}{3} \times A \times 15$ ✓
$= 5A$
$A = 90 \div 5$
$= 18$
The area of the base is 18 cm². ✓ (2 marks)

6 $DC = 4$ m, $BC = 1.6$ m

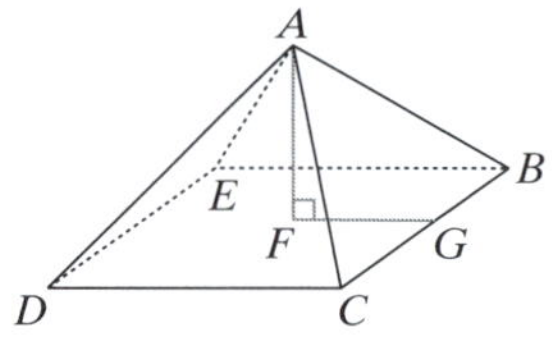

a $AG = 2.5$ m, $FG = 2$ m
Let the perpendicular height be h m.
$h^2 = 2.5^2 - 2^2$
$= 6.25 - 4$
$= 2.25$
$h = \sqrt{2.25} \quad (h > 0)$
$= 1.5$
The perpendicular height is 1.5 m. ✓ (1 mark)

b $V = \frac{1}{3}Ah$
$= \frac{1}{3} \times 4 \times 1.6 \times 1.5$
$= 3.2$
The volume of the pyramid is 3.2 m³. ✓ (1 mark)

7 $V = \frac{2}{3}\pi r^3$
When $r = 5$,
$V = \frac{2}{3} \times \pi \times 5^3$ ✓
$= 261.799\,3878\ldots$
∴ the volume is 262 cm³ to the nearest cubic centimetre. ✓
∴ the capacity is 262 mL to the nearest millilitre ✓ (3 marks)

8

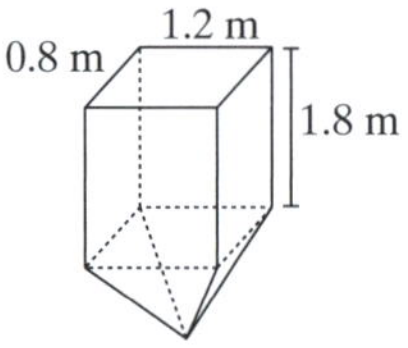

The bin consists of a rectangular prism and rectangular pyramid.
Rectangular prism:
$V = lbh$
$= 1.2 \times 0.8 \times 1.8$
$= 1.728$
The volume of the rectangular prism is 1.728 m³. ✓
Height of pyramid
$= 2.6 \text{ m} - 1.8 \text{ m}$
$= 0.8 \text{ m}$ ✓
Pyramid:
$V = \frac{1}{3}AH$
$= \frac{1}{3} \times 1.2 \times 0.8 \times 0.8$
$= 0.256$
The volume of the rectangular pyramid is 0.256 m³. ✓
Total volume
$= 1.728 \text{ m}^3 + 0.256 \text{ m}^3$
$= 1.984 \text{ m}^3$ ✓ (4 marks)

9 $V = \frac{4}{3}\pi r^3$
$523.6 = \frac{4}{3}\pi r^3$ ✓
$r^3 = \dfrac{523.6}{\frac{4}{3}\pi}$
$= 125.000\,2923\ldots$
$r = \sqrt[3]{125.000\,2923\ldots}$
$= 5.000\,003\,897\ldots$
The radius of the sphere is 5 cm. ✓ (2 marks)

10 **a** Ratio of volumes $= 27:64$
Ratio of heights $= \sqrt[3]{27} : \sqrt[3]{64}$
$= 3:4$ ✓
(1 mark)

b Ratio of areas $= 3^2:4^2$
$= 9:16$ ✓
(1 mark)

c Larger pyramid has
$V = 144, \; h = 9$
$V = \frac{1}{3}Ah$
$144 = \frac{1}{3} \times A \times 9$
$144 = 3A$
$A = 144 \div 3$
$= 48$
The area of the base of the larger pyramid is 48 cm². ✓
Let x cm² be the area of the base of the smaller pyramid.
$x:48 = 9:16$
$\frac{x}{48} = \frac{9}{16}$
$x = \frac{9}{16} \times 48$
$= 27$
The area of the base of the smaller pyramid is 27 cm². ✓ (2 marks)

(Total: 25 marks)

VOLUME OF PYRAMIDS, CONES AND SPHERES
ADVANCED TEST PAGE 56

1 **a** $V = \frac{1}{3}Ah$

$= \frac{1}{3} \times 18 \times 15$

$= 90\text{ cm}^3$ ✓ (1 mark)

b Area = 18 cm² ✓ (1 mark)
[The volume of a right pyramid is one-third the area of the base multiplied by the perpendicular height, regardless of what shape the base is.]

2

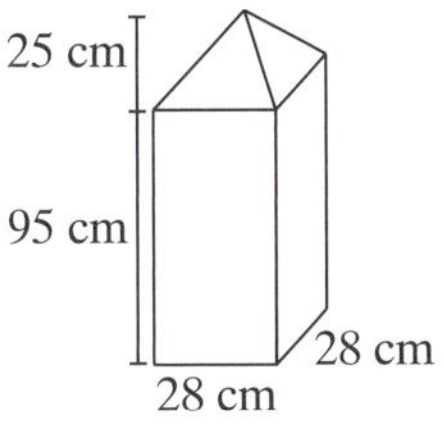

Total height = 1.2 m
= 120 cm
Height of pyramid
= (120 – 95) cm
= 25 cm

$V = 28^2 \times 95 + \frac{1}{3} \times 28^2 \times 25$ ✓

$= 81\,013.3333\ldots$

$= 81\,013\text{ cm}^3$ (nearest cm³) ✓
(2 marks)

3

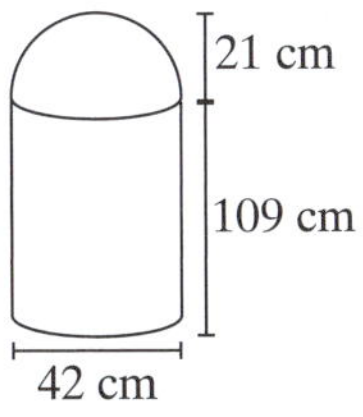

Radius = 21 cm
Height of cylinder
= 1.3 m – 21 cm
= 109 cm ✓

$V = \pi \times 21^2 \times 109 + \frac{2}{3} \times \pi \times 21^3$ ✓

$= 170\,409.4103\ldots$

$= 170\,409\text{ cm}^3$ (nearest cm³) ✓
(3 marks)

4 **a** $V = \frac{1}{3}\pi r^2 h$

$= \frac{1}{3} \times \pi \times 48^2 \times 81$

$= 195\,432.195\ldots$

$= 195\,432\text{ cm}^3$
(nearest cm³) ✓ (1 mark)

b $V = \frac{4}{3}\pi r^3$

$\frac{4}{3}\pi r^3 = 195\,432.195\ldots$

$r^3 = 195\,432.195\ldots \div \left(\frac{4}{3}\pi\right)$

$= 46\,656$ ✓

$r = \sqrt[3]{46\,656}$

$= 36$

The radius of the sphere is 36 cm. ✓ (2 marks)

5

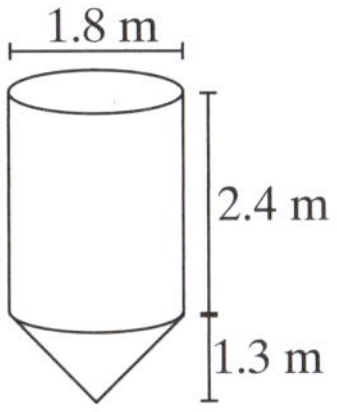

radius = 0.9 m
Height of cone = (3.7 – 2.4) m
= 1.3 m

$V = \pi r^2 H + \frac{1}{3}\pi r^2 h$

$= \pi \times 0.9^2 \times 2.4 + \frac{1}{3} \times \pi \times 0.9^2 \times 1.3$ ✓

$= 7.209\,955\ldots$

$= 7.2\text{ m}^3$ (1 d.p.) ✓

Capacity
= 7.2 × 1000 L
= 7200 L (nearest 100 L) ✓
(3 marks)

6 **a**

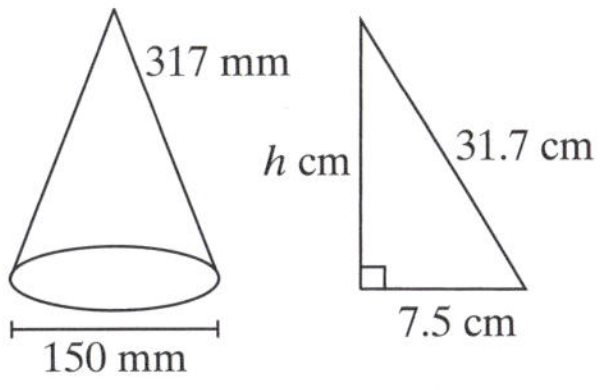

radius = 75 mm = 7.5 cm

Now $h^2 = 31.7^2 - 7.5^2$

$= 948.64$

$h = \sqrt{948.64}$ $(h > 0)$

$= 30.8$ ✓

$V = \frac{1}{3}\pi r^2 h$

$= \frac{1}{3} \times \pi \times 7.5^2 \times 30.8$ ✓

$= 1814.26975\ldots$

$= 1814\text{ cm}^3$ (nearest cm³) ✓
(3 marks)

b

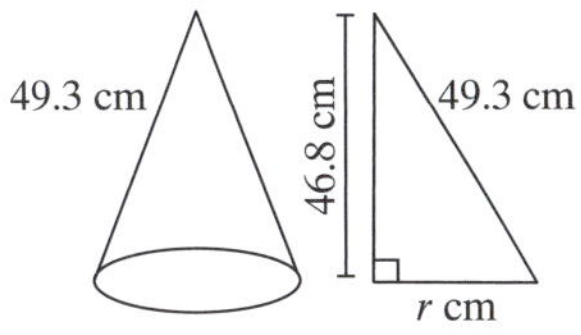

$r^2 = 49.3^2 - 46.8^2$

$= 240.25$

$r = \sqrt{240.25}$ $(r > 0)$

$= 15.5$ ✓

$V = \frac{1}{3}\pi r^2 h$

$= \frac{1}{3} \times \pi \times 15.5^2 \times 46.8$ ✓

$= 11\,774.3751\ldots$

$= 11\,774\text{ cm}^3$
(nearest cm³) ✓ (3 marks)

7 **a**

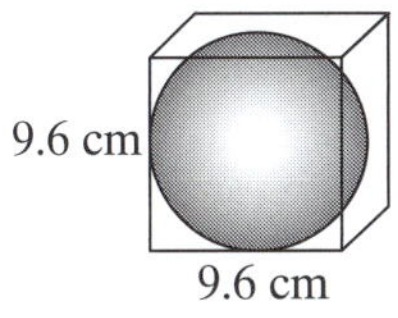

Radius of ball = 4.8 cm
Unused volume
= volume of box
– volume of ball

$= 9.6^3 - \frac{4}{3} \times \pi \times 4.8^3$ ✓

$= 421.489\,313\ldots$

$= 421.5\text{ cm}^3$ (1 d.p.) ✓
(2 marks)

b

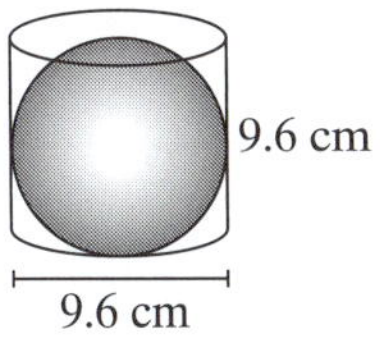

Unused volume
= volume of cylinder
– volume of ball

$= \pi \times 4.8^2 \times 9.6 - \frac{4}{3} \times \pi \times 4.8^3$

$= 231.623\,343\ldots$

$= 231.6\text{ cm}^3$ (1 d.p.) ✓

Difference
= 421.489313…
– 231.623343…
= 189.8659…
There would be around 190 cm³ less unused space if the ball was placed in the cylinder. ✓
[Or just find the difference in volume between the box and cylinder.] (2 marks)

8 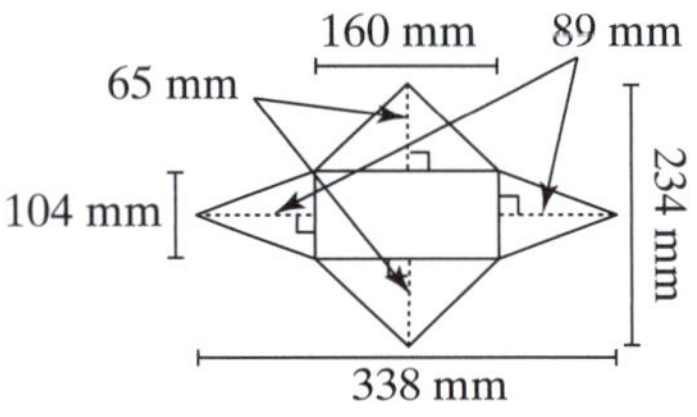

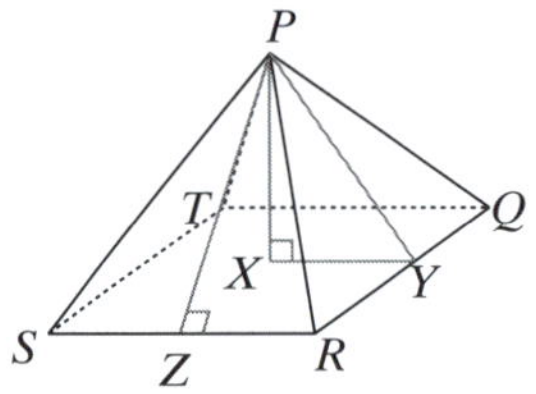

338 mm – 160 mm = 178 mm
(178 ÷ 2) mm = 89 mm ✓
So, PY = 89 mm
and XY = 80 mm

$PX^2 = 89^2 - 80^2$
$= 1521$
$PX = \sqrt{1521}$ $(PX > 0)$
$= 39$ mm ✓

$V = \frac{1}{3}Ah$
$= \frac{1}{3} \times 160 \times 104 \times 39$
$= 216\,320$ mm^3 ✓ (3 marks)

9 In ΔPQT, ΔSRT
$\angle QPT = \angle RST$ (both 90°)
$\angle PTQ = \angle STR$ (common)
$\therefore$ $\Delta PQT \;|||\; \Delta SRT$
(equiangular) ✓

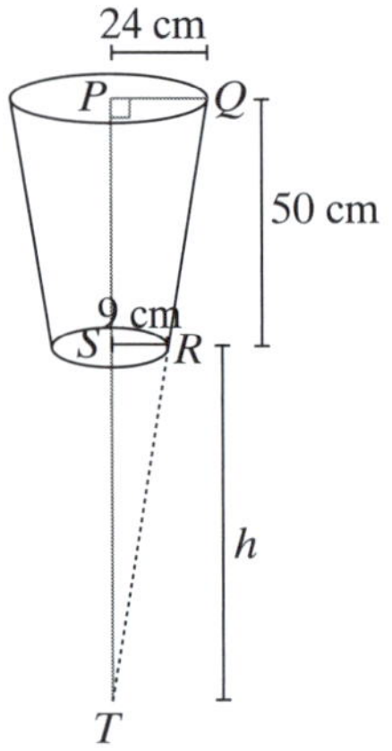

Let the length of ST be h cm.

Now $\frac{ST}{PT} = \frac{SR}{PQ}$ (corresponding sides of similar triangles)

$\frac{h}{h + 50} = \frac{9}{24}$
$= \frac{3}{8}$

$8h = 3h + 150$
$5h = 150$
$h = 30$ ✓

$V = \frac{1}{3}\pi R^2 H - \frac{1}{3}\pi r^2 h$
$= \frac{1}{3} \times \pi \times 24^2 \times 80$
$- \frac{1}{3} \times \pi \times 9^2 \times 30$ ✓
$= 45\,710.1731\ldots$
$= 45\,710$ cm^3 (nearest cm^3) ✓
(4 marks)

(Total: 30 marks)

SURFACE AREA OF PYRAMIDS, CONES AND SPHERES
SKILLS CHECK PAGE 58

1 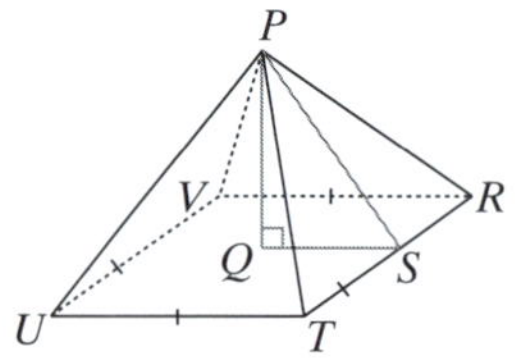

a the perpendicular height of the pyramid is PQ

b the perpendicular height of the triangular face PRT is PS.

2 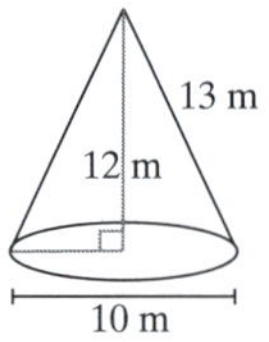

a diameter = 10 m
b perpendicular height = 12 m
c slant height = 13 m
d radius = 10 m ÷ 2
= 5 m

3 a 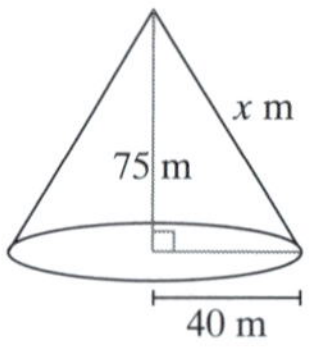

$x^2 = 40^2 + 75^2$
$= 1600 + 5625$
$= 7225$
$x = \sqrt{7225}$ $(x > 0)$
$= 85$

b 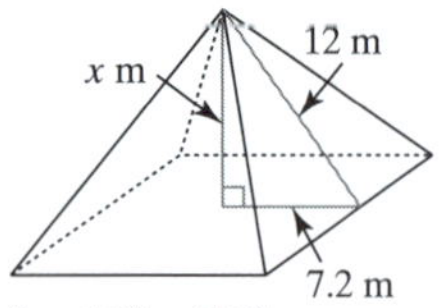

$x^2 = 12^2 - 7.2^2$
$= 144 - 51.84$
$= 92.16$
$x = \sqrt{92.16}$ $(x > 0)$
$= 9.6$

c Let the length of the side of the triangle be y mm.

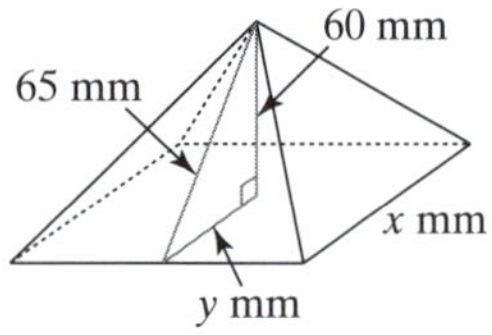

$y^2 = 65^2 - 60^2$
$= 4225 - 3600$
$= 625$
$y = \sqrt{625}$ $(y > 0)$
$= 25$
$x = 2y$
$= 50$

d 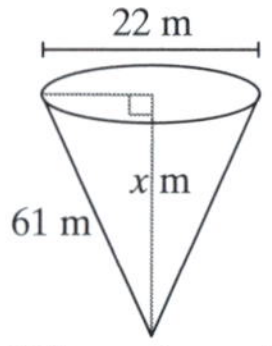

Diameter = 22 m
$\therefore$ radius = 11 m
$x^2 = 61^2 - 11^2$
$= 3721 - 121$
$= 3600$
$x = \sqrt{3600}$ $(x > 0)$
$= 60$

4 a $r = 1.8$
$A = \pi r^2$
$= \pi \times 1.8^2$
$= 10.1787602\ldots$
Area = 10.18 cm^2 (2 d.p.)

b $r = 3.3 \div 2 = 1.65$
$A = \pi r^2$
$= \pi \times 1.65^2$
$= 8.552985999\ldots$
Area = 8.55 m^2 (2 d.p.)

5 a $r = 48$, $l = 25$
$A = \pi rl$
$= \pi \times 48 \times 25$
$= 3769.911184\ldots$
Curved surface area
= 3769.9 cm^2 (1 d.p.)

b $r = 7.5,\ l = 35$
$A = \pi rl$
$= \pi \times 7.5 \times 35$
$= 824.6680716\ldots$
Curved surface area
$= 824.7\ \text{cm}^2$ (1 d.p.)

6 **a** $r = 12.6$
$A = 4\pi r^2$
$= 4 \times \pi \times 12.6^2$
$= 1995.036999\ldots$
Surface area $= 1995\ \text{cm}^2$ (nearest cm^2)

b diameter $= 784$ mm
$= 78.4$ cm
$r = 39.2$
$A = 4\pi r^2$
$= 4 \times \pi \times 39.2^2$
$= 19\,309.98774\ldots$
Surface area $= 19\,310\ \text{cm}^2$ (nearest cm^2)

7 Area of square base
$= 12\ \text{cm} \times 12\ \text{cm}$
$= 144\ \text{cm}^2$
Area of one triangle
$A = \frac{1}{2}bh$
$= \frac{1}{2} \times 12 \times 10$
$= 60$
Area $= 60\ \text{cm}^2$
Total surface area
$= 144\ \text{cm}^2 + 4 \times 60\ \text{cm}^2$
$= 384\ \text{cm}^2$

8 **a** $A = 122,\ l = 8$
$A = \pi rl$
$122 = \pi \times r \times 8$
$r = \frac{122}{8\pi}$
$= 4.854225764\ldots$
$= 4.9$ (1 d.p.)

b $A = 1195$
$A = 4\pi r^2$
$1195 = 4\pi r^2$
$r^2 = \frac{1195}{4\pi}$
$r = \sqrt{\frac{1195}{4\pi}}$ $(r > 0)$
$= 9.751670549\ldots$
$= 9.8$ (1 d.p.)

SURFACE AREA OF PYRAMIDS, CONES AND SPHERES
INTERMEDIATE TEST PAGE 59

1 $h^2 = 125^2 - 44^2$
$= 13\,689$
$h = 117$ $(h > 0)$

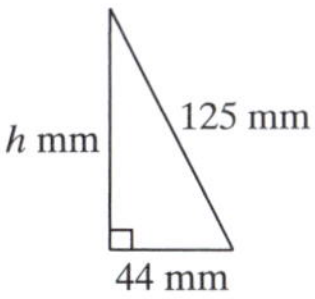

The perpendicular height is 117 mm.
$\therefore$ **[C]** ✓ (1 mark)

2 $A = 4\pi r^2$
$= 4 \times \pi \times 6400^2$
$= 514\,718\,540.4\ldots$
$= 500\,000\,000$ (1 sig. fig.)
$= 5 \times 10^8\ \text{km}^2$
$\therefore$ **[B]** ✓ (1 mark)

3 Ratio of areas $= 25:1$
$= 5^2:1^2$
Ratio of intervals $= 5:1$
Ratio of volumes $= 5^3:1^3$
$= 125:1$
$\therefore$ **[C]** ✓ (1 mark)

4

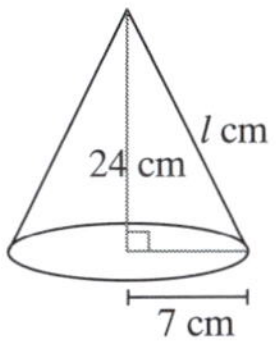

a $A = \pi r^2$
$= \pi \times 7^2$
$= 153.93804\ldots$
$= 153.9$ (1 d.p.)
Area is $153.9\ \text{cm}^2$ ✓(1 mark)

b $l^2 = 24^2 + 7^2$
$= 576 + 49$
$= 625$
$l = \sqrt{625}$ $(l > 0)$
$= 25$
The slant height is 25 cm ✓ (1 mark)

c $A = \pi rl$
$= \pi \times 7 \times 25$
$= 549.7787144\ldots$
The curved surface area
$\approx 549.8\ \text{cm}^2$ ✓
Total A
$= 153.93804\ldots$
$+ 549.7787144\ldots$
$= 703.71675\ldots$
The surface area is $703.7\ \text{cm}^2$ (1 d.p.) ✓
(2 marks)

5

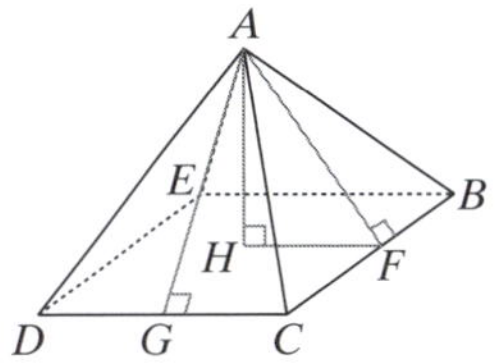

$AH = 24$ cm, $AF = 30$ cm, $AG = 26$ cm

a Let $HG = x$ cm
$x^2 = 26^2 - 24^2$
$= 676 - 576$
$= 100$
$x = 10$ $(x > 0)$
$HG = 10$ cm ✓
$\therefore BC = 20$ cm ✓ (2 marks)

b Let $HF = y$ cm
$y^2 = 30^2 - 24^2$
$= 900 - 576$
$= 324$
$y = \sqrt{324}$ $(y > 0)$
$= 18$
$HF = 18$ cm ✓
$\therefore DC = 36$ cm ✓
Area of base
$= 36\ \text{cm} \times 20\ \text{cm}$
$= 720\ \text{cm}^2$ ✓ (3 marks)

c Triangle ADC:
Area $= \frac{1}{2}bh$
$= \frac{1}{2} \times 36\ \text{cm} \times 26\ \text{cm}$
$= 468\ \text{cm}^2$ ✓
Triangle ABC:
Area $= \frac{1}{2}bh$
$= \frac{1}{2} \times 20\ \text{cm} \times 30\ \text{cm}$
$= 300\ \text{cm}^2$ ✓
Surface area
$= (2 \times 468 + 2 \times 300 + 720)\ \text{cm}^2$
$= 2256\ \text{cm}^2$ ✓ (3 marks)

6 Surface area $= 2463\ \text{cm}^2$
$4\pi r^2 = 2463$
$r^2 = \frac{2463}{4\pi}$ ✓
$r = \sqrt{\frac{2463}{4\pi}}$ $(r > 0)$
$= 13.99997544\ldots$
The radius is 14 cm. ✓
(2 marks)

7

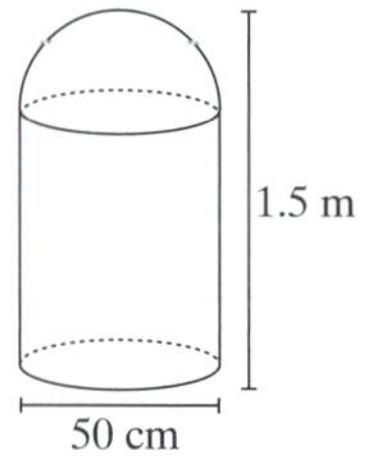

a $r = 25\text{ cm} = 0.25\text{ m}$

Height of cylinder
$= (1.5 - 0.25)\text{ m}$
$= 1.25\text{ m}$ ✓

Curved surface of cylinder:
$A = 2\pi rh$
$= 2 \times \pi \times 0.25 \times 1.25$
$= 1.963\,485\,408\ldots$ ✓

Hemisphere:
$A = 2\pi r^2$
$= 2 \times \pi \times 0.25^2$
$= 0.392699081\ldots$ ✓

Total $= 963\,485\,408\ldots$
$+ 0.392\,699\,081\ldots$
$= 2.356\,194\,49\ldots$

Surface area $= 2.36\text{ m}^2$ (2 d.p.) ✓ (4 marks)

b 25 bollards given 2 coats of paint
$= 50 \times$ surface area of 1 bollard
$\approx 50 \times 2.36$
$= 118\text{ m}^2$ ✓

1 litre covers 15 m^2

Required litres
$= 118 \div 15$
$= 7.8666\ldots$ ✓

Two 4-litre cans of paint will be required. ✓ (3 marks)

8

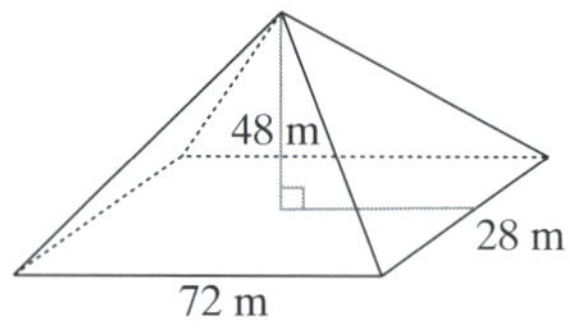

Area of base
$= 72\text{ m} \times 28\text{ m}$
$= 2016\text{ m}^2$ ✓

Let x m be the height of the side triangle.
$x^2 = 48^2 + 36^2$
$= 3600$
$x = 60 \quad (x > 0)$ ✓

Area $= \frac{1}{2} \times 28\text{ m} \times 60\text{ m}$
$= 840\text{ m}^2$ ✓

Let y m be the height of the front triangle.
$y^2 = 48^2 + 14^2$
$= 2500$
$y = 50 \quad (y > 0)$ ✓

Area $= \frac{1}{2} \times 72\text{ m} \times 50\text{ m}$
$= 1800\text{ m}^2$ ✓

Total $= (2016 + 2 \times 840 + 2 \times 1800)\text{ m}^2$
$= 7296\text{ m}^2$ ✓ (6 marks)

(Total: 30 marks)

SURFACE AREA OF PYRAMIDS, CONES AND SPHERES
ADVANCED TEST PAGE 60

1 **a** $A = 4\pi r^2$
$= 4 \times \pi \times 12^2$
$= 1809.557\,36\ldots$
$= 1809.6\text{ cm}^2$ (1 d.p.) ✓ (1 mark)

b $A = 4\pi r^2$
$= 4 \times \pi \times 24^2$
$= 7238.229\,47\ldots$
$= 7238.2\text{ cm}^2$ (1 d.p.) ✓ (1 mark)

c Percentage increase
$= \frac{7238.229\,47\ldots - 1809.557\,36\ldots}{1809.557\,36\ldots} \times 100\%$
$= 300\%$ ✓ (1 mark)

d Yes, if the radius of any sphere doubles the new area is 2^2 or 4 times greater than what it was before. This is a 300% increase. ✓ (1 mark)

e $A = \pi rl$
$= \pi \times 8 \times 20$
$= 502.654\,824\ldots$
$= 502.7\text{ cm}^2$ (1 d.p.) ✓ (1 mark)

f If both the radius and slant height double the new area will be 4 times greater than what it was before, so the percentage increase in area is 300%. ✓ (1 mark)

g Yes, if the radius of a circle doubles, the area is four times greater, so the total surface area will also be four times greater. ✓ (1 mark)

h If both the radius and slant height triple, the surface area will be 9 times greater. This is an 800% increase. ✓ (1 mark)

2 $V = \frac{4}{3}\pi r^3$

Now the volume is 11.5 cm^3.

So $\frac{4}{3}\pi r^3 = 11.5$
$r^3 = 11.5 \div \left(\frac{4}{3}\pi\right)$
$= 2.745\,422\,76\ldots$ ✓
$r = 1.400\,241\,92\ldots$
$= 1.4\text{ cm}$ (1 d.p.) ✓

$A = 4\pi r^2$
$= 4 \times \pi \times (1.400\,241\,92\ldots)^2$
$= 24.638\,599\,4\ldots$
$= 24.6\text{ cm}^2$ (1 d.p.) ✓ (3 marks)

3 **a**

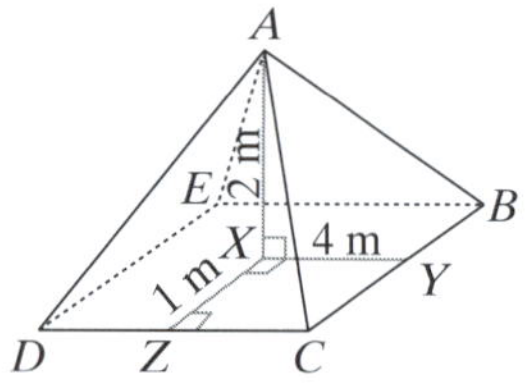

$AY^2 = 4^2 + 2^2$
$= 20$
$AY = \sqrt{20} \quad (AY > 0)$
$= 2\sqrt{5}\text{ m}$ ✓

Area of ΔABC:
$A = \frac{1}{2} \times 2 \times 2\sqrt{5}$
$= 2\sqrt{5}\text{ m}^2$ ✓ (2 marks)

b $AZ^2 = 2^2 + 1^2$
$= 5$
$AZ = \sqrt{5}\text{ m} \quad (AZ > 0)$ ✓

Area of ΔADC:
$A = \frac{1}{2} \times 8 \times \sqrt{5}$
$= 4\sqrt{5}\text{ m}^2$

$\therefore$ the area of ΔADC is twice that of ΔABC ✓ (2 marks)

c Total area
$= (8 \times 2 + 2 \times 4\sqrt{5} + 2 \times 2\sqrt{5})\text{ m}^2$
$= (16 + 12\sqrt{5})\text{ m}^2$ ✓ (1 mark)

4

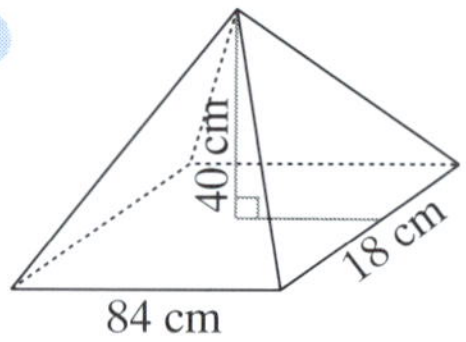

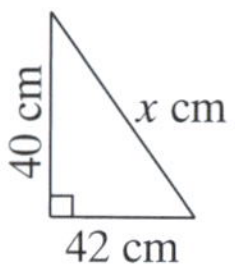

Side triangle:
$x^2 = 40^2 + 42^2$
$= 3364$
$x = 58 \quad (x > 0)$ ✓

$A = \frac{1}{2}bh$

$= \frac{1}{2} \times 18 \times 58$

$= 522 \text{ cm}^2$ ✓

Front triangle:

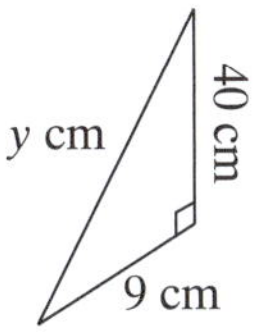

$y^2 = 40^2 + 9^2$

$= 1681$

$y = 41 \quad (y > 0)$ ✓

$A = \frac{1}{2}bh$

$= \frac{1}{2} \times 84 \times 41$

$= 1722 \text{ cm}^2$ ✓

Area of base $= (84 \times 18) \text{ cm}^2$

$= 1512 \text{ cm}^2$

Total

$= (1512 + 2 \times 522 + 2 \times 1722) \text{ cm}^2$

$= 6000 \text{ cm}^2$ ✓ (5 marks)

5 Each cone has perpendicular height 17.1 cm and radius 14 cm.

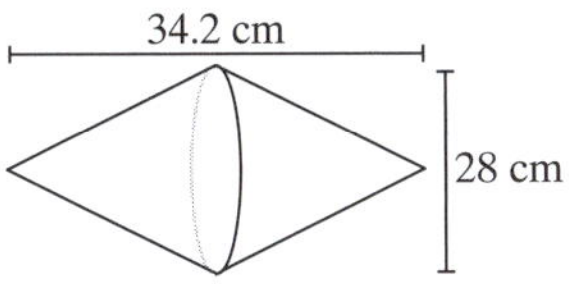

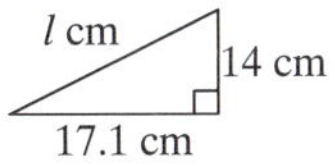

$l^2 = 14^2 + 17.1^2$

$= 488.41$

$l = \sqrt{488.41} \quad (l > 0)$

$= 22.1$ ✓

$A = 2\pi rl$

$= 2 \times \pi \times 14 \times 22.1$

$= 1944.0175\ldots$

$= 1944 \text{ cm}^2$ (nearest cm^2) ✓

(2 marks)

6 **a**

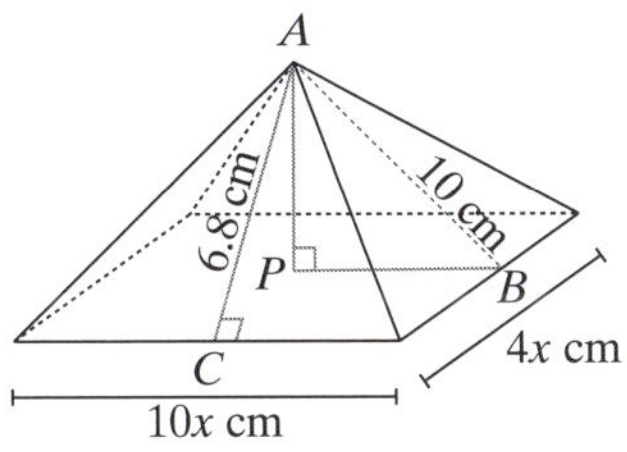

In ΔAPB,

$AP^2 = AB^2 - PB^2$

$h^2 = 10^2 - (5x)^2$

$= 100 - 25x^2$ ✓ (1 mark)

b In ΔAPC,

$AP^2 = AC^2 - PC^2$

$h^2 = 6.8^2 - (2x)^2$

$= 46.24 - 4x^2$ ✓

$\therefore 46.24 - 4x^2 = 100 - 25x^2$

$21x^2 = 53.76$

$x^2 = 2.56$

$x = 1.6 \quad (x > 0)$ ✓

(2 marks)

c $h^2 = 100 - 25x^2$

$= 100 - 25 \times 1.6^2$

$= 36$

$h = 6 \quad (h > 0)$

The perpendicular height is 6 cm. ✓ (1 mark)

d $4x = 4 \times 1.6$

$= 6.4$

$10x = 10 \times 1.6$

$= 16$

So the base is 16 cm by 6.4 cm ✓

$A = 16 \times 6.4 + 2 \times \frac{1}{2} \times 16 \times 6.8$

$+ 2 \times \frac{1}{2} \times 6.4 \times 10$

$= 275.2 \text{ cm}^2$ ✓ (2 marks)

7

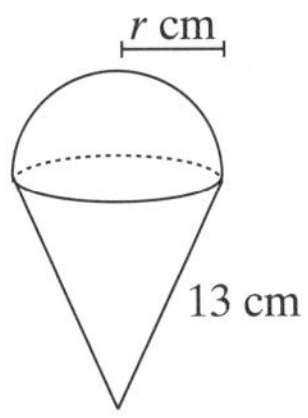

a $A = 2\pi r^2 + \pi rl$

$115\pi = 2\pi r^2 + \pi r \times 13$ ✓

$115 = 2r^2 + 13r$

$2r^2 + 13r - 115 = 0$ ✓

(2 marks)

b $2r^2 + 13r - 115 = 0$

$2r^2 + 23r - 10r - 115 = 0$

$r(2r + 23) - 5(2r + 23) = 0$

$(2r + 23)(r - 5) = 0$ ✓

$2r = -23 \quad \text{or} \quad r = 5$

$r = -11.5 \quad \text{or} \quad r = 5$

But $r > 0$,

$\therefore r = 5$ ✓ (2 marks)

c Let h cm be the height of the cone.

$h^2 = 13^2 - 5^2$

$= 144$

$h = 12 \quad (h > 0)$ ✓

Total height

$= (12 + 5)$ cm

$= 17$ cm ✓ (2 marks)

(Total: 35 marks)

GEOMETRY REVIEW
SKILLS CHECK PAGE 62

1 **a**

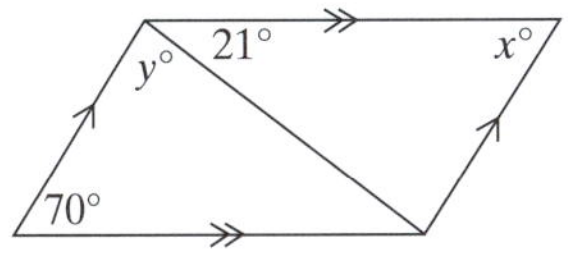

$x = 70$

(opposite ∠s of a parallelogram)

$70° + y° + 21° = 180°$

(co-interior angles, || lines)

$y + 91 = 180$

$y = 89$

b

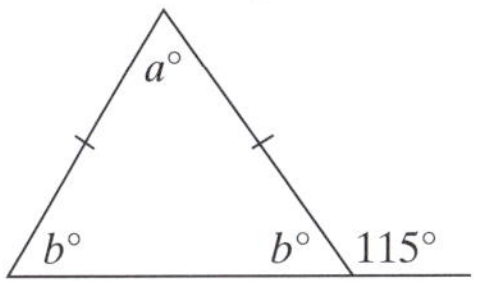

$b° + 115° = 180°$

(angles in a straight line)

$b = 65$

$a° + 2b° = 180°$

(angle sum of isosceles Δ)

$a + 130 = 180$

$a = 50$

c

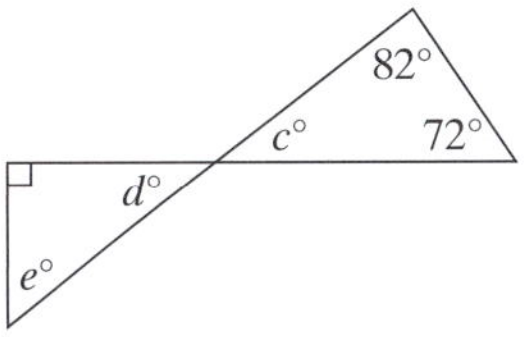

$c° + 82° + 72° = 180°$

(angle sum of a Δ)

$c + 154 = 180$

$c = 26$

$d = c$

(vertically opposite)

$d = 26$

$e° + d° + 90° = 180°$

(angle sum of Δ)

$e + 116 = 180$

$e = 64$

d

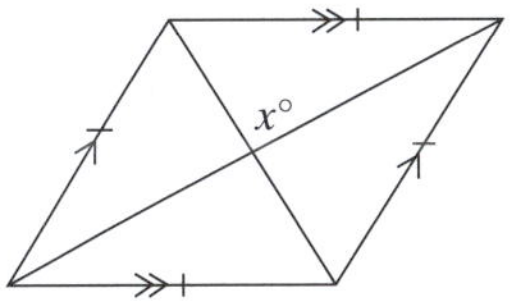

The quadrilateral is a rhombus because it is a parallelogram with a pair of adjacent sides equal.

$x = 90$

(diagonals of a rhombus meet at right angles)

e

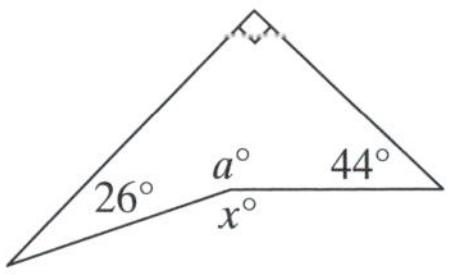

$a° + 44° + 90° + 26° = 360°$
($\angle$ sum of a quadrilateral)
$a + 160 = 360$
$a = 200$
$a° + x° = 360°$
($\angle$s at a point)
$x = 360 - 200$
$x = 160$

f

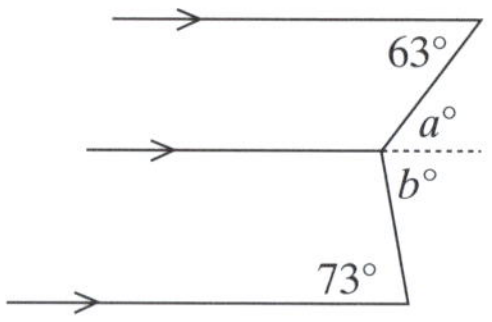

$a = 63$
(alternate $\angle$s, parallel lines)
$b = 73$
(alternate $\angle$s, parallel lines)
$x = a + b$
$= 63 + 73$
$= 136$

2 **a**

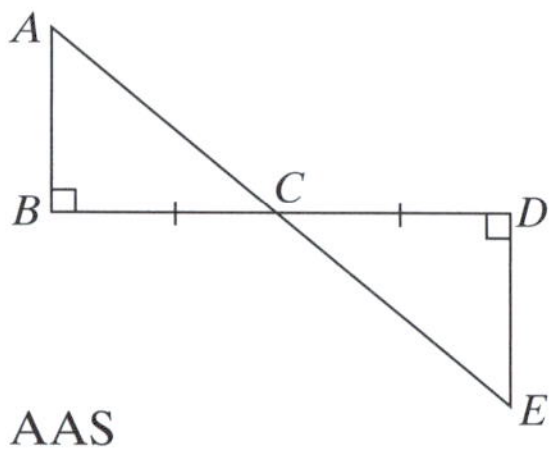

AAS
[$\angle B = \angle D$ (given)
$\angle ACB = \angle ECD$
(vertically opposite)
$BC = DC$ (given)]

b

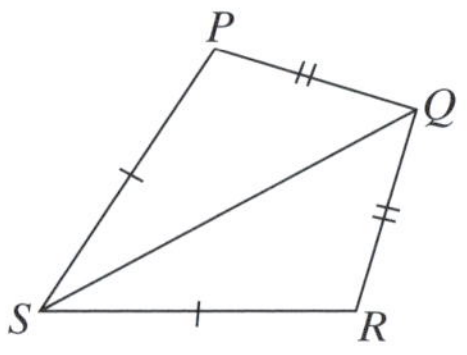

SSS
[$PQ = QR$ (given)
$PS = RS$ (given)
$QS = QS$ (common)]

c

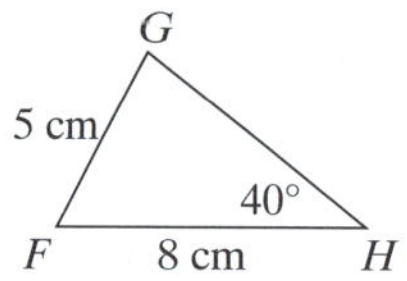

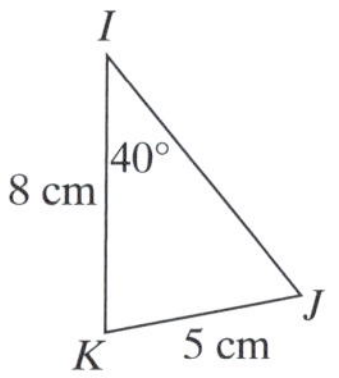

The triangles are not necessarily congruent.
[Both triangles have a pair of sides 8 cm and a pair 5 cm in length and an angle of 40°, but the angle must be included between the two sides for the triangles to be congruent.]

3 **a**

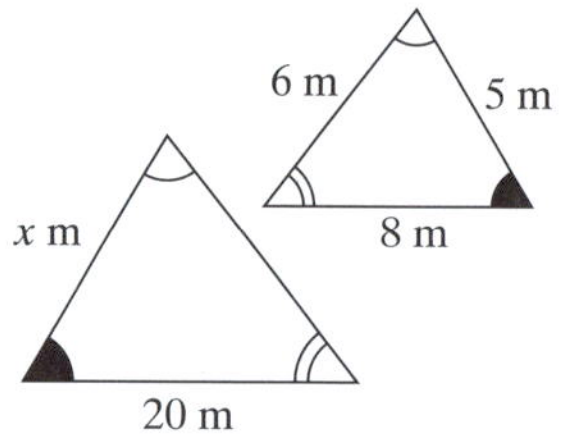

$\frac{x}{5} = \frac{20}{8}$
(corresponding sides of similar triangles)
$x = \frac{20}{8} \times 5$
$x = 12.5$

b

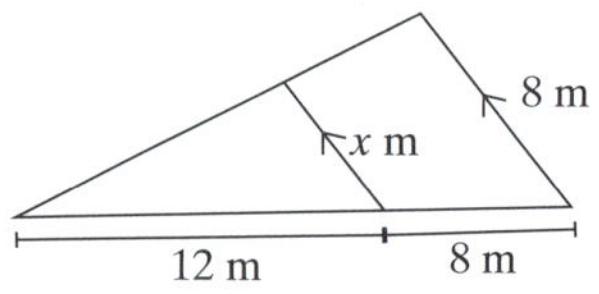

$\frac{x}{8} = \frac{12}{20}$
(corresponding sides of similar triangles)
$x = \frac{12}{20} \times 8$
$x = 4.8$

4 **a** The angles of a quadrilateral add up to 360°.

b The opposite angles of a parallelogram are equal.

c An isosceles triangle has two equal sides.

d The diagonals of a rectangle are equal in length.

e The diagonals of a rhombus bisect the vertex angles through which they pass.

f If two angles of a triangle are equal, the sides opposite those angles are equal.

g If all sides of a quadrilateral are equal the quadrilateral is a rhombus.

h If two angles of one triangle are equal to two angles of another triangle the two triangles are similar.

i If three sides of one triangle are equal to three sides of another triangle the two triangles are congruent.

GEOMETRY REVIEW
INTERMEDIATE TEST PAGE 63

1 All three triangles are congruent.
[All three triangles have angles of 50°, 60° and 70°. The side marked 7 cm is always opposite the angle of 60°.]
$\therefore$ **[D]** ✓ (1 mark)

2

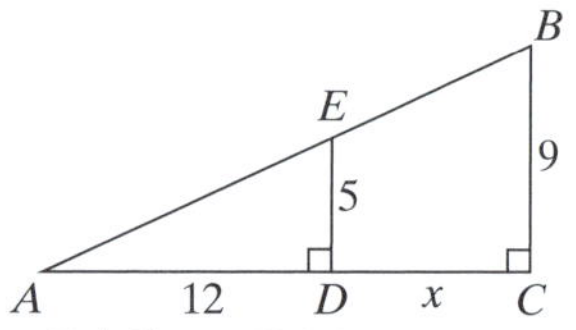

$\angle BAC = \angle EAD$
(common angle)
$\angle ACB = \angle ADE$
(both 90°, given)
$\therefore \; \Delta ABC \;|||\; \Delta AED$
(equiangular)
$\frac{AC}{AD} = \frac{BC}{ED}$
$\frac{x + 12}{12} = \frac{9}{5}$
$x + 12 = 21.6$
$x = 9.6$
$\therefore$ **[B]** ✓ (1 mark)

3

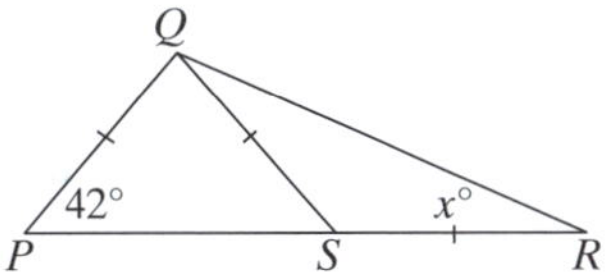

$\angle PSQ = 42°$
($= \angle QPS$, isosceles triangle)
$\angle SQR = x°$
($= \angle SRQ$, isosceles triangle)
$\angle PSQ = \angle SQR + \angle SRQ$
(exterior angle)
$2x = 42$
$x = 21$
$\therefore$ **[A]** ✓ (1 mark)

4

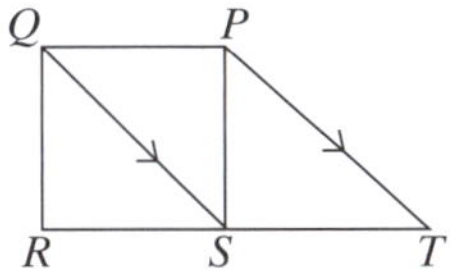

$PTSQ$ is a parallelogram (both pairs of opposite sides parallel) ✓
$\angle RQP = 90°$
($\angle$ of a square)
$\angle PQS = 45°$
(diagonals bisect $\angle$s) ✓
$\angle QPT = 180° - 45°$
$= 135°$ ✓
(co-interior $\angle$s, parallel lines)
(3 marks)

5

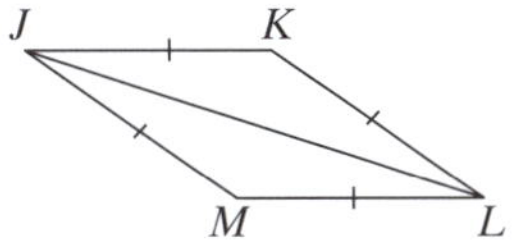

a In $\angle JKL$, $\angle JML$:
$JK = JM$ (given)
$KL = ML$ (given)
$JL = JL$ (common side) ✓
$\therefore \Delta JKL \equiv \Delta JML$ SSS ✓
$\therefore \angle KJL = \angle MJL$
(corresponding angles of congruent triangles)
$\therefore JL$ bisects $\angle KJM$ ✓
(3 marks)

b The diagonals of a rhombus bisect the vertex angles. ✓
(1 mark)

6

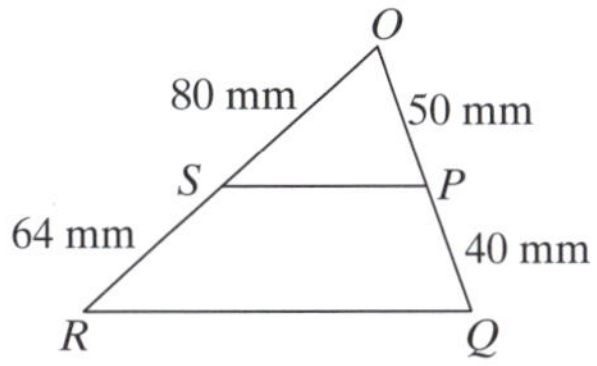

a In ΔOPS, ΔOQR:
$\angle SOP = \angle ROQ$
(common angle) ✓

$\frac{OS}{OR} = \frac{80}{144}$
$= \frac{5}{9}$

$\frac{OP}{OQ} = \frac{50}{90}$
$= \frac{5}{9}$

$\therefore \frac{OS}{OR} = \frac{OP}{OQ}$ ✓
$\therefore \Delta OPS ||| \Delta OQR$
(2 sides in proportion, included angles equal) ✓
(3 marks)

b Yes. SP is parallel to RQ. ✓
$\angle OSP = \angle ORQ$
(corresponding angles of similar triangles)
$\therefore SP \parallel RQ$
(corresponding angles are equal) ✓ (2 marks)

7

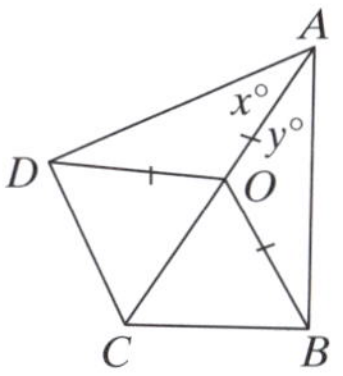

a $\angle ADO = \angle DAO$
(angles opposite equal sides, isosceles triangle)
$\therefore \angle ADO = x°$ ✓
$\angle DOC = \angle ADO + \angle DAO$
(exterior angle)
$\angle DOC = 2x°$ ✓ (2 marks)

b $\angle OCD + \angle ODC + \angle DOC$
$= 180°$
(angle sum triangle)
$\therefore \angle OCD + \angle ODC$
$= 180° - 2x°$
$\angle ODC = \angle OCD$
(angles opposite equal sides, isosceles triangle)
$\therefore \angle OCD = (180° - 2x°) \div 2$
$= 90° - x°$ ✓
Similarly,
$\angle COB = 2y$
(exterior $\angle$ of isosceles ΔAOB)
$\angle BCO = 90° - y°$
($\angle$ sum isosceles ΔBOC) ✓
$\angle DCB = \angle OCD + \angle BCO$
$= 90° - x° + 90° - y°$
$= 180° - (x + y)°$
$\angle DAB = x° + y°$
$\therefore \angle DAB + \angle DCB = 180°$
$\angle DAB$ and $\angle DCB$ are supplementary. ✓(3 marks)

(Total: 20 marks)

GEOMETRY REVIEW
ADVANCED TEST PAGE 64

1 In ΔPST, ΔPRQ
$\angle SPT = \angle RPQ$ (common $\angle$)
$\angle PTS = \angle PQR$
(both 90°, given)
$\therefore \Delta PST ||| \Delta PRQ$ equiangular ✓

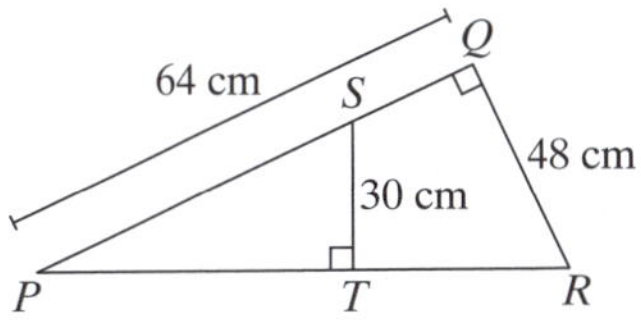

$\frac{PT}{PQ} = \frac{ST}{RQ}$
(corresp. sides similar triangles)
$\frac{PT}{64} = \frac{30}{48}$
$PT = 40$ ✓
$PR^2 = PQ^2 + QR^2$
$= 64^2 + 48^2$
$= 6400$
$PR = 80$ $(PR > 0)$
So $PR = 2 \times PT$
$\therefore T$ is the midpoint of PR. ✓
(3 marks)

2

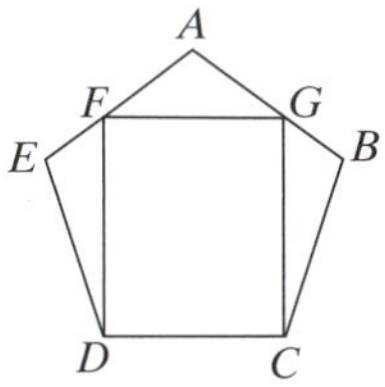

a Angle sum of pentagon
$= (5 - 2) \times 180°$
$= 540°$
Each angle of regular pentagon
$= 540° \div 5$
$= 108°$ ✓
$\therefore \angle EDC - 108°$
$\angle FDC = 90°$
(angle of rectangle)
$\angle FDE = 108° - 90°$
$= 18°$ ✓ (2 marks)

b $\angle DFG = 90°$
(angle of rectangle)
$\angle DEF = 108°$
(angle of reg. pentagon)
$\angle EFD + \angle DEF + \angle FDE = 180°$
(angle sum triangle)
$\angle EFD + 108° + 18° = 180°$
$\angle EFD + 126° = 180°$
$\angle EFD = 54°$ ✓
$\angle AFG + \angle DFG + \angle EFD = 180°$
(angles in straight line)
$\angle AFG + 90° + 54° = 180°$
$\angle AFG = 36°$
Now $36° = 2 \times 18°$
So $\angle AFG = 2\angle FDE$ ✓
(2 marks)

3

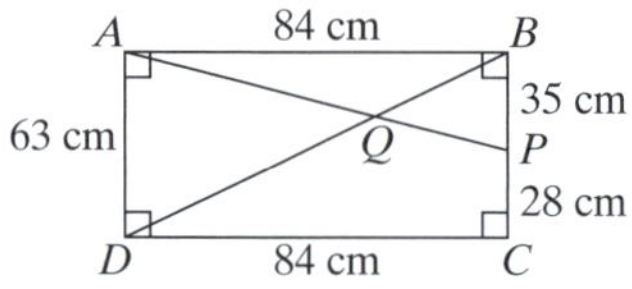

a In ΔAQD, ΔPQB
$\angle DAQ = \angle BPQ$
(alternate, $AD \parallel BC$) ✓
$\angle AQD = \angle PQB$
(vertically opposite)
$\therefore \Delta AQD \parallel\!| \Delta PQB$
equiangular ✓ (2 marks)

b $AP^2 = AB^2 + BP^2$
$= 84^2 + 35^2$
$= 8281$
$AP = 91$ cm $(AP > 0)$ ✓
$BD^2 = BC^2 + DC^2$
$= 63^2 + 84^2$
$= 11\,025$
$BD = 105$ cm $(BD > 0)$
Now $105 - 91 = 14$
So BD is 14 cm longer than AP. ✓ (2 marks)

c Let the length of BQ be x cm.
$$\frac{BQ}{DQ} = \frac{BP}{DA}$$
(corresp. sides similar Δs)
$$\frac{x}{105 - x} = \frac{35}{63} = \frac{5}{9}$$ ✓
$9x = 525 - 5x$
$14x = 525$
$x = 37.5$
The length of BQ is 37.5 cm ✓ (2 marks)

4

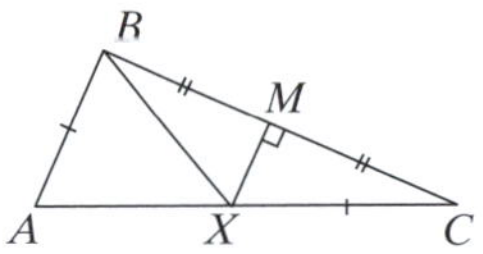

a In ΔBMX, ΔCMX
$BM = CM$
(M is midpoint of BC)
$\angle BMX = \angle CMX$
(both 90°, $XM \perp BC$) ✓
$MX = MX$ (common side)
$\therefore \Delta BMX \equiv \Delta CMX$ SAS ✓
(2 marks)

b $\angle XBM = \angle XCM$
(corresponding angles congruent triangles)
$\angle BXA = \angle XBC + \angle XCB$
(ext. angle ΔXBC)
$= 2\angle XCB$ ✓
$BX = CX$ (corresponding sides congruent triangles)
But $AB = CX$ (given)
$\therefore AB = BX$ ✓
$\angle BXA = \angle BAX$
(angles opp. equal sides isosceles triangle)
$\therefore \angle BAX = 2\angle XCB$
i.e. $\angle BAC = 2\angle BCA$ ✓
(3 marks)

5

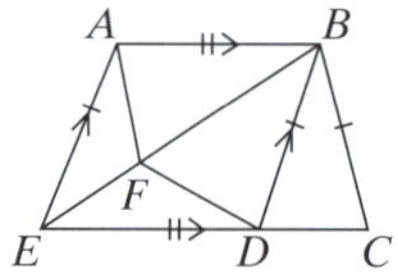

a In ΔAEF, ΔDEF
$AE = DE$
(sides of rhombus)
$\angle AEF = \angle DEF$
(diagonals of rhombus bisect angles) ✓
$EF = EF$ (common side)
$\therefore \Delta AEF \equiv \Delta DEF$ SAS ✓
(2 marks)

b $\angle AEF = 35°$
$\therefore \angle AED = 70°$
(diagonals bisect angles) ✓
$\angle BDC = 70°$
(corresponding, $AE \parallel BD$)
$\angle BCD = \angle BDC$
(angles opposite equal sides, isosceles ΔBDC)
$\therefore \angle BCD = 70°$ ✓
$\angle ABC + \angle BCD = 180°$
(co-interior, $AB \parallel EC$)
$\angle ABC + 70° = 180°$
$\angle ABC = 110°$ ✓
(3 marks)

6 In ΔPTQ, ΔSRQ
$\angle PTQ = \angle SRQ$ (given)
$\angle PQT = \angle SQR$
(vertically opposite)
$PT = SR$ (given)
$\therefore \Delta PTQ \equiv \Delta SRQ$ AAS ✓
Let the length of QR be x cm
$\therefore PQ = (50 - x)$ cm
$QT = QR$
(corresponding sides of congruent triangles)
So $QT = x$ cm ✓

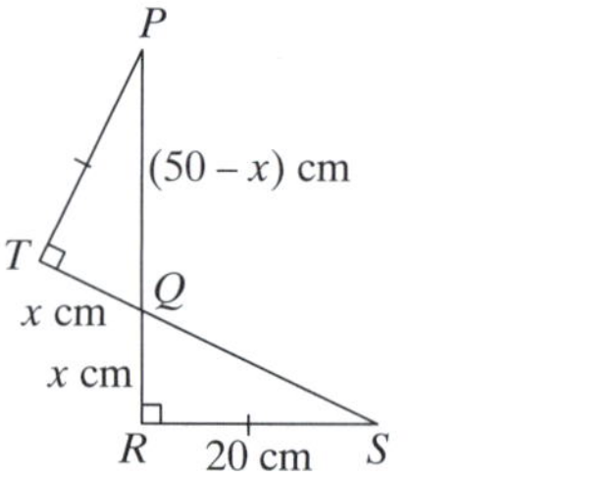

In ΔPTQ,
$PQ^2 = PT^2 + QT^2$
$(50 - x)^2 = 20^2 + x^2$ ✓
$2500 - 100x + x^2 = 400 + x^2$
$100x = 2100$
$x = 21$
$\therefore$ the length of QR is 21 cm. ✓
(4 marks)

7

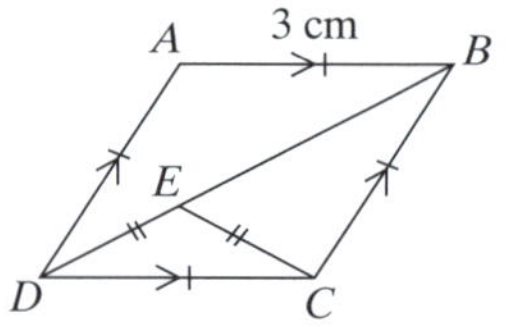

a In ΔEDC, ΔABD
$\angle EDC = \angle ABD$
(alternate, $AB \parallel DC$)
$\angle EDC = \angle ADB$
(diagonals bisect angles)
but $\angle ECD = \angle EDC$
(base angles isos. Δ)
$\therefore \angle ECD = \angle ADB$ ✓
$\therefore \Delta EDC \parallel\!| \Delta ABD$
equiangular ✓ (2 marks)

b $$\frac{BD}{DC} = \frac{AD}{EC}$$
(corresp. sides similar Δs)
$$\frac{BD}{3} = \frac{3}{\sqrt{3}}$$ ✓
$$BD = \frac{9}{\sqrt{3}} = \frac{9\sqrt{3}}{3} = 3\sqrt{3} \text{ cm}$$ ✓ (2 marks)

c $BC = 3$
$ED = EC = \sqrt{3}$
$BE = 3\sqrt{3} - \sqrt{3}$
$= 2\sqrt{3}$ ✓
$BE^2 = (2\sqrt{3})^2$
$= 12$
$BC^2 + EC^2 = 3^3 + (\sqrt{3})^2$
$= 9 + 3$
$= 12$
$= BE^2$
$\therefore \Delta BEC$ is right-angled at C. ✓ (2 marks)

d Let $\angle EDC = x^\circ$
$\therefore \angle ECD = x^\circ$
$\angle BEC = 2x^\circ$
(exterior angle ΔEDC)
$\angle EBC = \angle EDC$
(base angles isos. ΔBCD)
$\therefore \angle EBC = x^\circ$ ✓
$\angle EBC + \angle BCE + \angle BEC = 180^\circ$
(angle sum triangle)
$x^\circ + 90^\circ + 2x^\circ = 180^\circ$
$3x^\circ = 90^\circ$
$x = 30^\circ$ ✓
$\angle ADC = 2 \times 30^\circ$
(diagonals bisect angles)
$= 60^\circ$
$\angle BAD + \angle ADC = 180^\circ$
(co-interior, $AB \parallel DC$)
$\angle BAD + 60^\circ = 180^\circ$
$\angle BAD = 120^\circ$ ✓
(3 marks)

8

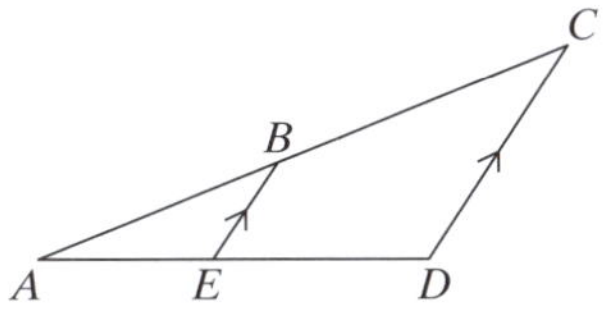

a In ΔABE, ΔACD
$\angle BAE = \angle CAD$
(common angle)
$\angle ABE = \angle ACD$
(corresponding angles, $BE \parallel CD$)
$\therefore \Delta ABE \, ||| \, \Delta ACD$
equiangular ✓ (1 mark)

b $\frac{BE}{CD} = \frac{AB}{AC}$
(corresp. sides similar Δs) ✓
$BE = CD \times \frac{AB}{AC}$
$= CD \times \frac{AC - BC}{AC}$ ✓
$= CD \times \frac{AC}{AC} - CD \times \frac{BC}{AC}$
$= CD \times 1 - \frac{CD \times BC}{AC}$
$= CD - \frac{BC \times CD}{AC}$ ✓
(3 marks)

(Total: 40 marks)

PARTS OF CIRCLES AND CHORD PROPERTIES
SKILLS CHECK PAGE 66

1 **a**

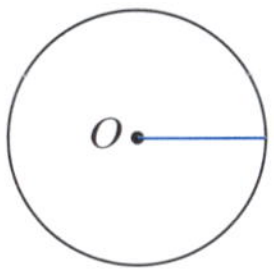

radius

b

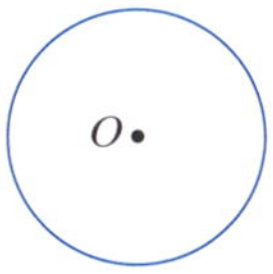

circumference

c

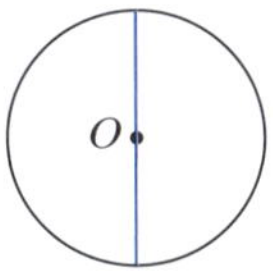

diameter

d

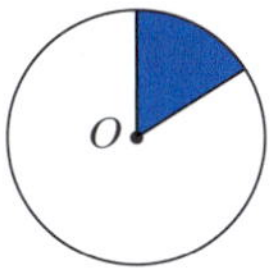

sector

e

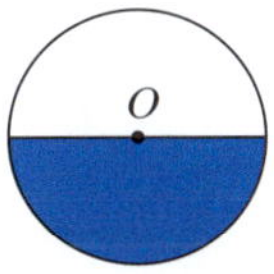

semi-circle

f

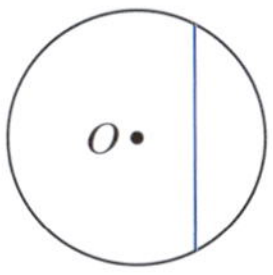

chord

g

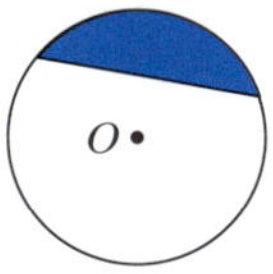

segment

h

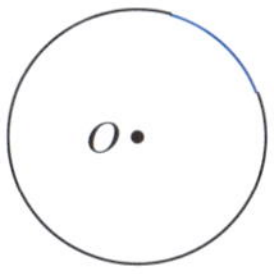

arc

i

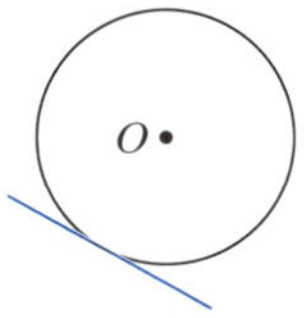

tangent

j

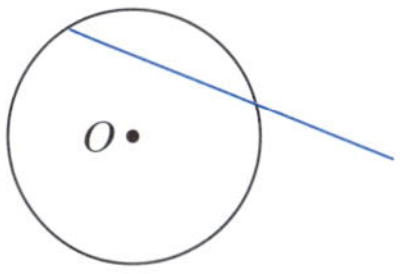

secant

2 **a**

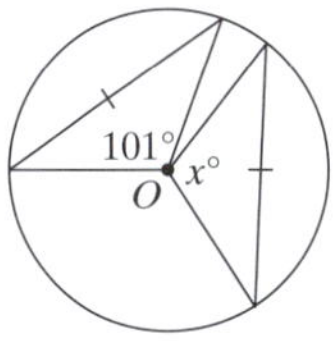

$x = 101$
(Equal chords subtend equal angles at the centre.)

b

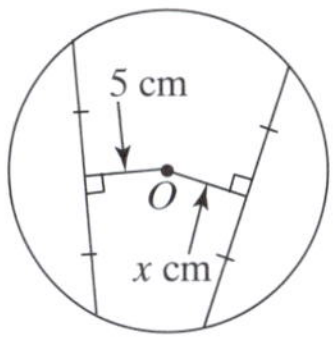

$x = 5$
(Chords of equal length are equidistant from the centre.)

c

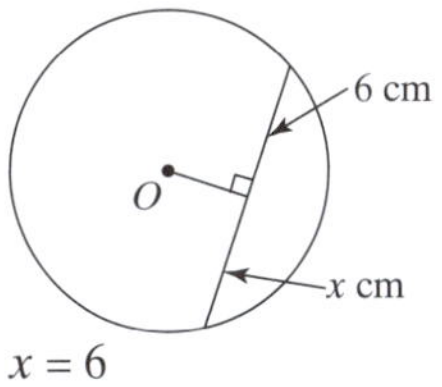

$x = 6$
(The perpendicular from the centre of a circle to a chord bisects the chord.)

d

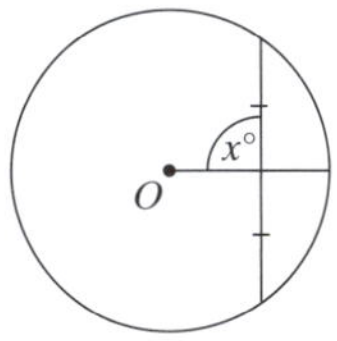

$x = 90$
(The line from the centre of the circle to the midpoint of a chord is perpendicular to the chord.)

e

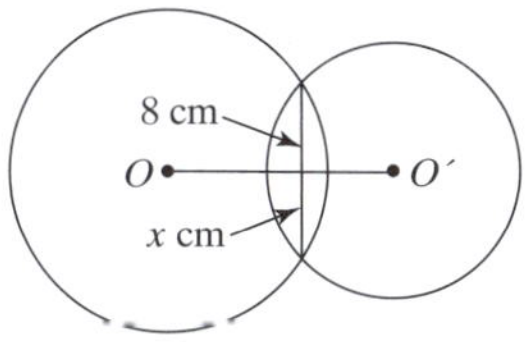

$x = 8$
(When two circles intersect, the line joining their centres bisects the common chord.)

f

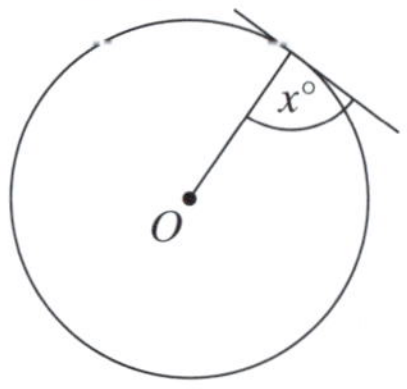

$x = 90$
(A tangent to a circle is perpendicular to the radius at the point of contact.)

3 **a** The perpendicular bisector of a chord passes through the centre of the circle.

b When two circles intersect, the line joining their centres bisects the common chord at right angles.

c When 3 points form a triangle, the point of intersection of the perpendicular bisectors of any 2 sides of the triangle is the centre of the circle which passes through all three points.

PARTS OF CIRCLES AND CHORD PROPERTIES
INTERMEDIATE TEST PAGE 67

1

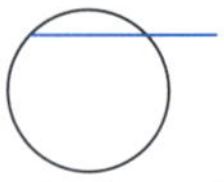

The blue line is called a secant.
$\therefore$ **[A]** ✓ (1 mark)

2 The perpendicular from the centre to the chord bisects the chord.
Let the length of the chord be $2x$ m.

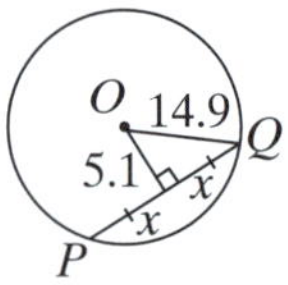

$x^2 = 14.9^2 - 5.1^2$
$= 196$
$x = \sqrt{196}$ $(x > 0)$
$= 14$
So the chord is 2×14 m or 28 m long.
$\therefore$ **[D]** ✓ (1 mark)

3

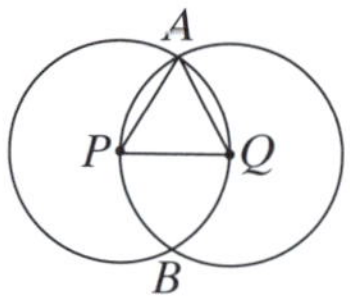

$PA = PQ$
(radii of circle, centre P)
$QA = QP$
(radii of circle, centre Q)
$\therefore AP = AQ = PQ$
The triangle is equilateral.
$\therefore$ **[C]** ✓ (1 mark)

4

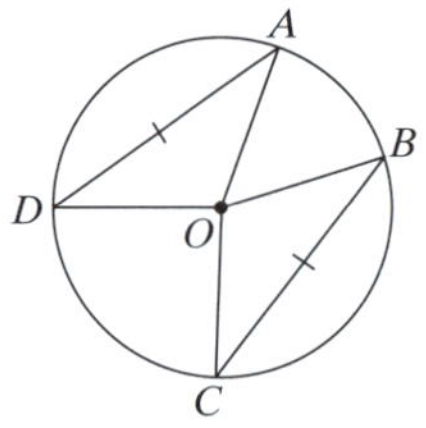

In ΔADO, ΔBCO
$AD = BC$ (given) ✓
$OA = OB$ (radii of circle)
$OD = OC$ (radii of circle) ✓
$\therefore \Delta ADO \equiv \Delta BCO$ SSS ✓
$\therefore \angle AOD = \angle BOC$
(corresponding angles of congruent Δs) ✓ (4 marks)

5 **a**

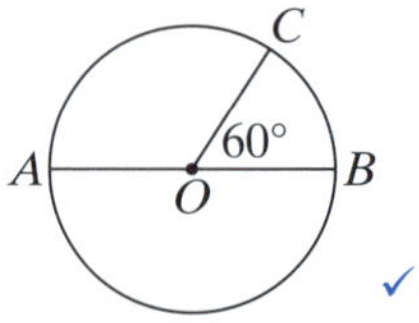

✓ (1 mark)

b $OC = OB$ (radii of circle)
$\angle COB = 60°$
So ΔCOB is equilateral ✓
$\therefore CB = OC = OB$
So CB has the same length as the radius. ✓ (2 marks)

c $\angle AOC + \angle COB = 180°$
(angles in straight line)
$\angle AOC + 60° = 180°$
$\angle AOC = 120°$ ✓
Now $120 = 2 \times 60$
The area of a sector is a fraction of the area of the circle. The particular fraction is determined by the angle at the centre.
Because the angle at the centre of sector AOC is twice that of sector COB, the area of sector AOC will be twice that of sector COB. ✓ (2 marks)

6

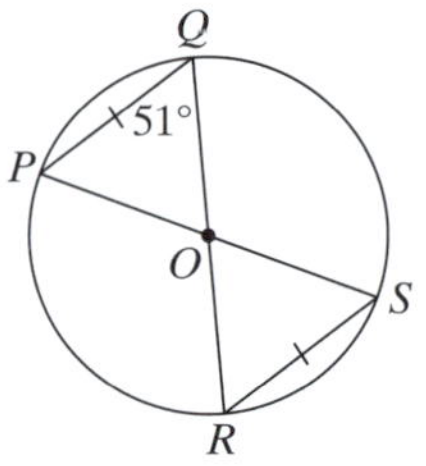

a $OP = OQ$ (radii of circle)
$\therefore \angle OQP = \angle OPQ$
(angles opposite equal sides, isosceles triangle)
$\therefore \angle OPQ = 51°$ ✓
$\angle POQ + \angle OPQ + \angle OQP = 180°$
(angle sum of a triangle)
$\angle POQ + 51° + 51° = 180°$
$\angle POQ = 78°$
$\therefore \angle ROS = 78°$
(vertically opp. $\angle POQ$) ✓
(2 marks)

b $OR = OS$ (radii of circle)
$\therefore \angle ORS = \angle OSR$
(angles opposite equal sides, isosceles triangle) ✓
$\angle ORS + \angle OSR + \angle ROS = 180°$
(angle sum of a triangle)
$2\angle ORS + 78° = 180°$
$2\angle ORS = 102°$
$\angle ORS = 51°$ ✓
$\therefore \angle ORS = \angle OQP$
$\therefore PQ \parallel RS$
(alternate $\angle$s are equal) ✓
(3 marks)

7

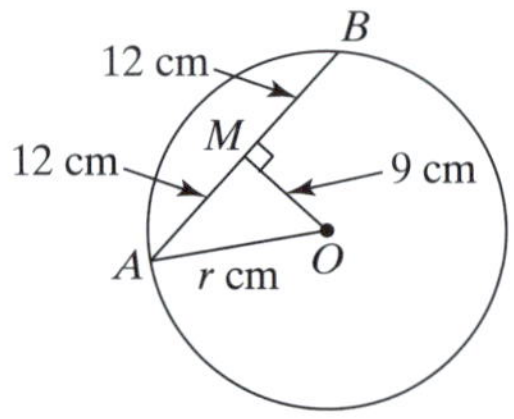

Let M be the foot of the perpendicular from O to AB.
$AM = MB$
(the perpendicular from the centre to the chord bisects the chord)
$AM = 12$ cm ✓
Let the length of OA be r cm.
$r^2 = 9^2 + 12^2$
$= 81 + 144$
$= 225$
$r = \sqrt{225}$ $(r > 0)$
$= 15$

The length of the radius of the circle is 15 cm. ✓

$C = 2\pi r$

$= 2 \times \pi \times 15$

$= 94.247\,779\,61\ldots$

$= 94.2 \quad (1\text{ d.p.})$

The circumference of the circle is 94.2 cm. ✓ (3 marks)

(Total: 20 marks)

PARTS OF CIRCLES AND CHORD PROPERTIES
ADVANCED TEST PAGE 68

1

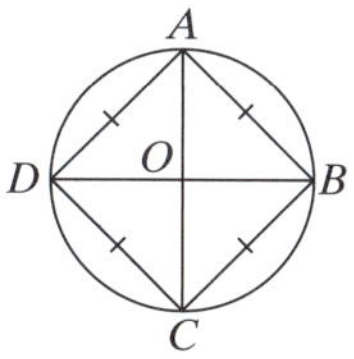

a $ABCD$ is a rhombus because all the sides are equal.
Now equal chords subtend equal angles at the centre.

$\therefore \angle AOB = \angle BOC$
$= \angle COD$
$= \angle DOA$

But $\angle AOB + \angle BOC + \angle COD + \angle DOA = 360°$ ($\angle$s at a point)

So $\angle AOB = \angle BOC$
$= \angle COD$
$= \angle DOA$
$= 90°$

$\angle COA = \angle COD + \angle DOA$
$= 180°$

$\angle ABC = 180° \div 2$
$= 90°$

(angle at centre is twice angle at circumference)

$\therefore$ $ABCD$ is a square because it is a rhombus with one angle 90°. ✓ (2 marks)

[In fact all the angles are 90°, but knowing that just one angle of a rhombus is 90° is enough to mean that it must be a square.]

b AC and BD are diameters of the circle (straight lines paassing through the centre)
Let the length of each chord be x cm.

$x^2 + x^2 = 12^2$
$2x^2 = 144$
$x^2 = 72$ ✓

$x = \sqrt{72} \quad (x > 0)$
$= 6\sqrt{2}$

The exact length of chord AB is $6\sqrt{2}$ cm. ✓ (2 marks)

2

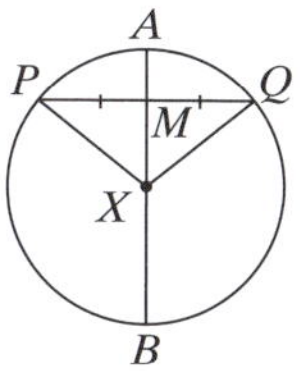

In ΔPMX, ΔQMX

$PM = QM$

(M is midpoint of PQ)

$\angle PMX = \angle QMX$

(both 90°, $XM \perp PQ$) ✓

$MX = MX$ (common side)

$\therefore \Delta PMX \equiv \Delta QMX$ SAS ✓

(2 marks)

b Corresponding sides of congruent triangles are equal. So $PX = QX$. ✓

(1 mark)

c No, Sophie is not correct. AB must be a diameter of the circle, but X could be any point on the diameter.✓

(1 mark)

3

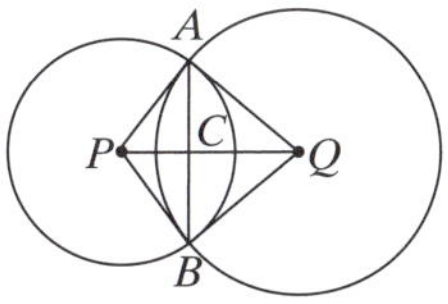

a In ΔPAQ, ΔPBQ

$PA = PB$

(radii of circle, centre P)

$QA = QB$

(radii of circle, centre Q) ✓

$PQ = PQ$

(common side)

$\therefore \Delta PAQ \equiv \Delta PBQ$ SSS ✓

(2 marks)

b In ΔPAC, ΔPBC

$PA = PB$

(radii of circle, centre P)

$\angle APC = \angle BPC$

(corresponding $\angle$s of congruent Δs PAQ, PBQ) ✓

$PC = PC$

(common side)

$\therefore \Delta PAC \equiv \Delta PBC$ SAS ✓

$\therefore AC = BC$

(corresponding sides of congruent triangles) ✓

(3 marks)

c $\angle ACP = \angle BCP$

(corresp. $\angle$s of cong. Δs) ✓

but $\angle ACP + \angle BCP = 180°$

(angles in a straight line)

$\therefore \angle ACP = \angle BCP = 90°$

$\therefore$ AB is perpendicular to PQ. ✓ (2 marks)

4

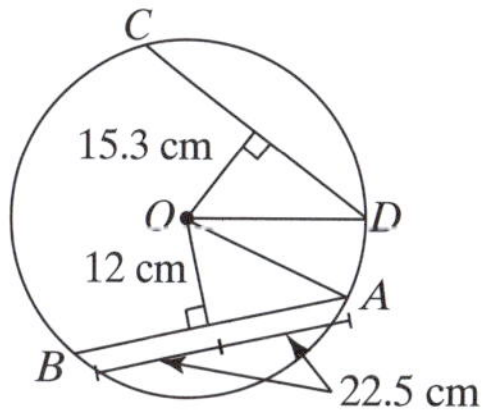

Let M be the centre of chord AB.

$\therefore AM = 22.5$ cm

In ΔOAM,

$OA^2 = 12^2 + 22.5^2$
$= 650.25$

$OA = \sqrt{650.25} \quad (OA > 0)$
$= 25.5$

$\therefore$ the radius of the circle is 25.5 cm ✓

So $OD = 25.5$ cm

Let N be midpoint of chord CD.

In ΔOND,

$ND^2 = 25.5^2 - 15.3^2$
$= 416.16$

$ND = \sqrt{416.16} \quad (ND > 0)$
$= 20.4$ ✓

Now $2 \times 20.4 = 40.8$

So chord $CD = 40.8$ cm long.

Difference $= 45 - 40.8$
$= 4.2$

Chord AB is 4.2 cm longer than chord CD. ✓ (3 marks)

5

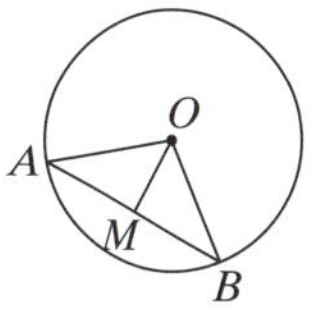

In ΔAOM, ΔBOM

$OA = OB$ (radii of circle) ✓

$OM = OM$ (common side)

$AM = BM$

(M is midpoint of AB) ✓

$\therefore \Delta AOM \equiv \Delta BOM$ SSS ✓

$\therefore \angle AMO = \angle BMO$

(corresponding angles of congruent triangles)

but $\angle AMO + \angle BMO = 180°$

(angles in a straight line)

$\therefore \angle AMO = 180° \div 2$
$= 90°$ ✓

(4 marks)

6

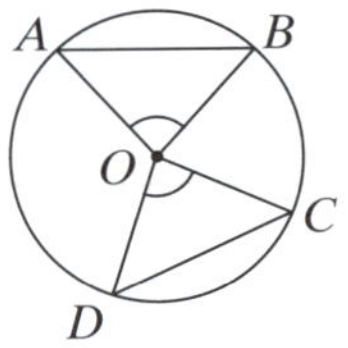

In ΔAOB, ΔDOC
$AO = DO$ (radii of circle)
$\angle AOB = \angle DOC$ (given) ✓
$OB = OC$ (radii of circle)
$\therefore \Delta AOB \equiv \Delta DOC$ SAS ✓
$\therefore AB = DC$
(corresponding sides of congruent triangles) ✓(3 marks)

7

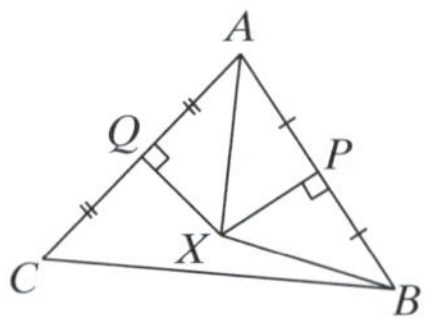

a In ΔAPX, ΔBPX
$AP = BP$ (XP bisects AB)
$\angle APX = \angle BPX$
(both 90°, $XP \perp AB$) ✓
$PX = PX$ (common side)
$\therefore \Delta APX \equiv \Delta BPX$ SAS ✓
(2 marks)

b $XA = XB$
(corresponding sides of congruent Δs) ✓
Similarly $\Delta AQX \equiv \Delta CQX$
and $XA = XC$. ✓
So $XA = XB = XC$
So X is the centre of the circle through A, B and C.✓
(3 marks)

(Total: 30 marks)

ANGLE PROPERTIES, TANGENTS AND SECANTS
SKILLS CHECK PAGE 70

1 **a**

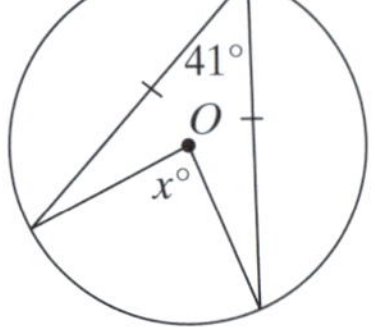

$x = 2 \times 41$
$= 82$
(The angle at the centre is twice the angle at the circumference.)

b

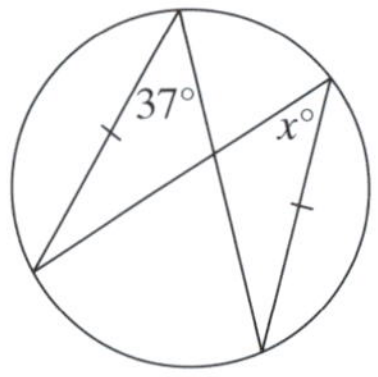

$x = 37$
(Angles in the same segment, standing on the same arc.)

c

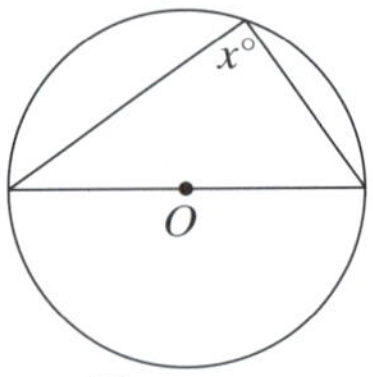

$x = 90$
(Angle in a semi-circle.)

d

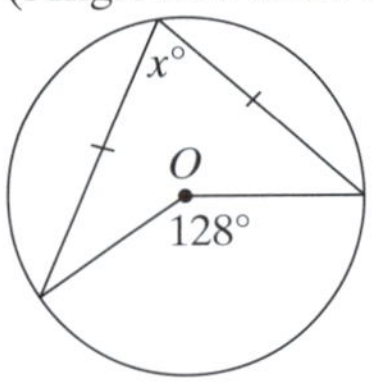

$x = 128 \div 2$
$= 64$
(The angle at the centre is twice the angle at the circumference.)

e

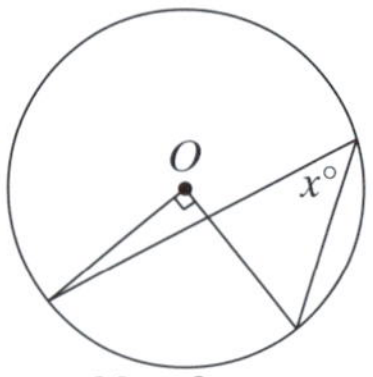

$x = 90 \div 2$
$= 45$
(The angle at the centre is twice the angle at the circumference.)

f

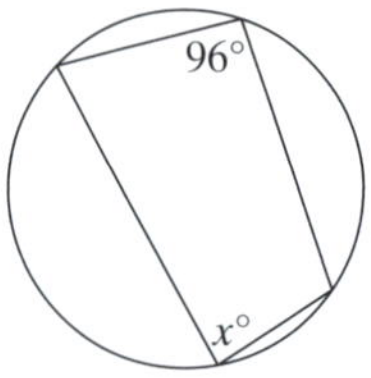

$x = 180 - 96$
$= 84$
(Opposite angles of cyclic quadrilaterals are supplementary.)

g

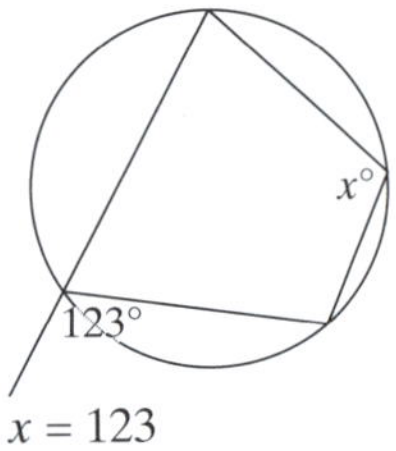

$x = 123$
(An exterior angle of a cyclic quadrilateral is equal to the interior opposite angle.)

h

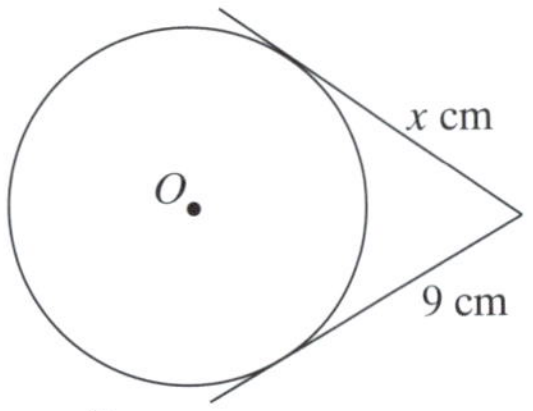

$x = 9$
(Tangents from the same point are equal.)

i

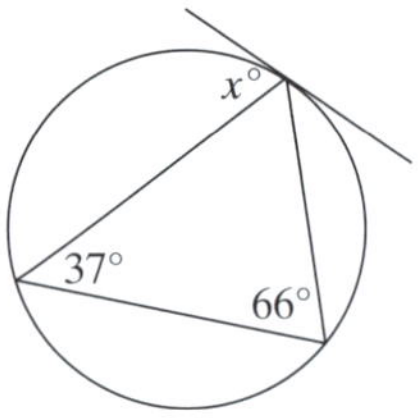

$x = 66$
(The angle between a tangent and a chord is equal to the angle in the alternate segment.)

j

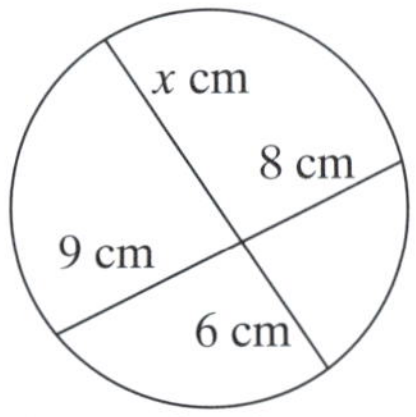

$6 \times x = 9 \times 8$
$6x = 72$
$x = 72 \div 6$
$x = 12$
(The products of the intercepts of two intersecting chords are equal.)

k

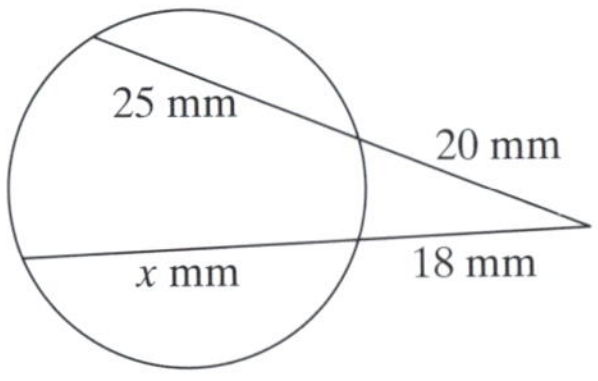

$18(x + 18) = 20 \times 45$
$18x + 324 = 900$
$18x = 576$
$x = 32$

(The products of the intercepts of two intersecting secants are equal.)

I

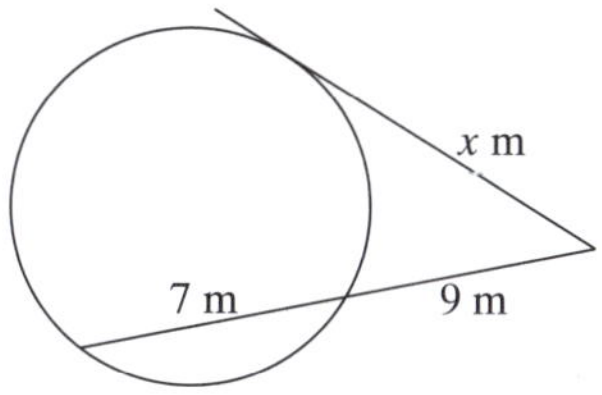

$x^2 = 9 \times (9 + 7)$
$x^2 = 144$
$x = 12 \quad (x > 0)$

(The square of a tangent to a circle from an external point is equal to the product of the intercepts of a secant from the same point.)

2

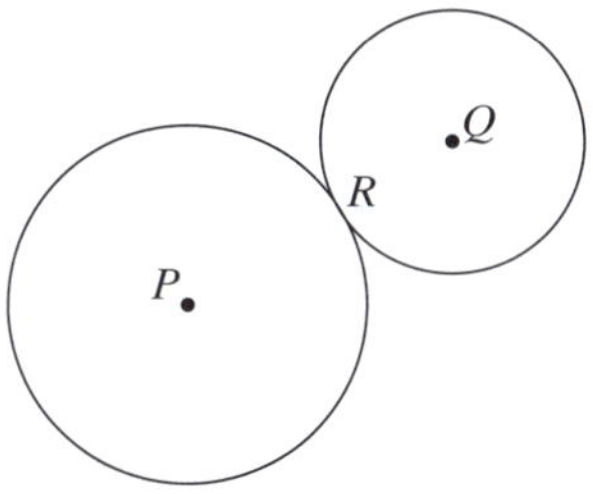

The points P, Q and R lie in a straight line. They are collinear.

ANGLE PROPERTIES, TANGENTS AND SECANTS
INTERMEDIATE TEST PAGE 71

1

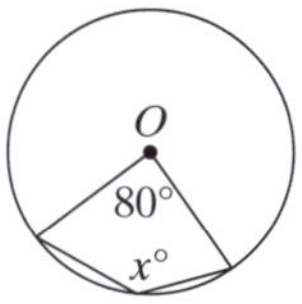

Exterior angle at centre
$= 360° - 80°$
$= 280°$
$2x = 280$
(angle at centre is twice angle at circumference)
$x = 140$
$\therefore$ **[C]** ✓ (1 mark)

2

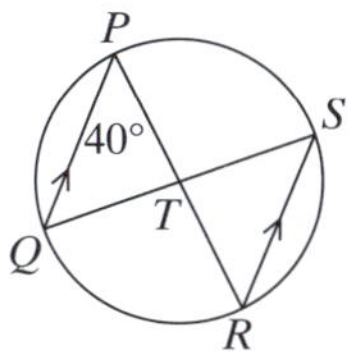

$\angle QSR = \angle QPR$
(angles in same segment standing on same arc)
So $\angle QSR = 40°$
$\angle SRT = \angle QPR$
(alternate, $PQ \parallel SR$)
So $\angle SRT = 40°$
$\angle STR + \angle SRT + \angle TSR = 180°$
(angle sum triangle)
$\angle STR + 40° + 40° = 180°$
$\angle STR = 100°$
$\therefore$ **[C]** ✓ (1 mark)

3

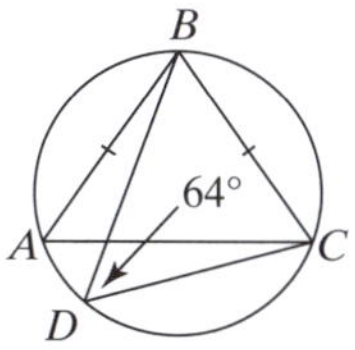

$\angle BAC = \angle BDC$
(angles in the same segment standing on the same arc)
$\angle BAC = 64°$
$\angle BCA = \angle BAC$
(base angles isosceles ΔABC)
$\angle BCA = 64°$
$\angle ABC + \angle BAC + \angle BCA = 180°$
(angle sum of a triangle)
$\angle ABC + 64° + 64° = 180°$
$\angle ABC + 128° = 180°$
$\angle ABC = 52°$
$\therefore$ **[A]** ✓ (1 mark)

4

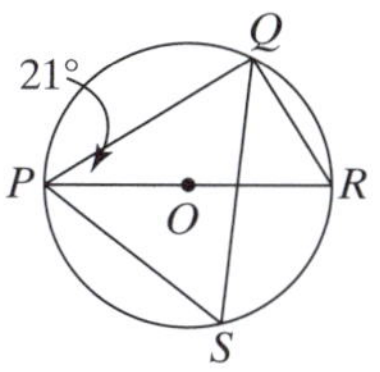

$\angle PQR = 90°$
(angle in a semi-circle) ✓
$\angle PRQ + \angle PQR + \angle QPR = 180°$
(angle sum of a triangle)
$\angle PRQ + 90° + 21° = 180°$
$\angle PRQ + 111° = 180°$
$\angle PRQ = 69°$ ✓
$\angle PSQ = \angle PRQ$
(angles in same segment)
$\therefore \angle PSQ = 69°$ ✓ (3 marks)

5

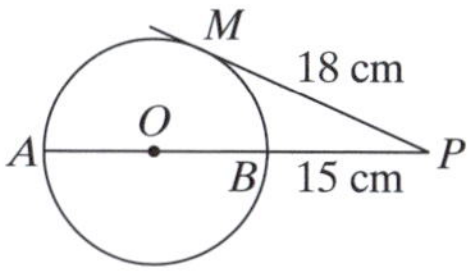

AP is a secant.
$AP \times BP = MP^2$
(product of intercepts of secant = square of tangent)
$AP \times 15 = 18^2$
$AP = \frac{18^2}{15}$
$AP = 21.6$ ✓
The length of AP is 21.6 cm
$AB = 21.6 - 15$
$= 6.6$
The diameter is 6.6 cm ✓
The radius is 3.3 cm. ✓
(3 marks)

6

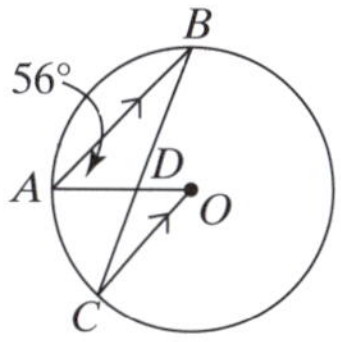

$\angle AOC = \angle BAO$
(alternate, $AB \parallel CO$)
$\angle AOC = 56°$ ✓
$\angle ABC = \frac{1}{2}\angle AOC$
(angle at circumference)
$\therefore \angle ABC = 28°$ ✓
$\angle DCO = \angle ABC$
(alternate, $AB \parallel CO$)
$\therefore \angle DCO = 28°$ ✓ (3 marks)

7

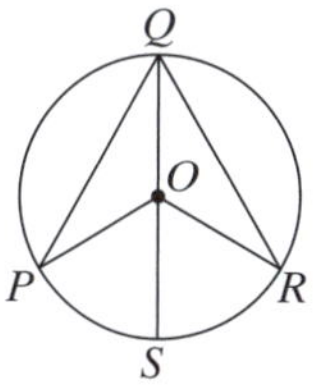

a $\angle PQO = x°$
$OP = OQ$
(radii of circle)
$\therefore \angle OPQ = \angle OQP$
(angles opposite equal sides, isosceles triangle)
$\therefore \angle OPQ = x°$ ✓
$\angle POS = \angle OPQ + \angle POQ$
(exterior angle of a triangle)
$= x° + x°$
$= 2x°$ ✓ (2 marks)

b Similarly if $\angle OQR = y°$
then $\angle ORQ = y°$
and $\angle ROS = 2y°$ ✓
$\angle POS = 2x° + 2y°$
$= 2(x° + y°)$
$= 2\angle POR$ ✓
(2 marks)

c $\angle QPS = 90°$
(angle in a semi-circle) ✓
(1 mark)

8

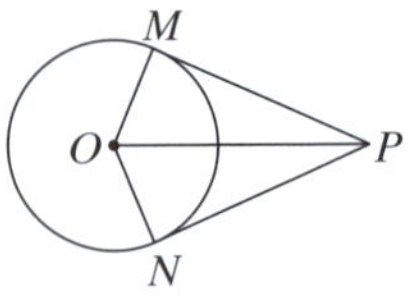

In ΔOMP, ΔONP
$\angle OMP = \angle ONP = 90°$
(angle between tangent and radius) ✓
$OP = OP$ (common side)
$OM = ON$ (radii of circle)
$\therefore \Delta OMP \equiv \Delta ONP$ RHS ✓
$\therefore MP = NP$
(corresponding sides of congruent triangles) ✓
(3 marks)

9

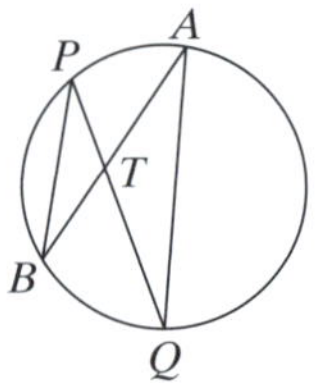

a In ΔAQT, ΔPBT
$\angle QAT = \angle BPT$
($\angle$s in the same segment)
$\angle AQT = \angle PBT$
($\angle$s in the same segment)
$\therefore \Delta AQT \parallel\!| \Delta PBT$
(equiangular) ✓
$\dfrac{AT}{PT} = \dfrac{TQ}{TB}$
(corresponding sides of similar triangles) ✓
$\therefore AT \times TB = PT \times TQ$ ✓
(3 marks)

b $AB = 14$ cm, $PT = 4$ cm, $QT = 12$ cm
Let the length of AT be x cm.
$\therefore TB = (14 - x)$ cm
$AT \times TB = PT \times TQ$
$x(14 - x) = 4 \times 12$ ✓
$14x - x^2 = 48$
$x^2 - 14x + 48 = 0$
$(x - 6)(x - 8) = 0$
$x - 6 = 0$ or $x - 8 = 0$
$x = 6$ or $x = 8$
Now $AT < TB$
$\therefore AT = 6$ cm and
$TB = 8$ cm ✓ (2 marks)

(Total: 25 marks)

ANGLE PROPERTIES, TANGENTS AND SECANTS
ADVANCED TEST PAGE 72

1

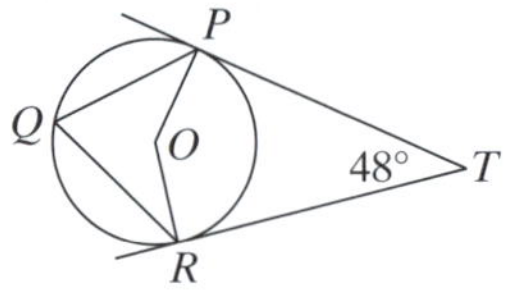

a Let O be the centre of the circle. Join OP and OR.
$\angle OPT = 90°$
(tangent $\perp$ radius)
$\angle ORT = 90°$
(tangent $\perp$ radius) ✓
$\angle POR + \angle OPR + \angle PTR + \angle ORT = 360°$
($\angle$ sum of a quadrilateral)
$\angle POR + 90° + 48° + 90° = 360°$
$\angle POR = 132°$ ✓
$\angle PQR = 132° \div 2$
(angle at centre is twice angle at circumference)
$= 66°$ ✓ (3 marks)

b No, it is not possible to draw a circle through P, Q, R and T.
$\angle PQR + \angle PTR \neq 180°$
So $PTRQ$ is not a cyclic quadrilateral. ✓ (1 mark)

2

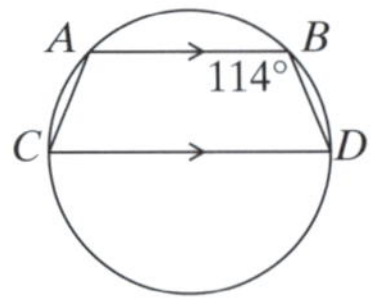

$\angle CBD = 90°$
(angle in semi-circle) ✓
$\angle ABC = 114° - 90°$
$= 24°$
$\angle BCD = \angle ABC$
(alternate, $AB \parallel CD$)
So $\angle BCD = 24°$ ✓
$\angle ACD + \angle ABD = 180°$
(opposite angles of cyclic quadrilateral)
$\angle ACD + 114° = 180°$
$\angle ACD = 66°$ ✓
$\angle ACB = 66° - 24°$
$= 42°$ ✓ (4 marks)

3

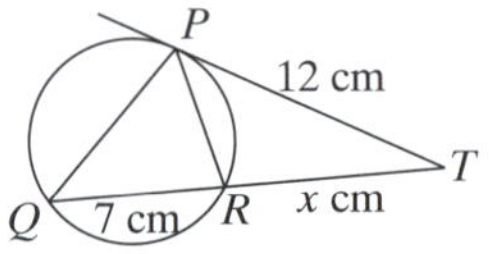

a $\angle TPR = \angle TQP$ because the angle between a tangent and a chord is equal to the angle in the alternate segment. ✓ (1 mark)

b Triangles TPR and TQP are similar because they are equiangular.
$\angle T$ is common to both triangles and, from part **a**, $\angle TPR = \angle TQP$. ✓ (1 mark)

c Let the length of RT be x cm.
Now, the square of the tangent from an external point equals the product of the intercepts of the secant from that point.
So $x(x + 7) = 12^2$ ✓
$x^2 + 7x = 144$
$x^2 + 7x - 144 = 0$
$(x + 16)(x - 9) = 0$
$x = -16$ or $x = 9$ ✓
But $x > 0$
So the length of RT is 9 cm. ✓ (3 marks)

4

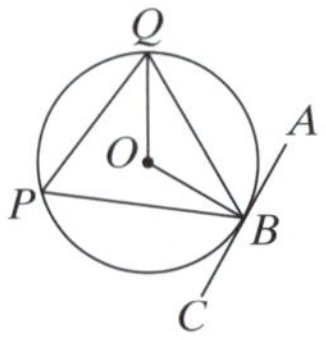

Let $\angle QPB = x°$
$\angle QOB = 2x°$
(angle at centre) ✓
$OQ = OB$
(radii of circle)
$\angle OBQ = \angle OQB$
(angles opposite equal sides, isosceles triangle)
$\angle OBQ + \angle OQB + \angle QOB = 180°$
(angle sum of a triangle)
$2\angle OBQ + 2x° = 180°$
$2\angle OBQ = 180° - 2x°$
$\angle OBQ = 90° - x°$ ✓

$\angle OBA = 90°$
(angle between tangent and radius)
$\angle QBA = 90° - (90° - x°)$
$= 90° - 90° + x°$
$= x°$
$\therefore \angle QBA = \angle QPB$ ✓ (3 marks)

5

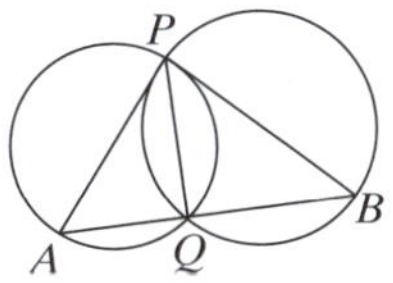

Join Q to A, B and P.
$\angle AQP = 90°$
(angle in a semicircle)
$\angle PQB = 90°$
(angle in a semicircle) ✓
$\angle AQB = \angle AQP + \angle PQB$
$= 90° + 90°$
$= 180°$
$\therefore AQB$ is a straight line
$\therefore Q$ lies on AB. ✓ (2 marks)

6

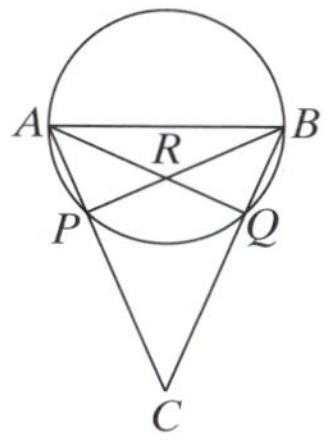

a $\angle APB = 90°$
(angle in a semicircle) ✓
$\angle APC = 180°$
(AC is a straight line)
$\angle BPC = 180° - 90°$
$= 90°$ ✓ (2 marks)

b Similarly $\angle AQB = \angle AQC$
$= 90°$
$\angle BPC + \angle AQD = 90° + 90°$
$= 180°$
$\therefore PRQC$ is a cyclic quadrilateral
(opposite angles are supplementary) ✓ (1 mark)

7

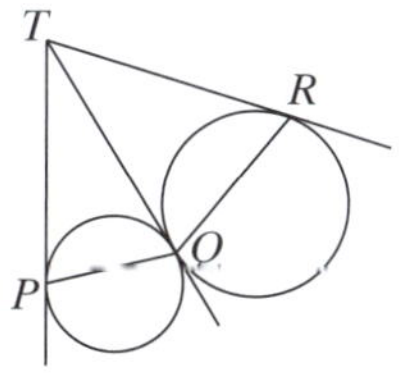

a $TR = TQ$
(tangents from same point are equal) ✓
But $TP = TQ$
(tangents from same point are equal)
$\therefore TR = TP$ ✓ (2 marks)

b $\angle PTR = 90°$
Let $\angle PTQ = x°$
$\therefore \angle QTR = 90° - x°$
$\angle TPQ = \angle TQP$
(base angles isos ΔTPQ) ✓
$\angle TPQ + \angle TQP + \angle PTQ$
$= 180°$
(angle sum of ΔPTQ)
$2\angle TQP + x° = 180°$
$2\angle TQP = 180° - x°$
$\angle TQP = 90° - \frac{x°}{2}$ ✓
$\angle TRQ = \angle TQR$
(base angles isos. ΔTRQ)
$\angle TRQ + \angle TQR$
$+ \angle QTR = 180°$
(angle sum ΔTRQ)
$2\angle TQR + 90° - x° = 180°$
$2\angle TQR = 90° + x°$
$\angle TQR = 45° + \frac{x°}{2}$ ✓
$\angle PQR = 90° - \frac{x°}{2} + 45° + \frac{x°}{2}$
$= 135°$ ✓ (4 marks)

8

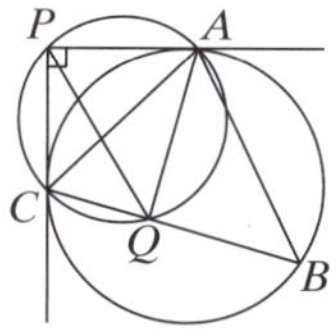

a $\angle PAC = \angle ABC$
(angle between tangent and chord = angle in alt. segment) ✓
But $\angle PAC = \angle PQC$
(angles at circumference standing on same arc)
$\therefore \angle ABC = \angle PQC$ ✓
(2 marks)

b $\angle APC + \angle AQC = 180°$
(opposite angles of cyclic quadrilateral)
but $\angle APC = 90°$ ($AP \perp PC$)
$\angle AQC = 90°$
$\therefore AQ \perp CB$ ✓ (1 mark)

9

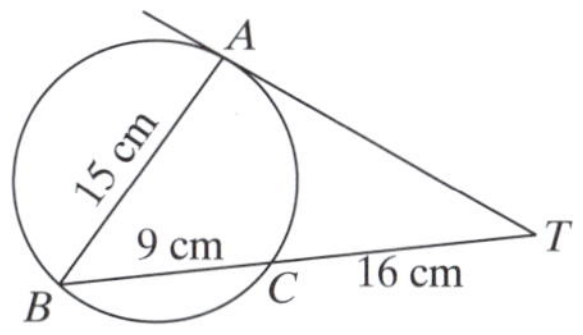

a The square of the tangent from an external point = the products of the intercepts of the secant from that point.
So $AT^2 = 16 \times (16 + 9)$
$= 400$
$AT = 20$ $(AT > 0)$ ✓
$BT^2 = (9 + 16)^2$
$= 625$
$AB^2 + AT^2 = 15^2 + 20^2$
$= 625$
$= BT^2$
$\therefore \angle ABT$ is right-angled at A. ✓
$\therefore AT \perp AB$
$\therefore AB$ is a diameter of the circle. ✓ (3 marks)

b $\angle ACB = 90°$
(angle in semicircle) ✓
$AC^2 = 15^2 - 9^2$
$= 144$
$\therefore AC = 12$ cm $(AC > 0)$ ✓
(2 marks)

(Total: 35 marks)

PROBLEMS IN TWO AND THREE DIMENSIONS

SKILLS CHECK PAGE 74

1 **a** $x^2 = 5^2 + 9^2$
$= 106$
$x = \sqrt{106}$ $(x > 0)$
$= 10.295\,6301\ldots$
$= 10.3$ (1 d.p.)

b $x^2 = 23^2 - 8^2$
$= 465$
$x = \sqrt{465}$ $(x > 0)$
$= 21.563\,858\,65\ldots$
$= 21.6$ (1 d.p.)

c $x^2 = 72^2 - 56^2$
$= 2048$
$x = \sqrt{2048}$ $(x > 0)$
$= 45.254\,83\ldots$
$= 45.3$ (1 d.p.)

d $x^2 = 32^2 + 34^2$
$= 2180$
$x = \sqrt{2180}$ $(x > 0)$
$= 46.690\,4701\ldots$
$= 46.7$ (1 d.p.)

2 **a** $\tan A = \frac{a}{b}$

b $\sin B = \frac{b}{c}$

c $\cos B = \frac{a}{c}$

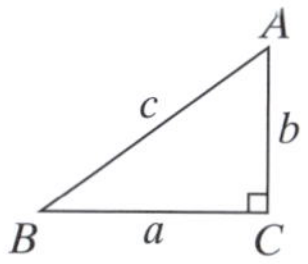

d $\cos A = \frac{b}{c}$

e $\sin A = \frac{a}{c}$

f $\tan B = \frac{b}{a}$

3 **a** $\tan 58° = 1.600\,334\,529\ldots$
$= 1.600$ (3 d.p.)

b $\sin 24° = 0.406\,736\,643\ldots$
$= 0.407$ (3 d.p.)

c $\cos 72°35' = 0.299\,318\,356\ldots$
$= 0.299$ (3 d.p.)

4 $\sin A = 0.375$
$A = 22.024\,312\,84\ldots°$
$= 22°$ (nearest degree)
[using 'shift sin 0.375']

5 **a** $\cos B = 0.2197$
$B = 77.308\,5868\ldots°$
$= 77°19'$
(nearest minute)
[There are 60 seconds in a minute, so if there are 30 seconds or more, the minutes must be rounded up.]

b $\tan B = 3$
$B = 71.565\,051\,18\ldots°$
$= 71°34'$
(nearest minute)

6 **a**

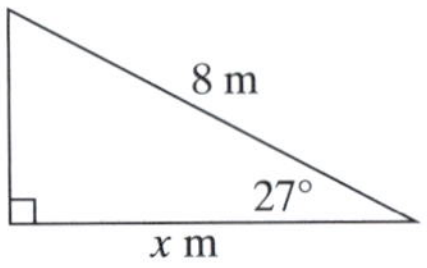

$\cos 27° = \frac{x}{8}$
$x = 8\cos 27°$
$= 7.128\,052\,194\ldots$
The side length is 7.1 m, correct to one decimal place.

b

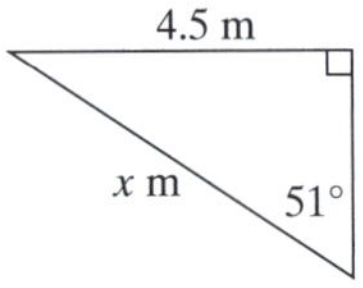

$\sin 51° = \frac{4.5}{x}$
$x = \frac{4.5}{\sin 51°}$
$= 5.790\,418\,046\ldots$
The length of the side is 5.79 m correct to two decimal places.

7

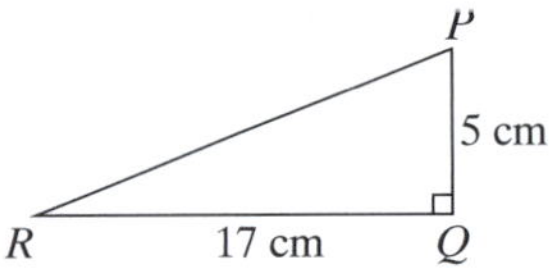

$\tan P = \frac{17}{5}$
$P = 73.610\,459\,67\ldots°$
$= 74°$ (nearest degree)

8

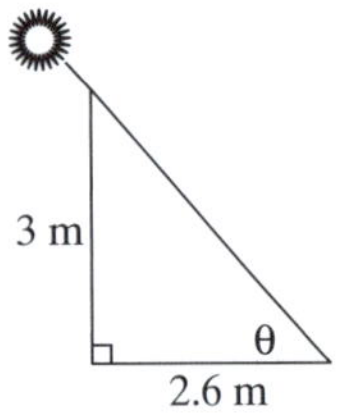

$\tan\theta = \frac{3}{2.6}$
$\theta = 49.085\,616\,78\ldots°$
$= 49°$ (nearest degree)
The angle of elevation of the sun is 49°.

9 **a** 090° is due east
[90° from north in a clockwise direction]

b 270° is due west

c 180° is due south

d 045° is north-east
[45° is half of 90°, halfway between north and east]

e 135° is south-east
[halfway between south and east]

10 **a**

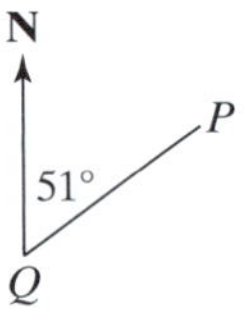

051°
[51° from north in a clockwise direction]

b

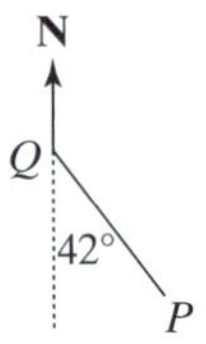

Bearing $= (180 - 42)°$
$= 138°$

c

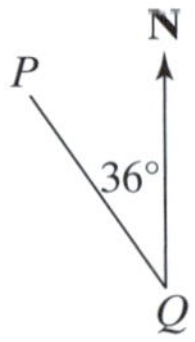

Bearing $= (360 - 36)°$
$= 324°$

PROBLEMS IN TWO AND THREE DIMENSIONS
INTERMEDIATE TEST PAGE 75

1 All the triangles have angles of 40°, 50° and 90°.
In B:

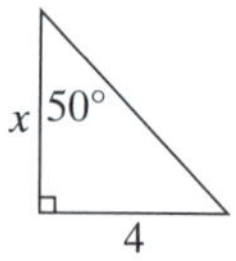

$\tan 40° = \frac{x}{4}$
$x = 4\tan 40°$
[In A and C, $\tan 40° = \frac{4}{x}$, so $x = \frac{4}{\tan 40°}$. In D, $\sin 40° = \frac{x}{4}$, so $x = 4\sin 40°$.]
∴ [B] ✓ (1 mark)

2

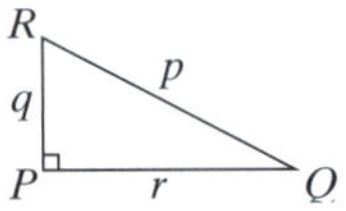

$\cos R = \frac{q}{p}$
∴ [C] ✓ (1 mark)

3

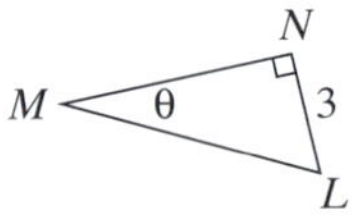

NL is opposite the angle.
LM is the hypotenuse.
$\sin\theta = \frac{3}{LM}$
$LM = \frac{3}{\sin\theta}$
∴ [C] ✓ (1 mark)

4 **a**

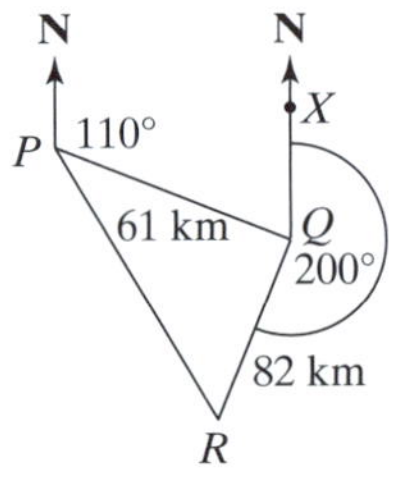

✓ (1 mark)

b Let X be a point due north of Q.

$\angle XQP = 180° - 110°$

(co-interior $\angle$s, $\parallel$ lines)

$\angle XQP = 70°$

$\angle PQR = 360° - 200° - 70°$

(angles in a revolution)

$\angle PQR = 90°$ ✓ (1 mark)

c $\tan P = \dfrac{82}{61}$ ✓

$P = 53.354\,350\,84\ldots°$

$= 53°$

(nearest degree)

$\angle RPQ = 53°$ to the nearest whole degree. ✓ (2 marks)

d

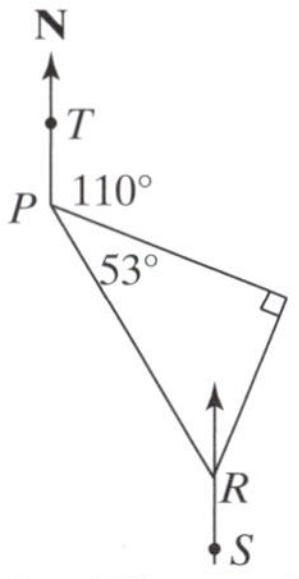

Let T be a point due north of P and let S be a point due south of R.

$\angle TPR = 110° + 53°$

$= 163°$

$\angle PRS = 163°$

(alternate angles, $\parallel$ lines)

The bearing of P from R is $(180 + 163)° = 343°$ ✓

(1 mark)

5

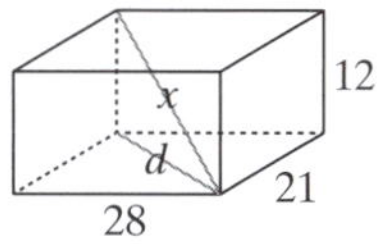

Let d cm be the length of the diagonal of the base of the box.

$d^2 = 28^2 + 21^2$

$= 1225$

$d = 35 \quad (d > 0)$ ✓

Let x cm be the length of the diagonal of the box.

$x^2 = 35^2 + 12^2$

$= 1369$

$= 37 \quad (x > 0)$ ✓

No, the cane will not fit as it is 2 cm longer than the diagonal of the box. ✓ (3 marks)

6 Let θ be the angle between the base and the side face.

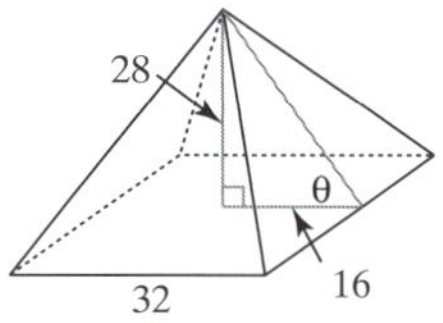

$\tan\theta = \dfrac{28}{16}$ ✓

$\theta = 60.255\,1187\ldots°$

∴ the angle between the base and the triangular face is 60° to the nearest degree. ✓

(2 marks)

7 a

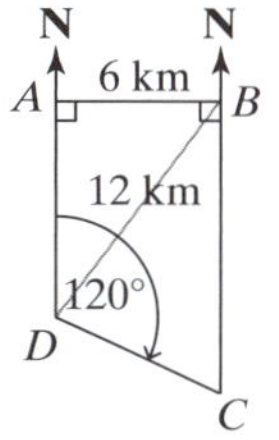

✓ (1 mark)

b In ΔABD,

$\sin D = \dfrac{6}{12}$

$D = \sin^{-1}\left(\dfrac{6}{12}\right)$

$= 30°$ ✓

$\angle BDC = 120° - 30°$

$= 90°$ ✓ (2 marks)

c $\angle DBC = \angle ADB$

(alternate, parallel lines)

∴ $\angle DBC = 30°$ ✓ (1 mark)

d In ΔBDC,

$\cos 30° = \dfrac{12}{BC}$

$BC = \dfrac{12}{\cos 30°}$ ✓

$= 13.856\,406\ldots$

$= 13.9$ km

(nearest 100 m) ✓

(2 marks)

e Required bearing

$= (180 + 30)°$

$= 210°$ ✓ (1 mark)

8

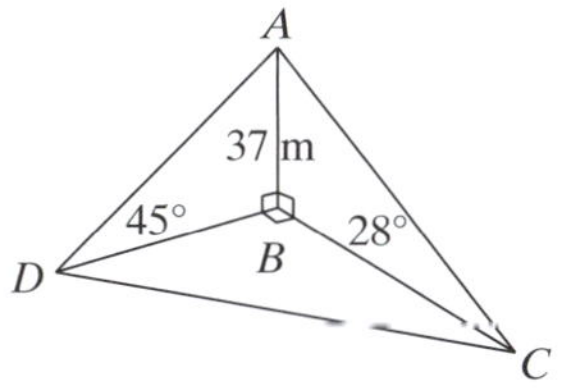

a ΔABD is a right-angled isosceles triangle.

So $BD = AB$

$= 37$ m ✓ (1 mark)

b In ΔABC,

$\tan 28° = \dfrac{37}{BC}$

$BC = \dfrac{37}{\tan 28°}$ ✓

$= 69.586879\ldots$

$= 69.6$ m (1 d.p.) ✓

(2 marks)

c $CD^2 = BD^2 + BC^2$

$= 37^2 + (69.586\,879\ldots)^2$

$= 6211.333\,75\ldots$ ✓

$CD = 78.812\,015\ldots \quad (CD > 0)$

$= 78.8$ m (1 d.p.) ✓

(2 marks)

(Total: 25 marks)

PROBLEMS IN TWO AND THREE DIMENSIONS
ADVANCED TEST

PAGE 76

1

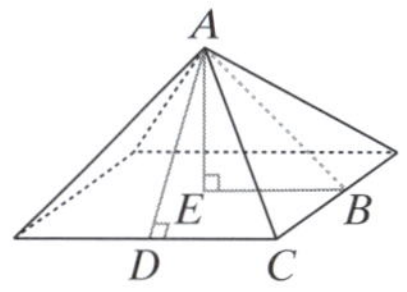

a $EB^2 = AB^2 - AE^2$

$= 25^2 - 24^2$

$= 49$

$EB = 7$ cm ✓ (1 mark)

b $AD^2 = ED^2 + AE^2$

$= 10^2 + 24^2$

$= 676$

$AD = 26$ cm ✓ (1 mark)

c $AC^2 = AD^2 + DC^2$

$= AD^2 + EB^2$

$= 676 + 49$

$= 725$ ✓

$AC = \sqrt{725} \quad (AC > 0)$

$= 5\sqrt{29}$ cm ✓ (2 marks)

2

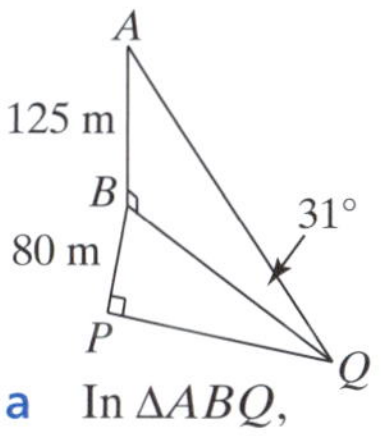

a In ΔABQ,

$\tan 31° = \dfrac{125}{BQ}$

$BQ = \dfrac{125}{\tan 31°}$ ✓

$= 208.034\,93\,161\ldots$

$= 208$ m

(nearest metre) ✓

(2 marks)

b $PQ^2 - (208.034\,93\ldots)^2 - 80^2$
$= 36\,878.534\ldots$
$PQ = 192.037\,84\ldots\ (PQ > 0)$
$= 192$ m
(nearest metre) ✓
(1 mark)

c In ΔBPQ,
$\sin Q = \dfrac{80}{208.034\,93\ldots}$
$Q = 22.615\,8559\ldots°$
$= 23°$
(nearest degree) ✓
Now P is due west of Q.
So the bearing of P from Q is 270°. ✓
The bearing of B from Q is $(270 + 23)° = 293°$ ✓
(3 marks)

3 **a**

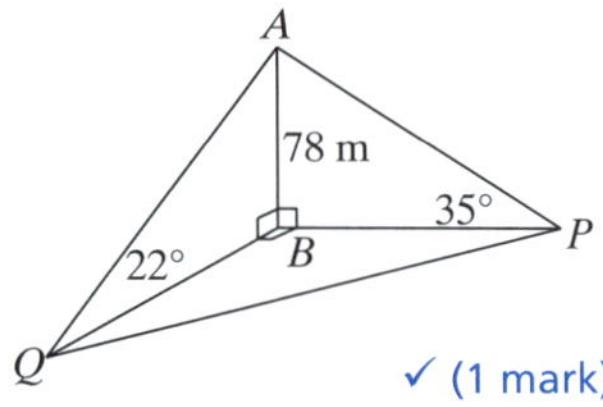

✓ (1 mark)

b In ΔABP,
$\tan 35° = \dfrac{78}{BP}$
$BP = \dfrac{78}{\tan 35°}$ ✓
$= 111.395\,544\ldots$
$= 111.4$ (1 d.p)
The length of BP is 111.4 m to one decimal place. ✓
(2 marks)

c In ΔABQ,
$\tan 22° = \dfrac{78}{BQ}$
$BQ = \dfrac{78}{\tan 22°}$ ✓
$= 193.056774\ldots$
$= 193.1$ (1 d.p)
The length of BQ is 193.1 m to one decimal place. ✓
(2 marks)

d In ΔQBP,
$PQ^2 = BP^2 + BQ^2$
$= (111.395\,544\ldots)^2 + (193.056\,774\ldots)^2$
$= 49\,679.885\,55\ldots$
$PQ = 222.8898\ldots\ (PQ > 0)$
$= 223$ (nearest unit)
∴ the distance from P to Q is 223 m to the nearest metre. ✓
(1 mark)

4

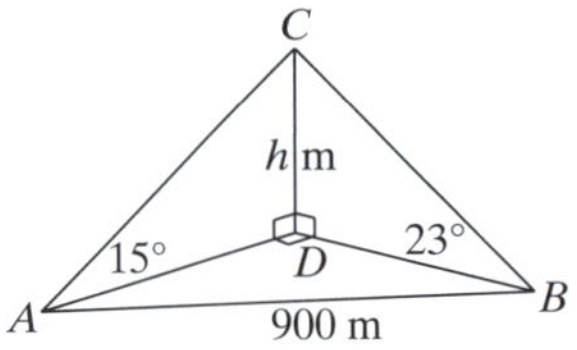

a In ΔCBD,
$\tan 23° = \dfrac{h}{BD}$
$BD = \dfrac{h}{\tan 23°}$ ✓ (1 mark)

b In ΔCAD,
$\tan 15° = \dfrac{h}{AD}$
$AD = \dfrac{h}{\tan 15°}$ ✓ (1 mark)

c In ΔABD,
$AB^2 = AD^2 + BD^2$
$900^2 = \left(\dfrac{h}{\tan 15°}\right)^2 + \left(\dfrac{h}{\tan 23°}\right)^2$ ✓
$= \dfrac{h^2}{(\tan 15°)^2} + \dfrac{h^2}{(\tan 23°)^2}$
$= h^2\left(\dfrac{1}{(\tan 15°)^2} + \dfrac{1}{(\tan 23°)^2}\right)$
$h^2 = 900^2 \div \left(\dfrac{1}{(\tan 15°)^2} + \dfrac{1}{(\tan 23°)^2}\right)$ ✓
$= 41\,584.8583\ldots$
$h = 203.923\,658\ldots\ (h > 0)$
$= 204$ (nearest unit) ✓
(3 marks)

5

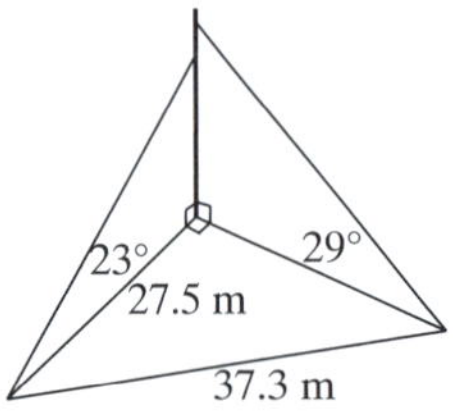

a Let x m be the length of the lower wire.
$\cos 23° = \dfrac{27.5}{x}$
$x = \dfrac{27.5}{\cos 23°}$ ✓
$= 29.874\,9103\ldots$
The lower wire is 29.9 m long, to one decimal place. ✓
(2 marks)

b Let h m be the height at which the lower wire is attached.
$\tan 23° = \dfrac{h}{27.5}$
$h = 27.5 \times \tan 23°$ ✓
$= 11.673\,057\ldots$
The lower wire is attached to the post at a height of 11.7 m above the ground, to one decimal place. ✓
(2 marks)

c Let y m be the distance from the base of the pole to the bottom of the higher wire.
$y^2 = 37.3^2 - 27.5^2$
$= 635.04$
$y = 25.2\quad (y > 0)$
The higher wire is attached to the ground at a point 25.2 m from the base of the pole. ✓ (1 mark)

d Let H m be the height at which the higher wire is attached to the pole.
$\tan 29° = \dfrac{H}{25.2}$
$H = 25.2 \times \tan 29°$
$= 13.968\,588\ldots$
The second wire is attached at a height of 14.0 m above the ground, to one decimal place. ✓
Difference
$= 13.968\,588\ldots - 11.673\,057\ldots$
$= 2.295\,530\,65\ldots$
$= 2.3$ m (1 d.p.)
The higher wire is 2.3 m higher on the pole. ✓
(2 marks)

e Let z m be the length of the higher wire.
$\cos 29° = \dfrac{25.2}{z}$
$z = \dfrac{25.2}{\cos 29°}$
$= 28.812\,5225\ldots$
The higher wire is 28.8 m long to one decimal place. ✓
Difference in lengths
$= 29.8749103\ldots - 28.8125225\ldots$
$= 1.0623878\ldots$
$= 1.06$ (2 d.p.)
The lower wire is 1.06 m, to the nearest centimetre, longer than the higher one. ✓ (2 marks)

(Total: 30 marks)

COMPLEMENTARY RATIOS, EXACT VALUES & THE UNIT CIRCLE
SKILLS CHECK PAGE 78

1 **a**

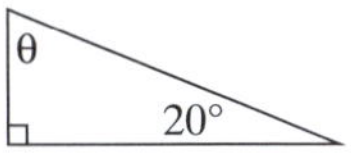

$\theta = 90° - 20°$
$= 70°$
$[\theta + 20° + 90° = 180°$ angle sum of a Δ]

b

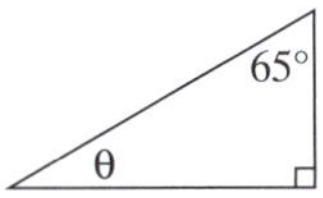

$\theta = 90° - 65°$
$= 25°$

c

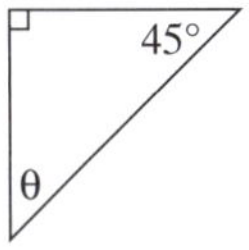

$\theta = 90° - 45°$
$= 45°$

2 **a** The complement of 20°
$= 90° - 20°$
$= 70°$

b The complement of 75°
$= 90° - 75°$
$= 15°$

c The complement of 42°
$= 90° - 42°$
$= 48°$

3 **a** $\cos 20° = 0.939\,692\,62\ldots$
$= 0.940$ (3 d.p.)

b $\sin 70° = 0.939\,692\,62$
$= 0.940$ (3 d.p.)

c $\sin 38° = 0.615\,661\,475\ldots$
$= 0.616$ (3 d.p.)

d $\cos 52° = 0.615\,661\,475\ldots$
$= 0.616$ (3 d.p.)

4 **a** $\sin 30° = \cos(90 - 30)°$
$= \cos 60°$

b $\cos 18° = \sin(90 - 18)°$
$= \sin 72°$

c $\sin 45° = \cos(90 - 45)°$
$= \cos 45°$

5 $\sin\theta = 0.6, \cos\theta = 0.8$
$\tan\theta = \dfrac{\sin\theta}{\cos\theta}$
$= \dfrac{0.6}{0.8}$
$= 0.75$

6 **a** $\sin 30° = \dfrac{1}{2}$

b $\tan 45° = 1$

c $\cos 60° = \dfrac{1}{2}$

d $\sin 60° = \dfrac{\sqrt{3}}{2}$

e $\tan 30° = \dfrac{1}{\sqrt{3}}$

f $\cos 45° = \dfrac{1}{\sqrt{2}}$

g $\tan 60° = \sqrt{3}$

h $\sin 45° = \dfrac{1}{\sqrt{2}}$

i $\cos 30° = \dfrac{\sqrt{3}}{2}$

7 **a** The statement is true.
$\sin 130° = \sin(180 - 130)°$
$= \sin 50°$

b The statement is false.
$\cos 110° = -\cos(180 - 110)°$
$= -\cos 70°$

c The statement is true.
$\tan 175° = -\tan(180 - 175)°$
$= -\tan 5°$

8 **a** $\tan 102° = -4.704\,630\,11\ldots$
$= -4.7046$ (4 d.p.)

b $\sin 156° = 0.406\,736\,643\ldots$
$= 0.4067$ (4 d.p.)

c $\cos 173° = -0.992\,546\,151\ldots$
$= -0.9925$ (4 d.p.)

9 **a** $\sin A = 0.237$
$A = 13.709\,545\ldots°$
[by calculator]
$= 14°$ (nearest degree)
But $\angle A$ is obtuse
$\therefore \angle A = 180° - 14°$
$= 166°$

b $\cos A = -0.5$
$A = 120°$ [by calculator]

c If $\tan\theta = 1.2345$
$\theta = 50.9909861\ldots°$
[by calculator]
$= 51°$
(nearest degree)
But $\angle A$ is obtuse,
$A = (180 - 51)°$
$= 129°$ (nearest degree)

10 **a** $\cos\theta = \dfrac{1}{\sqrt{2}}$
$\theta = 45°$

b $\tan\theta - 1 = 0$
$\tan\theta = 1$
$\theta = 45°$

c $2\sin\theta = \sqrt{3}$
$\sin\theta = \dfrac{\sqrt{3}}{2}$
$\theta = 60°$

11 **a** $\sin\theta = 0.5$
$\theta = 30°$ or $180° - 30°$
$\theta = 30°$ or $150°$

b $\sin\theta = 0.7654$
$\theta = 49°57'$ or $180 - 49°57'$
$\theta = 49°57'$ or $130°3'$
[by calculator to the nearest minute]

COMPLEMENTARY RATIOS, EXACT VALUES & THE UNIT CIRCLE
INTERMEDIATE TEST PAGE 79

1

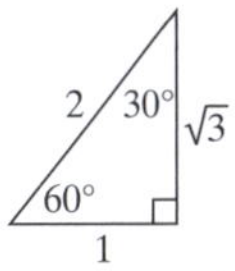

$\cos 60° = \dfrac{1}{2}$
$\therefore$ **[A]** ✓ (1 mark)

2 $\sin 55° = \sin(180 - 55)°$
$= \sin 125°$
Statement I is true.
$\sin 55° = \cos(90 - 55)°$
$= \cos 35°$
Statement II is true.
Both I and II.
$\therefore$ **[C]** ✓ (1 mark)

3 $\sin 30° = 0.5$
$\sin 150° = 0.5$
Either statement I or II is correct.
$\therefore$ **[C]** ✓ (1 mark)

4 $\tan 150° = \tan(180 - 150)°$
$= -\tan 30°$
$= -\dfrac{1}{\sqrt{3}}$

$\therefore$ **[D]** ✓ (1 mark)

5 $\sin\theta = 0.66, \cos\theta = 0.75$

$\tan\theta = \dfrac{\sin\theta}{\cos\theta}$
$= \dfrac{0.66}{0.75}$
$= 0.88$

$\therefore$ **[C]** ✓ (1 mark)

6 $\sin\theta$ and $\cos\theta$ are both negative in the third quadrant.
[Both x and y are negative.]
$\therefore$ **[C]** ✓ (1 mark)

7 Let the length of side BC be x m.

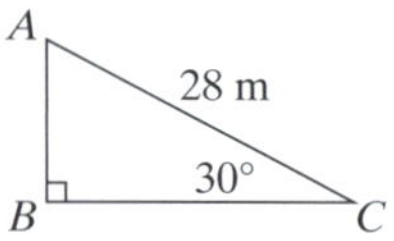

$\cos 30° = \dfrac{x}{28}$ ✓

$x = 28\cos 30°$
$= 28 \times \dfrac{\sqrt{3}}{2}$ ✓
$= 14\sqrt{3}$

The exact length of BC is $14\sqrt{3}$ m. ✓ (3 marks)

8 $\sin A = \dfrac{7}{9}$

If A is acute:
$A = 51.057\,558\,73\ldots°$
$= 51°$ (nearest degree) ✓
but A is obtuse
$A = 180° - 51°$ ✓
$= 129°$ (nearest degree) ✓
(3 marks)

9 **a**

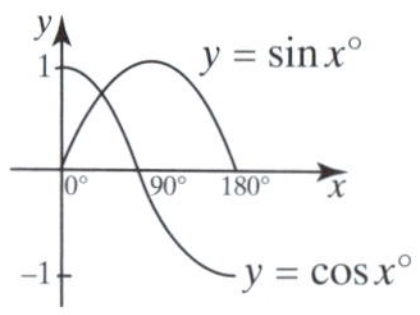

✓✓✓✓ (4 marks)

b From the graph,
$\sin x = \cos x$ when $x = 45°$. ✓
(1 mark)

10

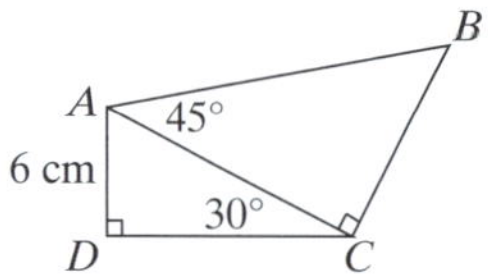

a In ΔADC

$\sin 30° = \dfrac{6}{AC}$ ✓

$AC = \dfrac{6}{\sin 30°}$
$= \dfrac{6}{\frac{1}{2}}$
$= 12$

The length of AC is 12 cm. ✓ (2 marks)

b In ΔABC, $\cos 45° = \dfrac{AC}{AB}$

$AB = \dfrac{AC}{\cos 45°}$ ✓
$= \dfrac{12}{\frac{1}{\sqrt{2}}}$
$= 12 \times \dfrac{\sqrt{2}}{1}$
$= 12\sqrt{2}$

The length of AB is $12\sqrt{2}$ cm. ✓ (2 marks)

11 $\sqrt{2}\sin x° = 1$

$\sin x° = \dfrac{1}{\sqrt{2}}$

Now $\sin 45° = \dfrac{1}{\sqrt{2}}$

sin is positive in first and second quadrants
$\therefore x = 45$ or $(180 - 45)$
$x = 45$ or 135 ✓✓
(2 marks)

12 $y = 5 - x$
$= -x + 5$
gradient: $m = -1$, $m = \tan\theta$

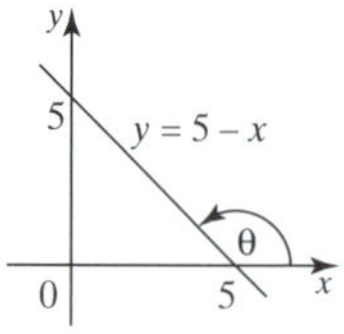

$\therefore \tan\theta = -1$ ✓
Now $\tan 45° = 1$
But θ is an obtuse angle.
$\therefore \theta = (180 - 45)°$
$\theta = 135°$ ✓ (2 marks)

(Total: 25 marks)

COMPLEMENTARY RATIOS, EXACT VALUES & THE UNIT CIRCLE

ADVANCED TEST PAGE 80

1 $\cos 210° = \cos(180 + 30)°$
$= -\cos 30°$ ✓
$= -\dfrac{\sqrt{3}}{2}$ ✓ (2 marks)

2 **a** $\sin A = 0.654$
$A = \sin^{-1} 0.654$
$= 40.843\,868\ldots°$
$= 40°51'$
(nearest minute) ✓
(1 mark)

b If A is obtuse,
$A = 180° - 40°51'$
$= 139°9'$ ✓ (1 mark)

3 **a** $\sin(180° - \theta) = \sin\theta$
$= y$ ✓ (1 mark)

b $\cos(180° - \theta) = -\cos\theta$
$= -x$ ✓
(1 mark)

c $\tan\theta = \dfrac{\sin\theta}{\cos\theta}$
$= \dfrac{y}{x}$ ✓ (1 mark)

d $\cos(90° - \theta) = \sin\theta$
$= y$ ✓ (1 mark)

e $\sin(360° - \theta) = -\sin\theta$
$= -y$ ✓
(1 mark)

f $\cos(180° + \theta) = -\cos\theta$
$= -x$ ✓
(1 mark)

4

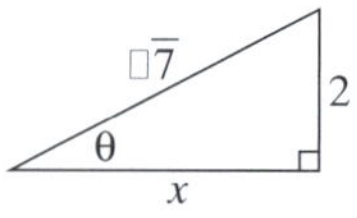

$\sin\theta = \dfrac{2}{\sqrt{7}}$

$x^2 = (\sqrt{7})^2 - 2^2$
$= 3$
$x = \sqrt{3}$ $(x > 0)$ ✓

$\tan\theta = \dfrac{2}{\sqrt{3}}$ ✓ (2 marks)

5 $3\cos\theta = \cos\theta - 1$
$2\cos\theta = -1$

$\cos\theta = -\dfrac{1}{2}$ ✓

Now $\cos 60° = \dfrac{1}{2}$

cos is negative in 2nd and 3rd quadrants

$\therefore$ Either $\theta = 180° - 60°$

$\theta = 120°$ ✓

Or $\theta = 180° + 60°$

$\theta = 240°$ ✓ (3 marks)

6

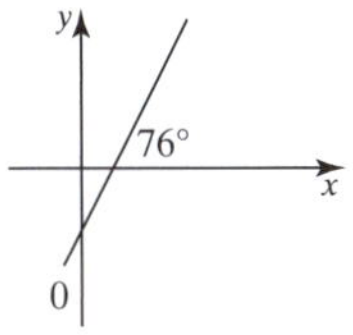

$m = \tan\theta$

$= \tan 76°$

$= 4.010\,78\ldots$

≈ 4 ✓ (1 mark)

7 $\tan\theta = -\frac{1}{4}$

Now $\tan^{-1}\left(\frac{1}{4}\right) = 14.036\,24\ldots°$

$= 14°$

(nearest degree)

If θ is obtuse,

$\theta = 180° - 14°$

$= 166°$ ✓ (1 mark)

8

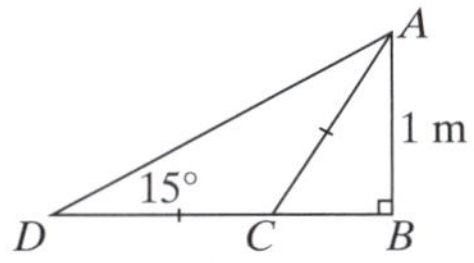

a $\angle DAC = \angle ADC$

(base angles isos. ΔADC)

$\therefore \angle DAC = 15°$

$\angle ACB = \angle DAC + \angle ADC$

(exterior $\angle$ of ΔADC)

$= 15° + 15°$

$= 30°$ ✓ (1 mark)

b In ΔACB,

$\sin 30° = \frac{1}{AC}$

$AC = \frac{1}{\sin 30°}$

$= 2$ m ✓ (1 mark)

c $CB^2 = 2^2 - 1^2$

$= 3$

$CB = \sqrt{3}$ m $(CB > 0)$ ✓ (1 mark)

d $DC = AC$ (given)

So $DC = 2$ m

$DB = (2 + \sqrt{3})$ m ✓ (1 mark)

e In ΔADB,

$\tan 15° = \frac{1}{DB}$

$= \frac{1}{2+\sqrt{3}}$ ✓

$= \frac{1}{2+\sqrt{3}} \times \frac{2-\sqrt{3}}{2-\sqrt{3}}$

$= \frac{2-\sqrt{3}}{4-3}$

$= 2 - \sqrt{3}$ ✓ (2 marks)

f $\angle DAB + \angle ABD + \angle BDA = 180°$

(angle sum triangle)

So $\angle DAB = 75°$

$\tan 75° = \frac{2+\sqrt{3}}{1}$

$= 2 + \sqrt{3}$ ✓ (1 mark)

9 **a** When $x = 60$,

$\sin 2x° = \sin 120°$

$= \sin(180 - 60)°$

$= \sin 60°$

$= \sin x°$ ✓ (1 mark)

b

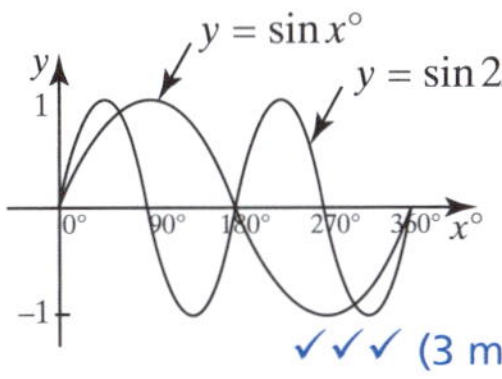

✓✓✓ (3 marks)

c From the graph (and using the result of part **a**)

$\sin 2x° = \sin x°$ when

$x = 0, 60, 180, 300$ and 360

for $0° \le x° \le 360°$. ✓✓ (2 marks)

(Total: 30 marks)

SINE RULE, COSINE RULE AND AREA OF A TRIANGLE

SKILLS CHECK PAGE 82

1

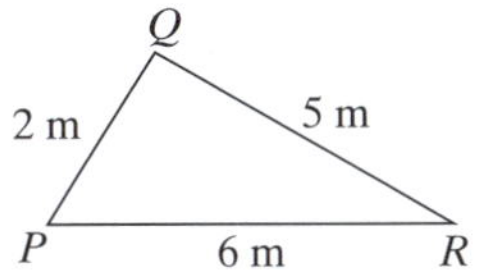

a 5 m

[side p is opposite angle P]

b 6 m

[side q is opposite angle Q]

c 2 m

2

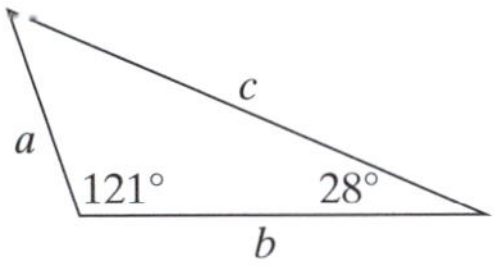

a $\angle A = 28°$

[angle A is opposite side a]

b $\angle B = 180° - (121° + 28°)$

[angle sum Δ]

$= 180° - 149°$

$= 31°$

c $\angle C = 121°$

3 $b = 5$, $\angle A = 70°$, $\angle B = 56°$

$\frac{a}{\sin A} = \frac{b}{\sin B}$

$\frac{a}{\sin 70°} = \frac{5}{\sin 56°}$

$a = \frac{5 \sin 70°}{\sin 56°}$

$= 5.667\,370\,526\ldots$

$= 5.7$ (1 d.p.)

4 $\sin C = 0.376$

$C = 22.086\,132\,42\ldots°$

[by calculator]

$= 22°5'$ (nearest minute)

But $\angle C$ is obtuse

$\angle C = 180° - 22°5'$

$= 157°55'$

5 $b = 4.2$, $c = 5.7$, $\angle C = 64°$

$\frac{\sin B}{b} = \frac{\sin C}{c}$

$\frac{\sin B}{4.2} = \frac{\sin 64°}{5.7}$

$\sin B = \frac{4.2 \sin 64°}{5.7}$

$[= 0.662\,269\,297\ldots]$

$B = 41.473\,1728\ldots°$

$\angle B = 41°$ (nearest degree)

6 **a** $a = 3.7$, $b = 2.8$, $\angle C = 60°$

$c^2 = a^2 + b^2 - 2ab\cos C$

$c^2 = 3.7^2 + 2.8^2$

$- 2 \times 3.7 \times 2.8 \times \cos 60°$

$= 11.17$

$c = \sqrt{11.17}$ $(c > 0)$

$= 3.342\,154\,993\ldots$

$= 3.34$ (2 d.p.)

b $a = 65$, $b = 21.1$, $\angle C = 146°$

$c^2 = a^2 + b^2 - 2ab\cos C$

$c^2 = 65^2 + 21.1^2$

$- 2 \times 65 \times 21.1 \times \cos 146°$

$= 6944.260\,062\ldots$

$c = \sqrt{6944.260\,062\ldots}$ $(c > 0)$

$= 83.332\,227\,03\ldots$

$= 83.33$ (2 d.p.)

7 $a = 7, b – 5, c – 8$

$\cos A = \dfrac{b^2 + c^2 - a^2}{2bc}$

$= \dfrac{5^2 + 8^2 - 7^2}{2 \times 5 \times 8}$

$= 0.5$

$\angle A = 60°$

8 **a** cosine rule
[Including the side that needs to be found, there are 3 sides and just one angle.]

b sine rule
[Including the side that needs to be found, there are 2 sides and 2 angles.]

9 $a = 8, b = 4, \angle C = 30°$

Area $= \dfrac{1}{2}ab\sin C$

$= \dfrac{1}{2} \times 8 \times 4 \times \sin 30°$

$= 8$

The area of the triangle is 8 units^2.

SINE RULE, COSINE RULE AND AREA OF A TRIANGLE
INTERMEDIATE TEST PAGE 83

1

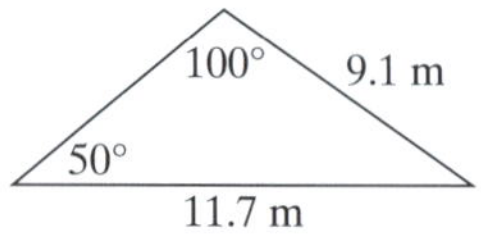

Third angle of triangle = 30°

Area $= \dfrac{1}{2}ab\sin C$

$= \dfrac{1}{2} \times 9.1 \times 11.7 \times \sin 30°$

$= 26.6175$ m^2

≈ 27 m^2

∴ **[A]** ✓ (1 mark)

2

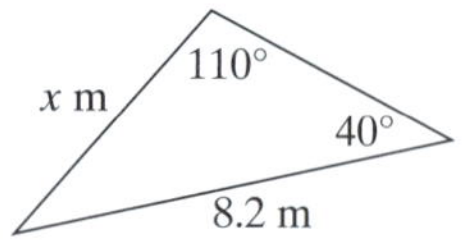

$\dfrac{a}{\sin A} = \dfrac{b}{\sin B}$

$\dfrac{x}{\sin 40°} = \dfrac{8.2}{\sin 110°}$

$x = \dfrac{8.2\sin 40°}{\sin 110°}$

$= 5.609\,13\ldots$

$= 5.6$ (1 d.p.)

∴ **[B]** ✓ (1 mark)

3

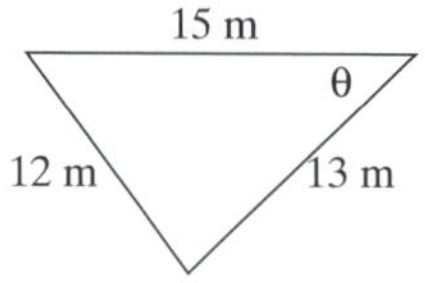

$\cos C = \dfrac{a^2 + b^2 - c^2}{2ab}$

$\cos\theta = \dfrac{15^2 + 13^2 - 12^2}{2 \times 15 \times 13}$

$= \dfrac{25}{39}$

$\theta = \cos^{-1}\left(\dfrac{25}{39}\right)$

$= 50.131\,658\ldots°$

$= 50°$ (nearest degree)

∴ **[A]** ✓ (1 mark)

4

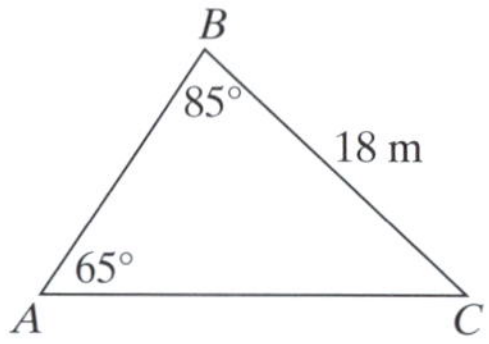

$\angle ACB + \angle ABC + \angle CAB = 180°$

(angle sum of a triangle)

$\angle ACB = 180° - (65 + 85)°$

$= 30°$ ✓

$\dfrac{c}{\sin C} = \dfrac{a}{\sin A}$

$\dfrac{c}{\sin 30°} = \dfrac{18}{\sin 65°}$

$c = \dfrac{18\sin 30°}{\sin 65°}$ ✓

$= 9.930\,401\,271\ldots$

$= 9.9$ (1 d.p.)

The length of AB is 9.9 m to one decimal place. ✓ (3 marks)

5

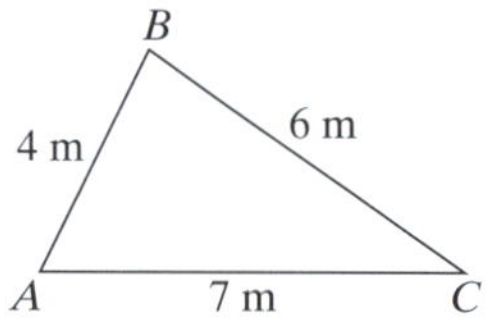

$\cos B = \dfrac{a^2 + c^2 - b^2}{2ac}$

$= \dfrac{6^2 + 4^2 - 7^2}{2 \times 6 \times 4}$ ✓

$= \dfrac{3}{48}$ $\left[= \dfrac{1}{16} = 0.0625\right]$ ✓

$B = 86.416\,6783\ldots°$

$= 86°$ (nearest degree)

The size of the largest angle is 86°, to the nearest whole degree. ✓ (3 marks)

6

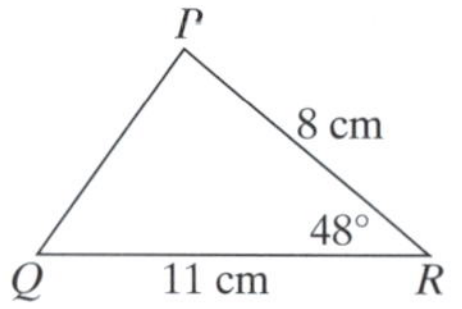

a $r^2 = p^2 + q^2 - 2pq\cos R$

$= 11^2 + 8^2 - 2 \times 11 \times 8 \times \cos 48°$ ✓

$= 67.233\,013\,28\ldots$

$r = 8.199\,573\,969\ldots$ $(r > 0)$ ✓

$= 8.2$ (1 d.p.)

The length of PQ is 8.2 cm, to the nearest mm. [or 82 mm] ✓ (3 marks)

b $A = \dfrac{1}{2}pq\sin R$

$= \dfrac{1}{2} \times 11 \times 8 \times \sin 48°$ ✓

$= 32.698\,372\,32\ldots$

$= 32.7$ (1 d.p.)

The area of the triangle is 32.7 cm^2, to one decimal place. ✓ (2 marks)

7

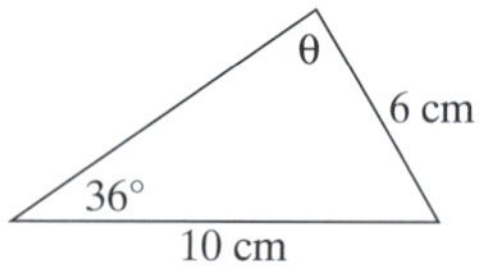

$\dfrac{\sin A}{a} = \dfrac{\sin B}{b}$

$\dfrac{\sin\theta}{10} = \dfrac{\sin 36°}{6}$

$\sin\theta = \dfrac{10\sin 36°}{6}$ ✓

$= 0.979\,642\,087\ldots$

If θ is acute,

$\theta = 78.419\,060\,35\ldots°$

$= 78°$ (nearest degree)

But the angle could be acute or obtuse.

$\theta = 78°$ or $\theta = (180 - 78)°$

$\theta = 78°$ or $\theta = 102°$ ✓✓

(3 marks)

8 **a**

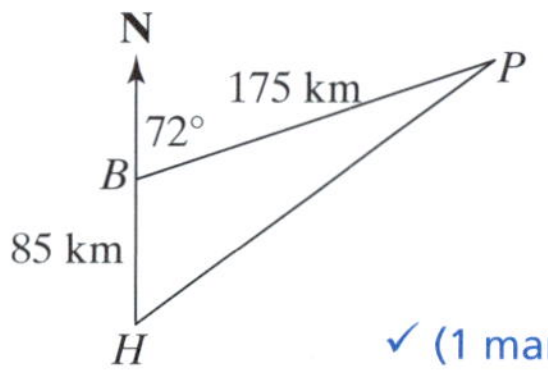

✓ (1 mark)

b $\angle PBH = 180° - 72°$

(angles in a straight line)

$= 108°$ ✓

$b^2 = h^2 + p^2 - 2hp\cos B$
$= 175^2 + 85^2 - 2 \times 175 \times 85 \times \cos 108°$
$= 47\,043.255\,58\ldots$ ✓
$b = \sqrt{47\,043.255\,58\ldots}$ $(b > 0)$
$= 216.894\,5725\ldots$
$= 217$ (near. whole number) ✓
Total distance
$\approx (175 + 217 + 85)$ km
$= 477$ km
The helicopter flew 477 km, to the nearest kilometre. ✓ (4 marks)

9

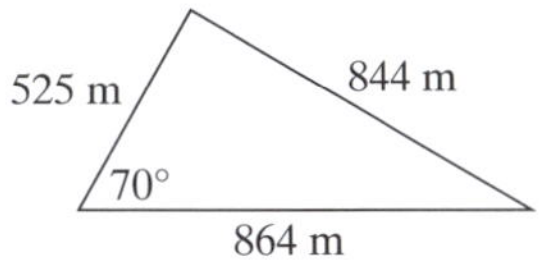

$A = \frac{1}{2}ab\sin C$
$= \frac{1}{2} \times 525 \times 864 \times \sin 70°$ ✓
$= 213\,122.2864\ldots$
Area $= 213\,122$ m^2 ✓ (nearest square metre)
$= 21.31$ ha (2 d.p.)
The area of the block of land is 21.31 ha, correct to two decimal places. ✓ (3 marks)

(Total: 25 marks)

SINE RULE, COSINE RULE AND AREA OF A TRIANGLE
ADVANCED TEST PAGE 84

1

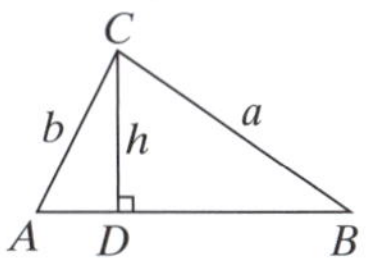

a In ΔACD,
$\sin A = \frac{h}{b}$
$h = b\sin A$ ✓ (1 mark)

b In ΔBCD,
$\sin B = \frac{h}{a}$
$h = a\sin B$ ✓ (1 mark)

c $a\sin B = b\sin A$
$\frac{a\sin B}{\sin A\sin B} = \frac{b\sin A}{\sin A\sin B}$
$\therefore \frac{a}{\sin A} = \frac{b}{\sin B}$ ✓ (1 mark)

2

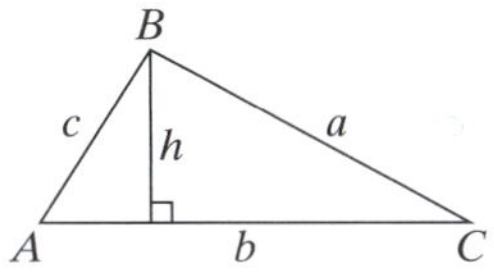

a In ΔBCD,
$\sin C = \frac{h}{a}$
$h = a\sin C$ ✓ (1 mark)

b Area $= \frac{1}{2}bh$
$= \frac{1}{2}b \times (a\sin C)$
$= \frac{1}{2}ab\sin C$ ✓ (1 mark)

3

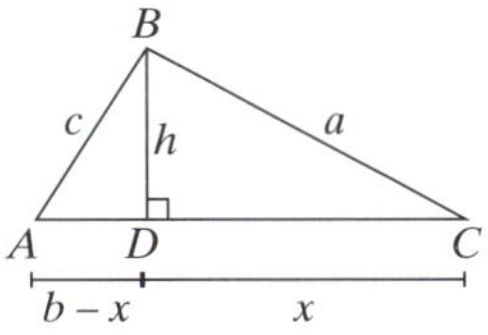

a In ΔABD,
$c^2 = h^2 + (b - x)^2$
$= h^2 + b^2 - 2bx + x^2$ ✓
In ΔBCD,
$a^2 = h^2 + x^2$
$\therefore h^2 = a^2 - x^2$ ✓
So $c^2 = a^2 - x^2 + b^2 - 2bx + x^2$
$c^2 = a^2 + b^2 - 2bx$ ✓ (3 marks)

b In ΔBCD,
$\cos C = \frac{x}{a}$
$x = a\cos C$
$\therefore c^2 = a^2 + b^2 - 2b(a\cos C)$
$= a^2 + b^2 - 2ab\cos C$ ✓ (1 mark)

4

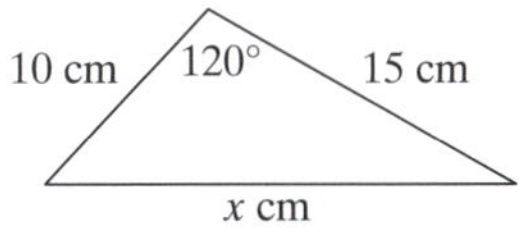

a Let the length of the third side be x cm.
$c^2 = a^2 + b^2 - 2ab\cos C$
$x^2 = 10^2 + 15^2 - 2 \times 10 \times 15 \times \cos 120°$
$= 325 - 300 \times \left(-\frac{1}{2}\right)$
$= 475$ ✓
$x = \sqrt{475}$ $(x > 0)$
$= \sqrt{25} \times \sqrt{19}$
$= 5\sqrt{19}$
The exact length of the third side of the triangle is $5\sqrt{19}$ cm. ✓ (2 marks)

b $A = \frac{1}{2}ab\sin C$
$= \frac{1}{2} \times 10 \times 15 \times \sin 120°$ ✓
$= 75 \times \frac{\sqrt{3}}{2}$
The exact area of the triangle is $\frac{75\sqrt{3}}{2}$ cm^2. ✓ (2 marks)

5 **a**

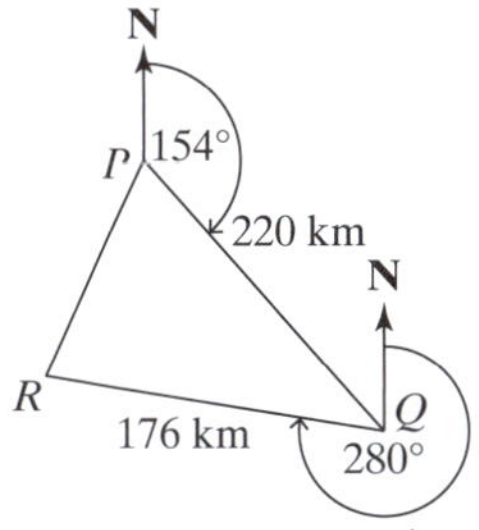

✓ (1 mark)

b Let X be a point due south of Q.
$\angle PQX = 154°$
(co-interior angles, || lines)
$\angle RQX = 280° - 180°$
$= 100°$
$\therefore \angle PQR = 54°$ ✓
$q^2 = 220^2 + 176^2 - 2 \times 220 \times 176 \times \cos 54°$
$= 33\,857.9100\ldots$
$q = \sqrt{33\,857.9100\ldots}$ $(q > 0)$ ✓
$= 184.005\,19\ldots$
$= 184$ km (nearest km) ✓
Total distance
$= (220 + 176 + 184)$ km
$= 580$ km ✓ (4 marks)

c $\frac{\sin P}{p} = \frac{\sin Q}{q}$
$\frac{\sin P}{176} = \frac{\sin 54°}{184.005\,19\ldots}$
$\sin P = \frac{176\sin 54°}{184.005\,19\ldots}$
$P = 50.698\,2183\ldots°$
$= 51°$ (nearest degree) ✓
[The obtuse ∠ will not work.]

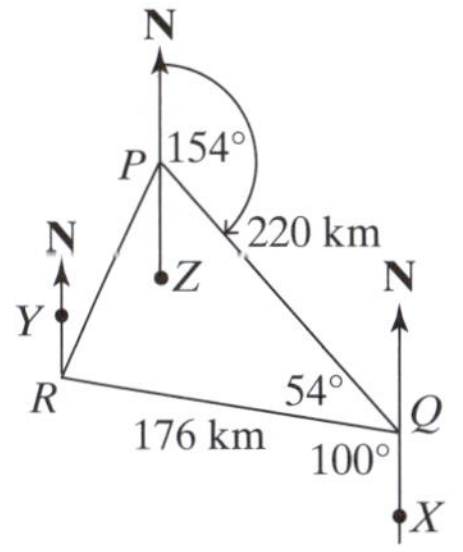

Let Y be a point due north of R and let Z be a point due south of P.

$\angle ZPQ = 180° - 154°$
$= 26°$

$\angle ZPR = 51° - 26°$
$= 25°$ ✓

$\angle YRP = \angle ZPR$
(alternate angles, || lines)

$\therefore\ \angle YRP = 25°$

$\therefore$ the bearing on which they must fly to return to P from Y is 025°. ✓ (3 marks)

6

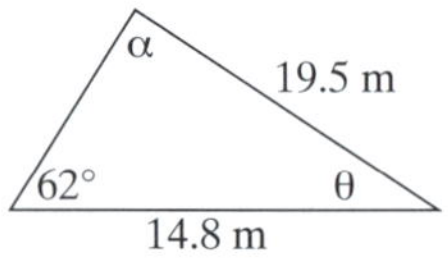

Let the third angle of the triangle be α.

$\frac{\sin\alpha}{14.8} = \frac{\sin 62°}{19.5}$

$\sin\alpha = \frac{14.8\sin 62°}{19.5}$

$\alpha = \sin^{-1}\left(\frac{14.8\sin 62°}{19.5}\right)$
$= 42.07745\ldots°$
$= 42°5'$
(nearest minute) ✓

Now $180° - 42°5' = 137°55'$ and $137°55' + 62° > 180°$

So α cannot be obtuse. ✓

$\theta + 42°5' + 62° = 180°$
(angle sum of Δ)

$\theta = 180° - 104°5'$
$= 75°55'$ ✓

(3 marks)

7 **a**

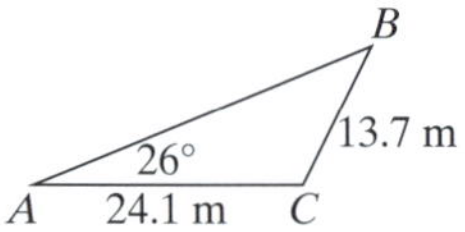

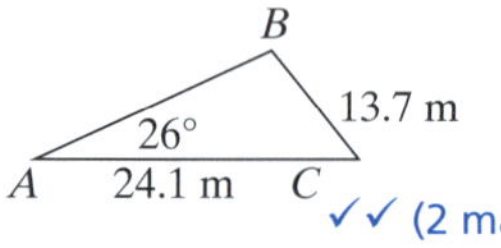

✓✓ (2 marks)

b $\frac{\sin B}{b} = \frac{\sin A}{a}$

$\frac{\sin B}{24.1} = \frac{\sin 26°}{13.7}$

$\sin B = \frac{24.1\sin 26°}{13.7}$ ✓

$B = \sin^{-1}\left(\frac{24.1\sin 26°}{13.7}\right)$
$= 50.457202\ldots°$
$= 50°27'$
(nearest minute) ✓

So the acute angle is $50°27'$

The obtuse angle is $180° - 50°27' = 129°33'$ ✓

(3 marks)

c $\angle A + \angle B + \angle C = 180°$
(angle sum Δ)

When $\angle B = 50°27'$

$\angle C = 180° - 26° - 50°27'$
$= 103°33'$ ✓

Area $= \frac{1}{2}ab\sin C$
$= \frac{1}{2} \times 13.7 \times 24.1 \times \sin 103°33'$
$= 160.4899\ldots\ \text{m}^2$ ✓

When $\angle B = 129°33'$

$\angle C = 180° - 26° - 129°33'$
$= 24°27'$ ✓

Area $= \frac{1}{2}ab\sin C$
$= \frac{1}{2} \times 13.7 \times 24.1 \times \sin 24°27'$
$= 68.3285\ldots.\ \text{m}^2$ ✓

Difference
$= 160.4899\ldots - 68.3285\ldots$
$= 92\ \text{m}^2$
(nearest square metre) ✓

(5 marks)

8

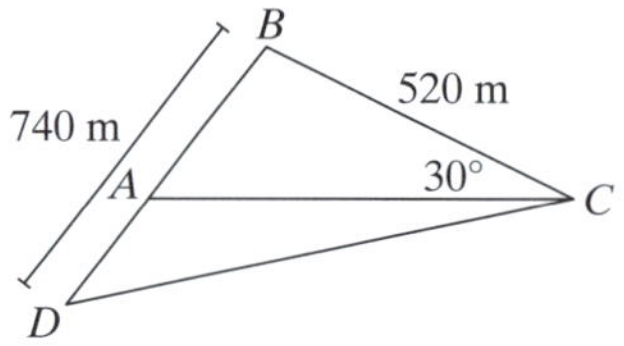

a Area of ΔABC
$= 10.4$ ha
$= 104\,000\ \text{m}^2$

$\frac{1}{2} \times 520 \times AC \times \sin 30°$
$= 104\,000$ ✓

$130 \times AC = 104\,000$

$AC = 800$

The length of AC is 800 m. ✓ (2 marks)

b Let the length of AB be x m.

$x^2 = 520^2 + 800^2 - 2 \times 520 \times 800 \times \cos 30°$
$= 189\,866.8641\ldots$ ✓

$x = 435.73715\ldots$ m ✓

$\cos B = \frac{189\,866.8641\ldots + 520^2 - 800^2}{2 \times 435.73715\ldots \times 520}$

$B = 113.36679\ldots.°$ ✓

$CD^2 = 520^2 + 740^2 - 2 \times 520 \times 740 \times \cos 113.36679\ldots°$
$= 1\,123\,235.669\ldots$

$CD = 1059.8281\ldots\ (CD > 0)$
$= 1060$ m
(nearest metre) ✓

(4 marks)

(Total: 40 marks)

MEAN AND STANDARD DEVIATION
SKILLS CHECK PAGE 86

1 **a** mean
$= \frac{1+2+2+5+6+7+7+7+8}{9}$
$= \frac{45}{9}$
$= 5$

b mode = 7
[There are more 7s than any other number.]

c median = 6
[There are 9 scores. The middle score is the 5th score, which is 6.]

d range $= 8 - 1$
$= 7$

2 11, 11, 12, 13, 13, 13, 14, 14, 14, 14, 15, 15, 16, 16, 16, 16, 16, 17, 17, 19

a mean $= 292 \div 20$
$= 14.6$

b median $= \frac{14 + 15}{2}$
$= 14.5$

c range $= 19 - 11$
$= 8$

d lower quartile = 13

e upper quartile = 16

f interquartile range $= 16 - 13$
$= 3$

3 **a** 5, 7, 9, 12, 12, 12, 15, 17, 18, 19

$\bar{x} = 12.6$

$\sigma_n = 4.45421149\ldots$
$= 4.5$ (1 d.p.)

b 56, 49, 32, 48, 91, 66, 78, 55

$\bar{x} = 59.375$
$= 59.4$ (1 d.p.)

$\sigma_n = 17.34890126\ldots$
$= 17.3$ (1 d.p.)

c

Score	5	6	7	8	9	10
Freq.	4	6	8	9	9	7

$\bar{x} = 7.790\,697\,674\ldots$
$= 7.8$ (1 d.p.)
$\sigma_n = 1.548\,564\,82\ldots$
$= 1.5$ (1 d.p.)

4 18, 28, 34, 35, 39, 41, 44, 46, 46, 49

a Mean = 380 ÷ 10 = 38

b $(18 - 38)^2 = (-20)^2 = 400$
$(28 - 38)^2 = (-10)^2 = 100$
$(34 - 38)^2 = (-4)^2 = 16$
$(35 - 38)^2 = (-3)^2 = 9$
$(39 - 38)^2 = 1^2 = 1$
$(41 - 38)^2 = 3^2 = 9$
$(44 - 38)^2 = 6^2 = 36$
$(46 - 38)^2 = 8^2 = 64$
$(46 - 38)^2 = 8^2 = 64$
$(49 - 38)^2 = 11^2 = 121$
$\Sigma(x - \bar{x})^2 = 400 + 100 + 16 + 9 + 1 + 9 + 36 + 64 + 64 + 121$
$= 820$

c $\sqrt{\dfrac{\sum(x-\bar{x})^2}{n}} = \sqrt{\dfrac{820}{10}}$
$= 9.055\,385\,13\ldots$
$= 9.06$ (2 d.p.)

5 59, 64, 67, 69, 71, 72, 73, 75, 76, 76, 79, 83

a mean = 72 [by calculator]

b standard deviation
$= 6.324\,555\,32\ldots$
$= 6.32$ (2 d.p.)

c New mean = 72
New standard deviation
$= 6.076\,4362\ldots$
$= 6.08$ (2.d.p.)
The score being added was equal to the mean so the mean did not change. Because a score close to the mean was added the standard deviation decreased.

d New mean = 77
Standard deviation
$= 6.32$ (2 d.p.)
Every score increased by 5 so the mean increased by 5. The spread of scores relative to the mean didn't change so the standard deviation didn't change.

e New mean = 144
New standard deviation
$= 12.649\,110\,64\ldots$
$= 12.65$ (2 d.p.)
Both the mean and standard deviation doubled.

6 **a** A and B have the same mean; 23.5

b A: standard deviation
$= 3.67$ (2 d.p.)
B: standard deviation
$= 4.22$ (2 d.p.)
So B has the greater standard deviation.

c Range = 12 for both sets.

d A has the more consistent scores, because it has the smaller standard deviation.

MEAN AND STANDARD DEVIATION
INTERMEDIATE TEST PAGE 87

1 The scores in B are spread further from the mean.
∴ **[B]** ✓ (1 mark)

2 4, 5, 5, 5, 6, 7, 7, 8, 9, 9, 10, 10
Mean = 7.083 333…
Median = 7
Stand. deviation = 2.019 006 907…
Lower quartile = 5
Upper quartile = 9
Interquartile range = 9 – 5 = 4
The mean is the largest.
∴ **[A]** ✓ (1 mark)

3

Score	11	12	13	14	15	16	17	18	19	20
Freq.	15	17	12	8	13	10	7	4	2	1

a Total frequency
$= 15 + 17 + 12 + 8 + 13 + 10 + 7 + 4 + 2 + 1$
$= 89$
89 people sat for the exam. ✓ (1 mark)

b Mean
$= 13.943\,820\,22\ldots$
$= 13.9$ (1 d.p.) ✓ (1 mark)

c Standard deviation
$= 2.324077928\ldots$
$= 2.3$ (1 d.p.) ✓ (1 mark)

d The mean would increase. ✓
The score being added is higher than the mean. ✓ (2 marks)

4 In English Josh scored 72%,
mean = 65%
standard deviation = 3.5
In History Josh scored 81%,
mean = 77%
standard deviation = 4
English was the better result. ✓
The History mark was one standard deviation above the mean while the English mark was two standard deviations above the mean. ✓ (2 marks)

5 **a** Theory: 35, 28, 30, 41, 32, 44, 36, 42, 37, 32, 38, 33
Mean = 35.6666…
$= 35.67$ (2 d.p.) ✓
Standard deviation
$= 4.749\,268\,95\ldots$
$= 4.75$ (2 d.p.) ✓ (2 marks)

b Practical: 33, 39, 29, 42, 37, 44, 33, 49, 38, 32, 37, 40
Mean = 37.75 ✓
Standard deviation
$= 5.371\,607\,456\ldots$
$= 5.37$ (2 d.p.) ✓ (2 marks)

c The theory marks are more consistent. ✓
The standard deviation is lower for the theory exam than for the practical exam. ✓ (2 marks)

d The students did better on the practical exam because the mean was higher. ✓ (1 mark)

e Elise did better in the theory exam. ✓
In the theory exam her mark of 41 was more than one standard deviation above the mean. In the practical exam, her score of 42 was less than one standard deviation above the mean. ✓ (2 marks)

6 In order the numbers are: 17, 18, 20, 21, 22, 24, 25, 29, 32, 35, 38, 46

a mean = 27.25 ✓

median $= \dfrac{24 + 25}{2}$

$= 24.5$ ✓

standard deviation

$= 8.554\,969\,316\ldots$

$= 8.55$ (2 d.p.) ✓

range $= 46 - 17$

$= 29$ ✓

lower quartile $= \dfrac{20 + 21}{2}$

$= 20.5$

upper quartile $= \dfrac{32 + 35}{2}$

$= 33.5$

Interquartile range

$= 33.5 - 20.5$

$= 13$ ✓ (5 marks)

b The mean is higher than the median. The mean has been affected by the one high score of 46. The median gives a better picture of the average monthly sales. ✓ The standard deviation is quite large so the data is not particularly consistent. The range gives the entire spread of sales but the interquartile range, showing the middle 50% of scores, is probably more helpful. ✓ (2 marks)

(Total: 25 marks)

MEAN AND STANDARD DEVIATION
ADVANCED TEST PAGE 88

1 **a** $\bar{x} = 73$ [by calculator] ✓ (1 mark)

b $\sigma_n = 7.08116\ldots$

$= 7.08$ (2 d.p.) ✓ (1 mark)

c The mean will stay the same because the mark being added is equal to the original mean. ✓ (1 mark)

d The standard deviation will decrease. ✓ The spread of scores from the mean is smaller because the mark being added is equal to the mean. ✓ (2 marks)

e The mean will increase by 1. Because the mark being added is 16 higher than the original mean and there are now 16 marks altogether. ✓ (1 mark)

f The standard deviation will increase. ✓ The mark being added is greater than all the previous marks so the spread will be greater. ✓ (2 marks)

2 **a** $\bar{x} = 12$ ✓ (1 mark)

b

x	$x - \bar{x}$	$(x - \bar{x})^2$
3	−9	81
5	−7	49
8	−4	16
9	−3	9
12	0	0
13	1	1
15	3	9
17	5	25
18	6	36
20	8	64

✓✓ (2 marks)

c $\Sigma(x - \bar{x}) = 0$ ✓ (1 mark)

d $\Sigma(x - \bar{x})^2 = 290$ ✓ (1 mark)

e $\sqrt{\dfrac{\Sigma(x - \bar{x})^2}{n}} = \sqrt{\dfrac{290}{10}}$

$= 5.385\,1648\ldots$

$= 5.385$ (3 d.p.)

✓ (1 mark)

f $\sigma_n = 5.385\,1648\ldots$

$= 5.385$ (3 d.p.)

Yes it is the same because the formula used in part **e** is the formula to find the standard deviation. ✓ (1 mark)

g No, Gwen is not correct. The sum of the differences from the mean $\Sigma(x - \bar{x})$ must always be 0 because the total below the mean must be equal to the total above the mean. So $\dfrac{\Sigma(x - \bar{x})}{n} = 0$. ✓

The sum of the squares is not the same as the square of the sum. So taking the square root is not the opposite operation. The differences are first squared and then summed. Squaring these differences means that there will be no negatives and hence we have a measure of how far the scores are from the mean regardless of whether they are above or below the mean. ✓ (2 marks)

3 **a** If every score in a data set increases by 4 the mean will also increase by 4. The location of every score increases by 4 and hence the mean, a measure of location, increases by 4. This is because the sum of all the scores will increase by 4 times the number of scores and when this increase is divided by the number of scores the increase is 4. ✓ The standard deviation will remain the same because the spread of scores relative to the mean will stay the same. ✓ (2 marks)

b If every score doubles the mean will also double because the sum of the scores will double, but the number of scores remains the same. ✓ The standard deviation will also double because the spread of scores will have doubled. ✓ (2 marks)

4

Day	Morning	Afternoon
Monday	26	19
Tuesday	39	29
Wednesday	41	25
Thursday	32	29
Friday	44	34
Monday	28	21
Tuesday	53	30
Wednesday	40	27
Thursday	37	24
Friday	35	32

a $\bar{x} = 37.5$ ✓
$\sigma_n = 7.5$ ✓ (2 marks)

b $\bar{x} = 27$ ✓
$\sigma_n = 4.5166359\ldots$
$= 4.52$ (2 d.p.) ✓
(2 marks)

c The morning times are longer than the afternoon ones by an average of around 10 minutes. The spread of times is also much greater in the morning than in the afternoon. The afternoon times are more consistent. ✓✓ (2 marks)

5 **a** The location of the data sets is about the same but the second set of data is more consistent than the first. The spread of scores is greater in the first set. ✓
(1 mark)

b A score of 72 would be equivalent if the mean of both sets was 72. ✓ (1 mark)

c If the mean was less than 72 then a mark of 72 would be better in the second set. If the mean was greater than 72, a score of 72 would be better in the first set. ✓
(1 mark)

(Total: 30 marks)

INVESTIGATIONS OF DATA SETS AND REPORTS
SKILLS CHECK PAGE 90

1

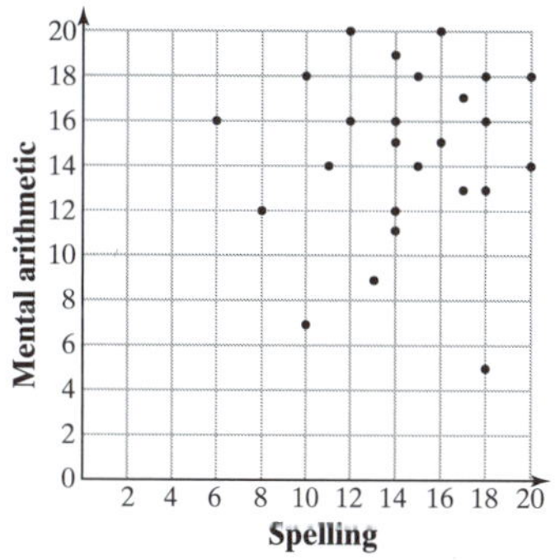

a 25 students are in the class. [There are 25 dots on the graph.]

b The student who scored 13 in spelling scored 9 in mental arithmetic.

c The student who scored 17 in mental arithmetic scored 17 in spelling.

d Of the students who scored 10 in spelling one scored 7 in mental arithmetic and the other scored 18.
Difference $= 18 - 7$
$= 11$

e There is a slight relationship. Those who did well in spelling generally did well in mental arithmetic and those who did not do so well in spelling generally did not do well in mental arithmetic either.

f One student scored 18 in spelling but only 5 in mental arithmetic.

2 **a**

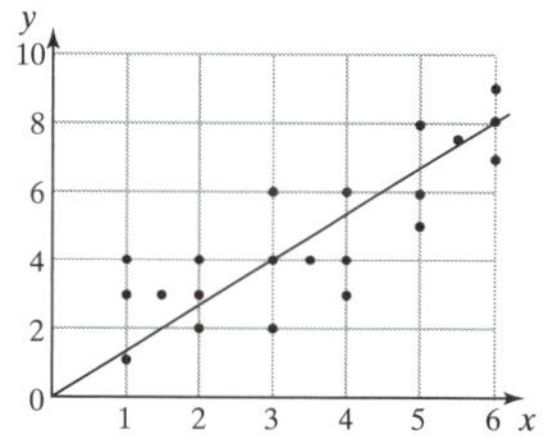

$m = \dfrac{4}{3}$

y-intercept $= 0$

$y = \dfrac{4}{3}x$

b

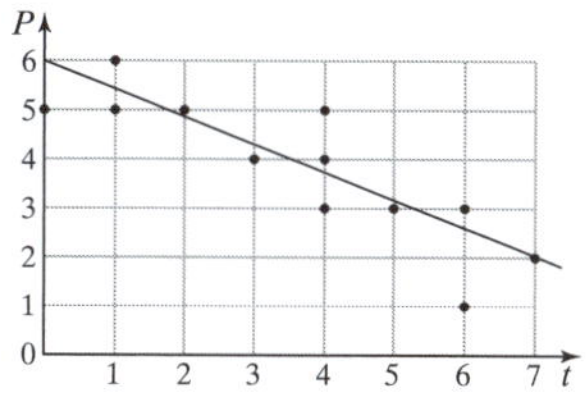

$m = \dfrac{-4}{7}$

vertical intercept $= 6$

$P = \dfrac{-4}{7}t + 6$

c

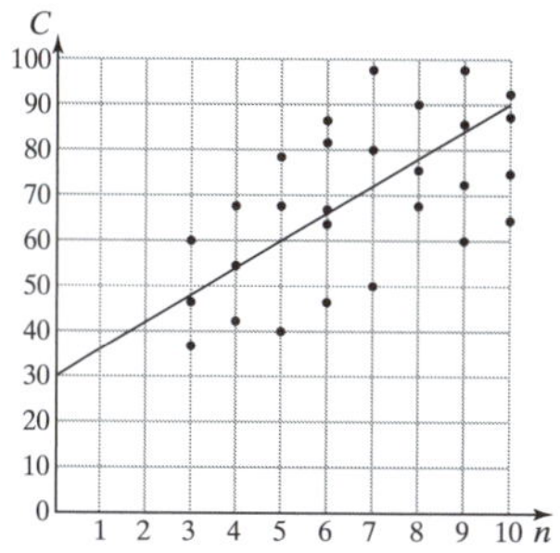

$m = \dfrac{90 - 30}{10 - 0}$
$= 6$

vertical intercept $= 30$

$C = 6n + 30$

3 $h = 9 - \dfrac{3}{4}t$

a When $t = 6$,
$h = 9 - \dfrac{3}{4} \times 6$
$= 4.5$

b When $h = 3$,
$3 = 9 - \dfrac{3}{4}t$
$\dfrac{3}{4}t = 6$
$t = 6 \div \dfrac{3}{4}$
$= 8$

c When $t = 20$,
$h = 9 - \dfrac{3}{4} \times 20$
$= -6$

d The height cannot be negative.

e $9 - \dfrac{3}{4}t \geq 0$
$\dfrac{3}{4}t \leq 9$
$t \leq 12$
The largest possible value of t is 12.

4 [Some possible reasons.]
The size of the sample is too small.
The sample is not likely to have the same demographics as the rest of Australia.
The age and sex of the respondents would not be representative.

5 **a** False; there can be more than one line of fit.

b True; the equation might be written in different forms but there can only be one equation for any line.

c False; the size must not be too small.

d True.

INVESTIGATIONS OF DATA SETS AND REPORTS
INTERMEDIATE TEST PAGE 91

[Note: some possible responses to questions that ask for a discussion of reports are given here, but there will be other perfectly valid responses.]

1

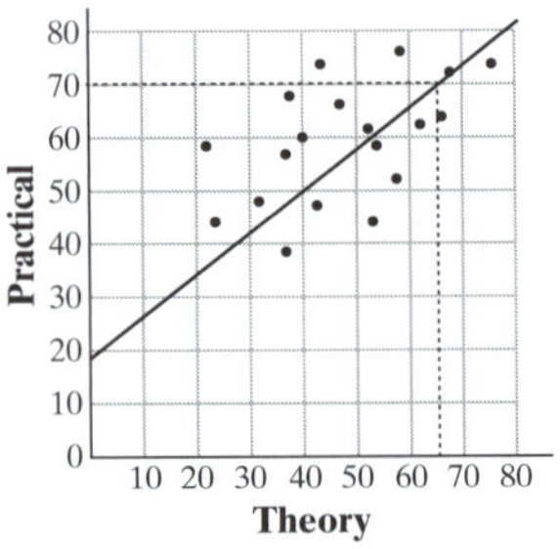

Josh's estimated mark should have been about 70.

∴ **[C]** ✓ (1 mark)

2 The lowest theory mark was 22. So Jess must have had a practical mark of 58. For a practical mark of 58 the line of fit suggests a theory mark of about 50.

Average $= \dfrac{22 + 50}{2}$

$= 36$

∴ **[A]** ✓ (1 mark)

3 For the line of fit:

y-intercept = 18.

$m = \dfrac{4}{5}$

So $y = \dfrac{4}{5}x + 18$

When $x = 90$,

$y = \dfrac{4}{5} \times 90 + 18$

$= 90$

∴ **[A]** ✓ (1 mark)

4 Percentage supporting X

$= (100 - 33)\%$

$= 67\%$

The only statement that must be correct is 'Around twice as many of those surveyed preferred X to Y.'

∴ **[C]** ✓ (1 mark)

5 a

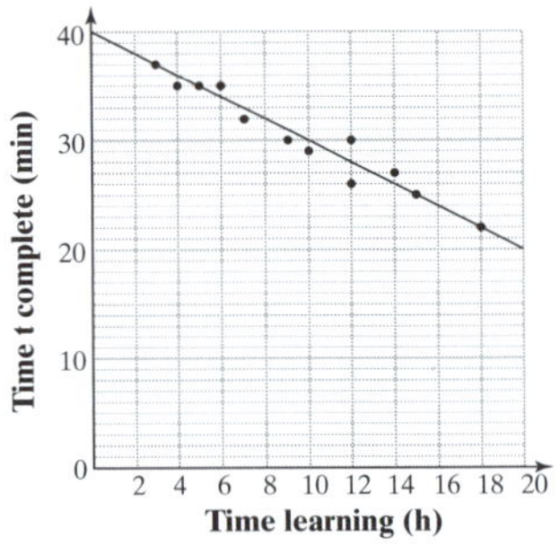

[The line of fit (and its equation) could be slightly different.] ✓✓✓ (3 marks)

b gradient = –1 ✓

y-intercept = 40

Equation: $y = 40 - x$ ✓

[or $y = -x + 40$] (2 marks)

c When $x = 24$,

$y = 40 - 24$

$= 16$

The prediction is that it would take 16 minutes to complete the procedure. ✓ (1 mark)

d When $y = 8$,

$8 = 40 - x$

$x = 40 - 8$

$= 32$

You would expect that she took 32 hours to learn the procedure. ✓ (1 mark)

e If $x = 48$,

$y = 40 - 48$

$= -8$

It makes no sense for a length of time to be negative so the equation of the line of fit cannot be used. ✓ (1 mark)

6

	Male	Female	Total
Plan A	84	68	152
Plan B	57	51	108
Undecided	19	21	40
Total	160	140	300

a Percentage that are male

$= \dfrac{160}{300} \times 100\%$

$= 53\frac{1}{3}\%$ ✓ (1 mark)

b Total not in favour of plan B

$= 152 + 40$

$= 192$

Percentage not in favour of plan B

$= \dfrac{192}{300} \times 100\%$

$= 64\%$ ✓ (1 mark)

c It is not a good sample of all Australians. The percentage of males in the sample is higher than for the actual population. The age and other demographics of the sample are also not likely to be an accurate representation of the Australian population. ✓✓ (2 marks)

d Total males

$= 84 + 57 + 19$

$= 160$

P(male prefers plan A)

$= \dfrac{84}{160}$ ✓

Total who prefer plan A

$= 84 + 68$

$= 152$

P(person who prefers plan A is male) $= \dfrac{84}{152}$

It is more likely that a person who prefers plan A is a male than that a male prefers plan A. ✓ (2 marks)

e No, it is not correct. ✓

The number of females who either prefer plan B or are undecided (51 + 21 = 72) is greater than the number of females who prefer plan A. So the majority of females did not prefer plan A. ✓

(2 marks)

(Total: 20 marks)

WORKED SOLUTIONS

CHECK YOUR SOLUTIONS

INVESTIGATIONS OF DATA SETS AND REPORTS
ADVANCED TEST PAGE 92

[Note: some possible responses to questions that ask for a discussion of reports are given here, but there will be other perfectly valid responses.]

1

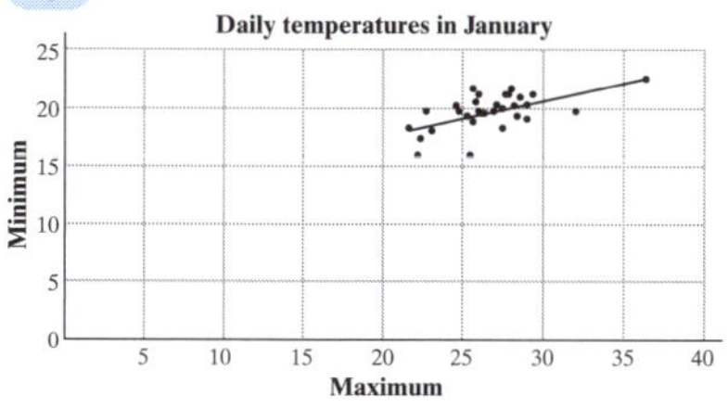

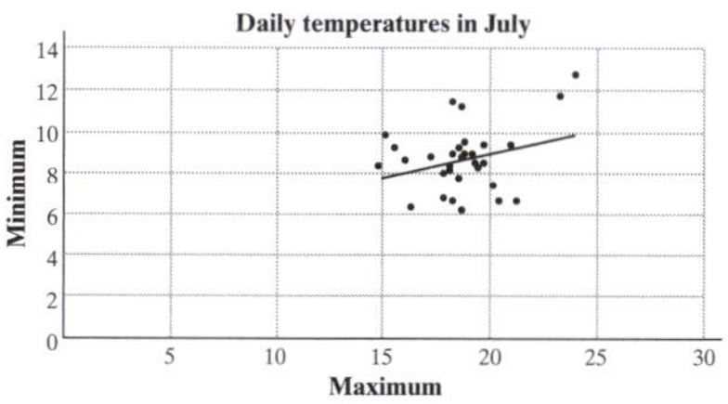

a The two lines look to have a similar gradient, but the scales are different on the vertical axis, so it is not possible to tell by looking alone. ✓
By calculation, the gradient for January is about 0.35 and that for July is about 0.25 so Charlotte is not correct. ✓ (2 marks)

b Because the scales are different on the two graphs, it is impossible to make an accurate judgement just by looking. ✓
However, the January temperatures appear to have the strongest correlation. ✓ (2 marks)

c In January, when the maximum is 23 °C the minimum is about 18 °C. In July, when the maximum is 23 °C the minimum is about 9.7 °C. So, the difference is about 8.3 °C. ✓ (1 mark)

d The lowest maximum in January was about 21.5 °C. On 2 days the maximum temperature in July was above 21.5 °C. ✓ (1 mark)

e A temperature of 27 °C is outside the range of temperatures for July so we shouldn't use the line of fit to make predictions. ✓
But, from the scatterplot we might expect that the minimum temperature could be around 11 °C. ✓
(2 marks)

2 It is not a valid response. ✓
This would be a valid response if it is the cars that are causing the drivers to be aggressive and reckless. It is more likely though that it is the type of people (young, inexperienced drivers) who are more likely to drive small cars that tend to be aggressive and reckless. ✓
(2 marks)

3 It is not a good question. ✓
It makes the assumption that there are problems with the policy. This introduces bias into the question. ✓ (2 marks)

4 It is a sensible approach if it provides better maintained and safer roads (although it is debatable whether the aim of council's policy is to provide a better service for their ratepayers or to limit their own liability). ✓
The council needs to consider how well the insurance companies guidelines apply to their area. For example, if the guidelines are based on statistics drawn from a metropolitan area they may not apply to a rural or regional area. ✓ (2 marks)

5 Because the study was done in Britain it won't necessarily mean that the same results would be found in Australia. It was also a controlled experiment and the results might have been different if a random sample of those giving up smoking was taken. ✓
Such media reports could discourage people from giving up smoking. It would be helpful if the media also made suggestions, for example, about seeking medical advice. ✓
(2 marks)

6 **a** The road was probably in poor condition but it is not possible to make any judgements without further information. ✓
It would be helpful to know how many people responded to the survey and how many of those voted for that particular road.
It could also be helpful to know who or what organisation set up and supported the survey and how it was publicised. ✓
(2 marks)

b Because it was an online survey, certain people would immediately be eliminated from participating, e.g. those who didn't have access to the internet. It might also not attract a representative cross-section of the population. ✓
The survey could perhaps be completed many times by the same person or that person could encourage others of similar mind to complete it. ✓ (2 marks)

(Total: 20 marks)

SAMPLE EXAM PAPER 1
Part A: Multiple Choice PAGE 95

1 $\sqrt{36x^{36}} = 6x^{18}$

$[\sqrt{36} = 6,\ \sqrt{x^{36}} = (x^{36})^{\frac{1}{2}} = x^{36\times\frac{1}{2}}]$

$\therefore$ **[B]** ✓ (1 mark)

2 The correct diagram is A.

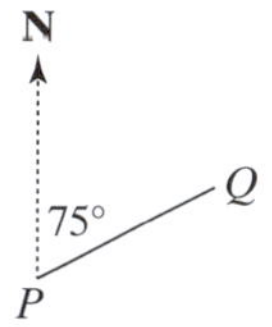

cont.

[Bearings are measured in a clockwise direction from north, so it is not C or D. The bearing is from P, so the angle must be at P.]

$\therefore$ **[A]** ✓ (1 mark)

3 $d = 6, \therefore r = 3;\ h = 4$

$V = \frac{1}{3}\pi r^2 h$

$= \frac{1}{3} \times \pi \times 3^2 \times 4$

$= 37.699\,111\,84\ldots$

The volume is closest to $38\ \text{cm}^3$.

$\therefore$ **[A]** ✓ (1 mark)

4 $\frac{x^2 - x}{x^2 - 1} = \frac{x(x-1)}{(x+1)(x-1)}$

$= \frac{x}{x+1}$

$\therefore$ **[C]** ✓ (1 mark)

5 $3n^0 - n^{-1} = 3 \times 1 - \frac{1}{n}$

$= 3 - \frac{1}{n}$

$\therefore$ **[C]** ✓ (1 mark)

6 $\frac{2}{\sqrt{2}} + \frac{\sqrt{2}}{2} = \frac{2}{\sqrt{2}} \times \frac{\sqrt{2}}{\sqrt{2}} + \frac{\sqrt{2}}{2}$

$= \frac{2\sqrt{2}}{2} + \frac{\sqrt{2}}{2}$

$= \frac{3\sqrt{2}}{2}$

$\therefore$ **[B]** ✓ (1 mark)

7 $y = x^2 - 6x + 8$

$a = 1, b = -6, c = 8$

Axis of symmetry:

$x = \frac{-b}{2a}$

$= \frac{-(-6)}{2 \times 1}$

$= 3$

$\therefore$ **[D]** ✓ (1 mark)

8 $\log_a 6 + \log_a 8 - \log_a 12$

$= \log_a \left(\frac{6 \times 8}{12}\right)$

$= \log_a 4$

$\therefore$ **[B]** ✓ (1 mark)

9

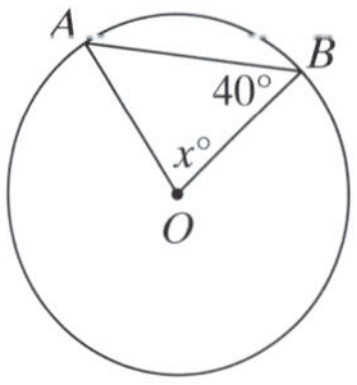

$OA = OB$ (radii of circle)

$\therefore \angle OAB = \angle OBA = 40°$ (isosceles Δ)

$x + 40 + 40 = 180$ ($\angle$ sum Δ)

$x + 80 = 180$

$x = 100$

$\therefore$ **[D]** ✓ (1 mark)

10 $xy - x - y + 1 = x(y-1) - 1(y-1)$

$= (y-1)(x-1)$

$\therefore$ **[A]** ✓ (1 mark)

11 $y = \frac{1}{3}x - 4$ has gradient $\frac{1}{3}$

Perpendicular lines have gradient -3.

A: $3x + y - 5 = 0$

$y = -3x + 5$

Gradient is -3.

[B has gradient 3, C has gradient $-\frac{1}{3}$ and D has gradient $\frac{1}{3}$.]

$\therefore$ **[A]** ✓ (1 mark)

12 $r = 8, l = 17$

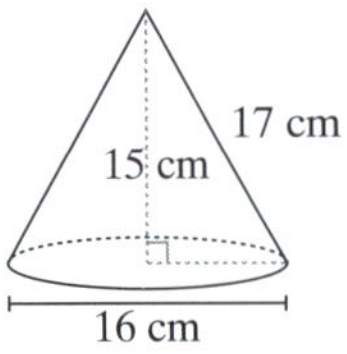

Curved surface area:

$A = \pi rl$

$= \pi \times 8 \times 17$

$= 427.256\,6009\ldots$

The curved surface area is $427\ \text{cm}^2$ to the nearest square centimetre.

$\therefore$ **[A]** ✓ (1 mark)

13 $3m^2 + m - 10$

$= 3m^2 + 6m - 5m - 10$

$= 3m(m+2) - 5(m+2)$

$= (m+2)(3m-5)$

$\therefore$ **[D]** ✓ (1 mark)

14

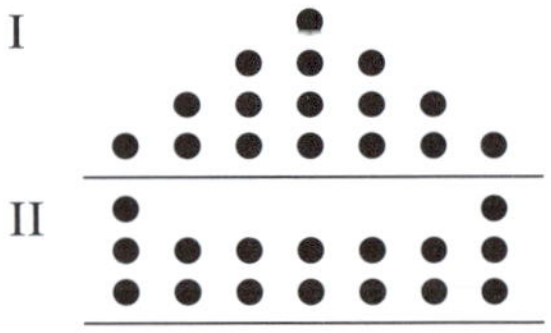

Both tests have the same mean. [Symmetrical about central position.] The scores in test II are spread further from the mean than those in test I.

Test II has the greater standard deviation.

$\therefore$ **[C]** ✓ (1 mark)

15 $P(x) = x^3 - 3x^2 + 6x - 9$

$P(-1) = (-1)^3 - 3 \times (-1)^2 + 6 \times -1 - 9$

$= -19$

$\therefore$ **[D]** ✓ (1 mark)

16

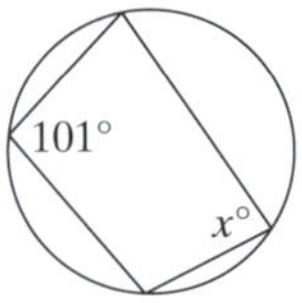

$x + 101 = 180$

(opposite $\angle$s of cyclic quad.)

$x = 79$

$\therefore$ **[A]** ✓ (1 mark)

17

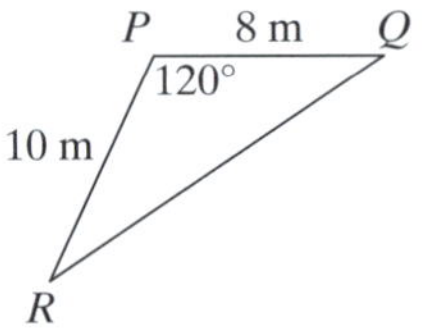

$p^2 = q^2 + r^2 - 2qr\cos P$

$= 10^2 + 8^2 - 2 \times 10 \times 8 \times \cos 120°$

$= 244$

$p = \sqrt{244}\quad (p > 0)$

$= 15.620\,499\,35\ldots$

$= 16$ (nearest whole number)

The length of QR is 16 m to the nearest metre.

$\therefore$ **[D]** ✓ (1 mark)

18 ratio of lengths $= 2:3$

ratio of volumes $= 2^3:3^3$

$= 8:27$

Let $x\ \text{m}^3$ be the volume of the smaller figure.

$\frac{x}{81} = \frac{8}{27}$

$x = \frac{8}{27} \times 81$

$= 24$

The volume of the smaller figure is 24 m³.
$\therefore$ **[B]** ✓ (1 mark)

19 The graph of $y = 2^{-x}$ is B.

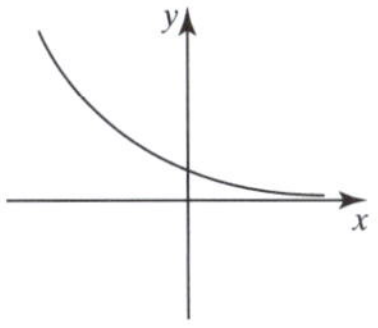

When $x = 0$, $y = 1$;
when $x = 1$, $y = \frac{1}{2}$
$\therefore$ **[B]** ✓ (1 mark)

20 Centre $(-1, 3)$, radius 2 units
$(x - (-1))^2 + (y - 3)^2 = 2^2$
$(x + 1)^2 + (y - 3)^2 = 4$
$\therefore$ **[D]** ✓ (1 mark)

21 $f(x) = x^2 - x + 1$
$f(-1) = (-1)^2 - (-1) + 1$
$= 1 + 1 + 1$
$= 3$
$\therefore$ **[D]** ✓ (1 mark)

22

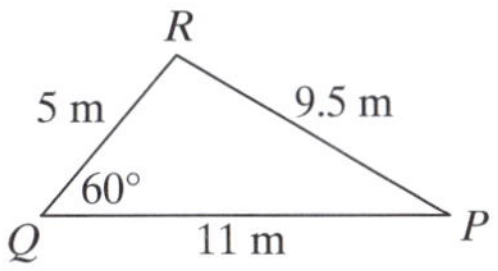

$\frac{\sin P}{p} = \frac{\sin Q}{q}$
$\frac{\sin P}{5} = \frac{\sin 60°}{9.5}$
$\sin P = \frac{5 \sin 60°}{9.5}$
$\therefore$ **[A]** ✓ (1 mark)

23 The mass is decreasing at a decreasing rate.

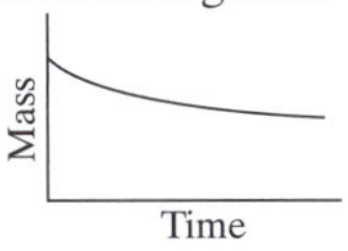

$\therefore$ **[D]** ✓ (1 mark)

24 $(3m - 2)(m - 4)$
$= 3m^2 - 12m - 2m + 8$
$= 3m^2 - 14m + 8$
$\therefore$ **[C]** ✓ (1 mark)

25 $9x^2 + 6x = (3x)^2 + 2 \times 3x \times 1$
Need to add $1^2 = 1$
$\therefore$ **[C]** ✓ (1 mark)
(Total Part A: 25 marks)

SAMPLE EXAM PAPER 1
Part B: Written Response PAGE 97

26 $\frac{8}{x^2 - 4} - \frac{1}{2x - 4}$
$= \frac{8}{(x + 2)(x - 2)} - \frac{1}{2(x - 2)}$ ✓
$= \frac{16 - (x + 2)}{2(x + 2)(x - 2)}$
$= \frac{14 - x}{2(x + 2)(x - 2)}$ ✓ (2 marks)

27 $8^x = 32$
$(2^3)^x = 2^5$
$2^{3x} = 2^5$ ✓
$3x = 5$
$x = 1\frac{2}{3}$ ✓ (2 marks)

28 $\sqrt{28} + \sqrt{63} = \sqrt{4} \times \sqrt{7} + \sqrt{9} \times \sqrt{7}$
$= 2\sqrt{7} + 3\sqrt{7}$
$= 5\sqrt{7}$ ✓ (1 mark)

29 $3x^2 - 7x - 1 = 0$
$a = 3$, $b = -7$, $c = -1$
$x = \frac{-b \pm \sqrt{b^2 - 4ac}}{2a}$
$= \frac{-(-7) \pm \sqrt{(-7)^2 - 4 \times 3 \times -1}}{2 \times 3}$
$= \frac{7 \pm \sqrt{61}}{6}$
$x = \frac{7 + \sqrt{61}}{6}$ or $x = \frac{7 - \sqrt{61}}{6}$
$= 2.468\,37\ldots$ or $= -0.135\,04\ldots$
$= 2.47$ or $= -0.14$
(2 d.p.) (2 d.p.)
✓✓ (2 marks)

30

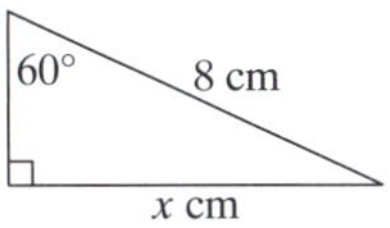

$\sin 60° = \frac{x}{8}$
$x = 8 \times \sin 60°$ ✓
$= 8 \times \frac{\sqrt{3}}{2}$
$= 4\sqrt{3}$ ✓ (2 marks)

31

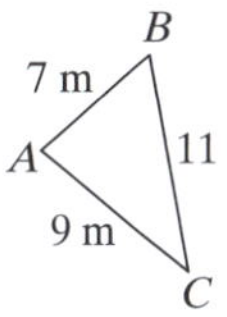

a $\cos A = \frac{b^2 + c^2 - a^2}{2bc}$
$= \frac{9^2 + 7^2 - 11^2}{2 \times 9 \times 7}$
$= 0.071\,428\,571\ldots$ ✓
$A = 85.903\,956\,24\ldots°$
$= 86°$
(nearest degree) ✓
(2 marks)

b $A = \frac{1}{2} bc \sin A$
$= \frac{1}{2} \times 9 \times 7 \times \sin 85.9039\ldots°$
$= 31.41954\ldots$
$= 31.4$ (1 d.p.)
The area of the triangle is 31.4 m², to one decimal place. ✓ (1 mark)

32

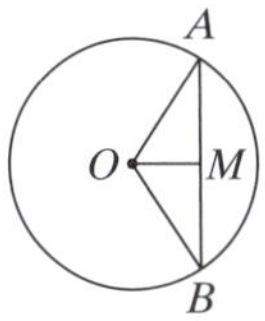

In ΔAOM, ΔBOM,
$OA = OB$ (radii of circle)
$AM = BM$
(M is the midpoint of AB)
$OM = OM$ (common side)
$\therefore \Delta AOM \equiv \Delta BOM$ SSS ✓
$\angle AMO = \angle BMO$
(corresponding sides congruent Δs) ✓
but $\angle AMO + \angle BMO = 180°$
(angles in a straight line)
$\therefore \angle AMO = \angle BMO = 90°$
$\therefore OM \perp AB$ ✓ (3 marks)

33

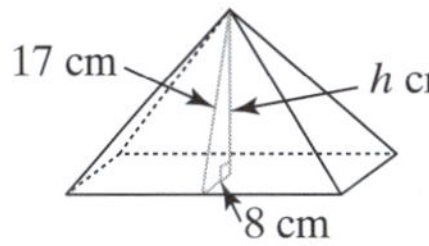

a $h^2 = 17^2 - 8^2$
$= 289 - 64$
$= 225$
$h = \sqrt{225}$ $(h > 0)$
$= 15$
The perpendicular height of the pyramid is 15 cm ✓
(1 mark)

b

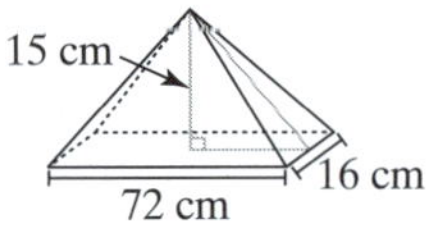

$V = \frac{1}{3}Ah$

$= \frac{1}{3} \times 72 \times 16 \times 15$

$= 5760$

The volume of the pyramid is 5760 cm³ ✓ (1 mark)

c Area of front (and back) triangular faces:

$A = \frac{1}{2}bh$

$= \frac{1}{2} \times 72 \times 17$

$= 612$

The area of the front triangle (and of the back triangle) is 612 cm². ✓

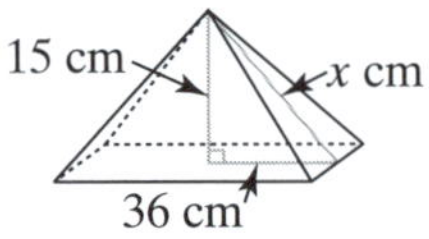

Let x cm be the perpendicular height of the side triangular face.

$x^2 = 36^2 + 15^2$

$= 1296 + 225$

$= 1521$

$x = \sqrt{1521} \quad (x > 0)$

$= 39$

The perpendicular height of the side triangular face is 39 cm. ✓

Area of the side triangular faces:

$A = \frac{1}{2}bh$

$= \frac{1}{2} \times 16 \times 39$

$= 312$

The area of each side triangular face is 312 cm². ✓

Area of rectangular base:

$A = lb$

$= 72 \times 16$

$= 1152$

The area of the base is 1152 cm².

Total surface area

$= 1152\text{ cm}^2 + 2 \times 612\text{ cm}^2 + 2 \times 312\text{ cm}^2$

$= 3000\text{ cm}^2$ ✓ (4 marks)

34 **a** $P(x) = x^3 - 3x^2 - 22x + 24$

$P(1) = 1^3 - 3 \times 1^2 - 22 \times 1 + 24$

$= 0$

$\therefore$ $(x - 1)$ is a factor of $P(x)$. ✓

$$\begin{array}{r}
x^2 - 2x - 24 \\
x - 1 \,\big)\, x^3 - 3x^2 - 22x + 24 \\
\underline{x^3 - x^2 \qquad\qquad} \\
-2x^2 - 22x \qquad \\
\underline{-2x^2 + 2x \qquad} \\
-24x + 24 \\
\underline{-24x + 24} \\
0
\end{array}$$ ✓

$P(x) = (x - 1)(x^2 - 2x - 24)$

$= (x - 1)(x + 4)(x - 6)$ ✓

(3 marks)

b $y = x^3 - 3x^2 - 22x + 24$

As $x \to \infty$, $y \to \infty$

The y-intercept is 24.

The x-intercepts are 1, 6 and –4. ✓ (1 mark)

(Total Part B: 25 marks)

SAMPLE EXAM PAPER 2

Part A: Multiple Choice PAGE 99

1 $a^2 - 25 = a^2 - 5^2$

$= (a + 5)(a - 5)$

$\therefore$ [C] ✓ (1 mark)

2 $(-2, -6)$ [Try each option.]

$y = x^2 + 5x + 13$

$= (-2)^2 + 5(-2) + 13$

$= 7$ (not –6)

$y = \frac{1}{2}x^3$

$= \frac{1}{2}(-2)^3$

$= -4$ (not –6)

$y = \frac{6}{x + 1}$

$= \frac{6}{-2 + 1}$

$= -6$

$y = -2^{-x}$

$= -2^{-(-2)}$

$= -2^2$

$= -4$ (not –6)

$\therefore$ [C] ✓ (1 mark)

3 Leading term $= 2x \times 4x^3$

$= 8x^4$

Leading coefficient = 8

$\therefore$ [D] ✓ (1 mark)

4 Try each option.

$\sin 60° = \frac{\sqrt{3}}{2}$

$\sin 120° = \sin (180 - 60)°$

$= \sin 60°$

$= \frac{\sqrt{3}}{2}$

$\cos 30° = \frac{\sqrt{3}}{2}$

$\cos 150° = \cos (180 - 30)°$

$= -\cos 30°$

$= -\frac{\sqrt{3}}{2}$

$\therefore$ [D] ✓ (1 mark)

5

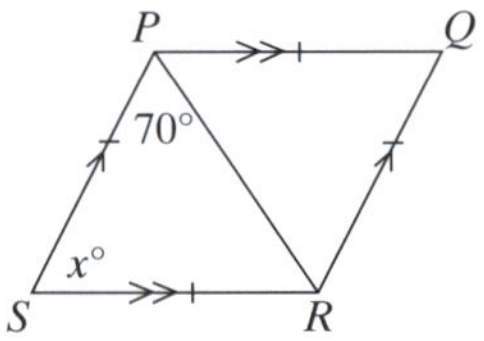

$PQRS$ is a rhombus.

(A parallelogram with a pair of adjacent sides equal.)

$\angle SPQ = 140°$

(diagonals bisect angles)

$\angle SPQ + \angle PSR = 180°$

(co-interior angles, || lines)

$140 + x = 180$

$x = 40$

$\therefore$ [A] ✓ (1 mark)

6 $y = x^2 - 6x + 11$

Axis of symmetry: $x = \frac{-b}{2a}$

$x = -\frac{-6}{2}$

$x = 3$

When $x = 3$,

$y = 3^2 - 6 \times 3 + 11$

$= 2$

Vertex is at (3, 2)

$\therefore$ [B] ✓ (1 mark)

7 $y = \frac{-2}{x}$ is the equation of a hyperbola.

A is a parabola. Its equation is of the form $y = -ax^2 + 2$ (or $y = 2 - ax^2$).

B is the straight line $y = \frac{-x}{2}$.

D is a parabola. Its equation is of the form $y = ax^2 - 2$.

$\therefore$ [C] ✓ (1 mark)

8 $2x^{\frac{1}{2}} \times 4x^{\frac{1}{2}} = 8x^{\frac{1}{2}+\frac{1}{2}}$
$= 8x^1$
$= 8x$
$\therefore$ **[A]** ✓ (1 mark)

9 $V = 36, h = 8$
$V = \frac{1}{3}Ah$
$36 = \frac{1}{3} \times A \times 8$
$108 = 8A$ [multiplying by 3]
$A = 108 \div 8$
$= 13.5$
The area is 13.5 cm^2
[Or $\frac{1}{3} \times A \times 8 = 36$
$\frac{8}{3}A = 36$
$A = 36 \div \frac{8}{3}$
$= 13.5$
It does not matter what type of pyramid it is.]
$\therefore$ **[D]** ✓ (1 mark)

10

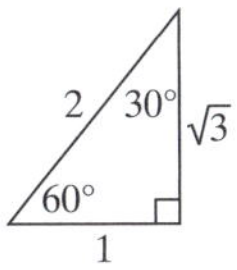

$\tan 30° = \frac{1}{\sqrt{3}}$
$\therefore$ **[B]** ✓ (1 mark)

11 $\sqrt{8} + 3\sqrt{2} = \sqrt{4} \times \sqrt{2} + 3\sqrt{2}$
$= 2\sqrt{2} + 3\sqrt{2}$
$= 5\sqrt{2}$
$\therefore$ **[A]** ✓ (1 mark)

12 $x = \frac{-b \pm \sqrt{b^2 - 4ac}}{2a}$
$\therefore$ **[A]** ✓ (1 mark)

13 Area of triangular base:
$A = \frac{1}{2}bh$
$= \frac{1}{2} \times 40 \times 9$
$= 180$
Volume:
$V = AH$
$= 180 \times 5$
$= 900$
The volume is 900 cm^3.

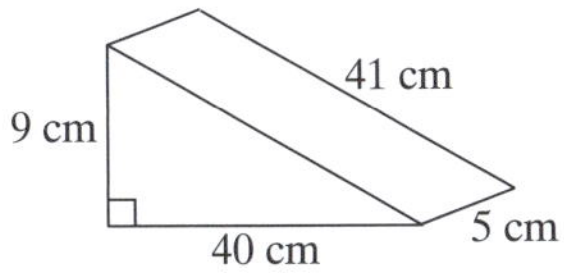

Surface area:
$A = 2 \times 180 + (9 + 40 + 41) \times 5$
$= 810$
The surface area is 810 cm^2.
$\therefore$ **[A]** ✓ (1 mark)

14

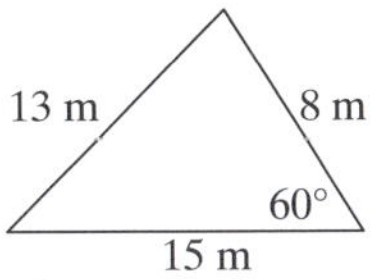

$A = \frac{1}{2}ab\sin C$
$= \frac{1}{2} \times 8 \times 15 \times \sin 60°$
$= 51.961\,524\,23\ldots$
The area is closest to 52 m^2.
$\therefore$ **[C]** ✓ (1 mark)

15 $(2x + 5)^2$
$= (2x)^2 + 2 \times 2x \times 5 + 5^2$
$= 4x^2 + 20x + 25$
The missing term is $20x$.
$\therefore$ **[C]** ✓ (1 mark)

16

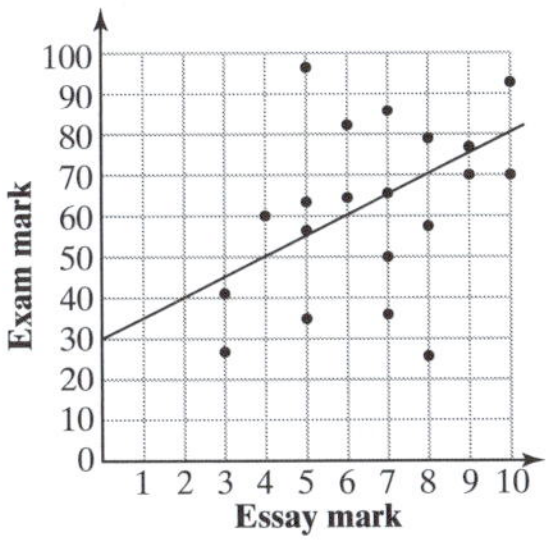

The lowest exam mark is about 25. The essay mark for this score was 8. For an essay mark of 8 the line of fit gives an exam mark of 70.
Difference = 70 – 25
= 45
$\therefore$ **[D]** ✓ (1 mark)

17

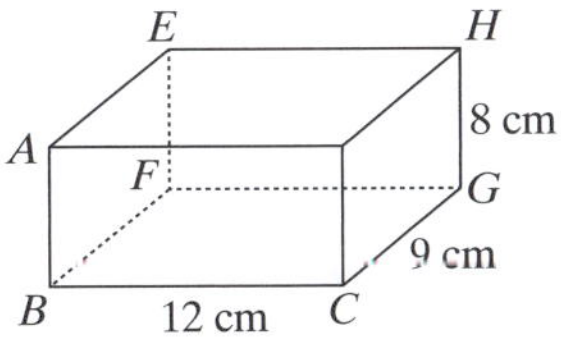

$BG^2 = 12^2 + 9^2$
$= 225$
$BG = 15 \quad (BG > 0)$
$AG^2 = 8^2 + 15^2$
$= 289$
$AG = 17 \quad (AG > 0)$
The length AG is 17 cm.
$\therefore$ **[B]** ✓ (1 mark)

18 $x^2 + 5x + 6 = 0$
$(x + 2)(x + 3) = 0$
$x + 2 = 0$ or $x + 3 = 0$
$x = -2$ or $x = -3$
$\therefore$ **[A]** ✓ (1 mark)

19 $m = 2, b = -1$
$y = mx + b$
$y = 2x - 1$
$2x - y - 1 = 0$
$\therefore$ **[A]** ✓ (1 mark)

20 $\log_8 2 = x$
$8^x = 2$
$8^{\frac{1}{3}} = 2 \quad [2 = \sqrt[3]{8}\,]$
$x = \frac{1}{3}$
$\therefore$ **[A]** ✓ (1 mark)

21 $(x + 4)^2 + (y - 2)^2 = 16$
$(x - (-4))^2 + (y - 2)^2 = 4^2$
Centre is $(-4, 2)$
$\therefore$ **[B]** ✓ (1 mark)

22 $P(x) = x^4 - 6x^3 - 4x^2 + 54x - 45$
$P(-5) = 960$
$P(-3) = 0$
$P(-2) = -105$
$P(-1) = -96$
$x + 3$ is a factor of $P(x)$
$\therefore$ **[B]** ✓ (1 mark)

23

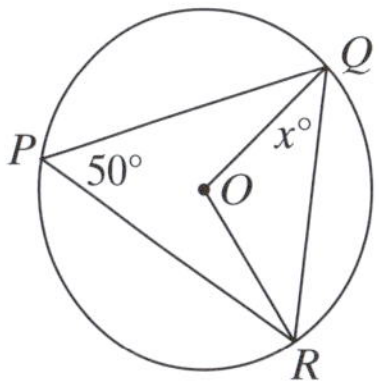

$\angle QOR = 100°$
(angle at centre)
$\angle OQR = \angle ORQ$
(angles opposite equal sides, $OQ = OR$ radii)
$2x + 100 = 180 \quad (\angle$ sum of $\Delta)$
$2x = 80$
$x = 40$
$\therefore$ **[C]** ✓ (1 mark)

24

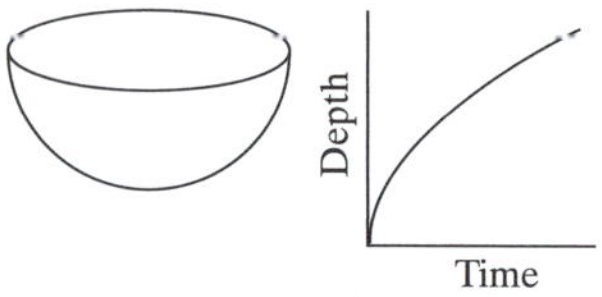

Water is flowing into the bowl at a constant rate but the depth will increase quickly at first and then more slowly.

$\therefore$ **[C]** ✓ (1 mark)

25

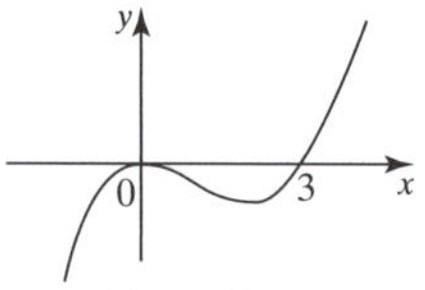

$y = x^2(x - 3)$ has a double root at $x = 0$ and a single root at $x = 3$.

As $x \to \infty$, $y \to \infty$

$\therefore$ **[D]** ✓ (1 mark)

(Total Part A: 25 marks)

SAMPLE EXAM PAPER 2

Part B: Written Response PAGE 101

26 **a** $4x^2 + 10x + 6$

$= 2(2x^2 + 5x + 3)$

$= 2(2x + 3)(x + 1)$ ✓ (1 mark)

b $x^2 + xy - x - y$

$= x(x + y) - 1(x + y)$ ✓

$= (x + y)(x - 1)$ ✓ (2 marks)

27 $n^2 - 6n = 18$

$n^2 - 6n + 9 = 18 + 9$

$(n - 3)^2 = 27$ ✓

$n - 3 = \pm\sqrt{27}$

$n = 3 \pm \sqrt{27}$

$= 3 \pm 3\sqrt{3}$ ✓ (2 marks)

28

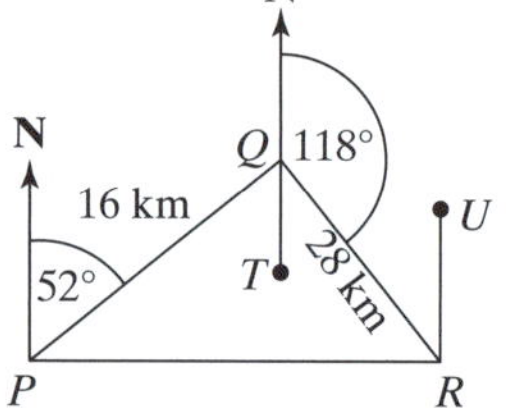

a Let T be a point due south of Q.

$\angle PQT = 52°$

(alternate $\angle$s, $\parallel$ lines)

$\angle RQT = 180° - 118°$

($\angle$s in a straight line)

$= 62°$

$\angle PQR = 52° + 62°$

$= 114°$ ✓ (1 mark)

b By the cosine rule,

$q^2 = p^2 + r^2 - 2pr\cos Q$

$= 28^2 + 16^2$

$\quad - 2 \times 28 \times 16 \times \cos 114°$ ✓

$= 1404.436\,032\ldots$

$q = \sqrt{1404.436\,032\ldots}$ $(q > 0)$

$= 37.475\,805\,96\ldots$

$= 37.5$ (1 d.p.)

The distance from P to R is 37.5 km. ✓ (2 marks)

c By the sine rule,

$\dfrac{\sin R}{r} = \dfrac{\sin Q}{q}$

$\dfrac{\sin R}{16} = \dfrac{\sin 114°}{37.4758\ldots}$

$\sin R = \dfrac{16\sin 114°}{37.4758\ldots}$

$= 0.390\,031\,033\ldots$

$R = 22.956\,4304\ldots°$

$\angle QRP = 23°$

(nearest degree) ✓

[$\angle PQR$ is obtuse so the other angles must be acute.]

Let U be a point due north of R.

$\angle QRU = 180° - 118°$

(co-interior $\angle$s, $\parallel$ lines)

$= 62°$

$\angle URP = 62° + 23°$

$= 85°$

Reflex angle URP

$= 360° - 85°$

$= 275°$

The bearing of P from R is 275°. ✓ (2 marks)

29 $3\log_3 6 - (\log_3 12 + \log_3 2)$

$= \log_3 6^3 - \log_3(12 \times 2)$

$= \log_3 216 - \log_3 24$ ✓

$= \log_3 \dfrac{216}{24}$

$= \log_3 9$

$= 2$ [since $3^2 = 9$] ✓ (2 marks)

30

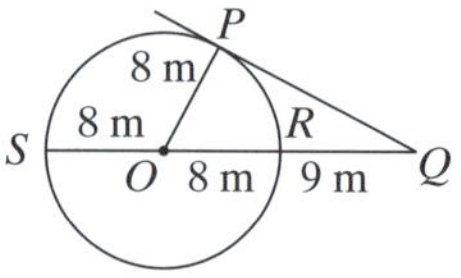

$PQ^2 = SQ \times RQ$

(product of intercepts on secant equals square of tangent) ✓

$= 25 \times 9$

$= 225$

$PQ = \sqrt{225}$ $(PQ > 0)$

$= 15$

The length of the tangent is 15 m. ✓

[Or use Pythagoras' theorem.

$\angle OPQ = 90°$

($\angle$ between tangent and radius)

$PQ^2 = 17^2 - 8^2 = 225$] (2 marks)

31 $\dfrac{4}{3+\sqrt{7}} = a + b\sqrt{7}$

$\dfrac{4}{3+\sqrt{7}} = \dfrac{4}{3+\sqrt{7}} \times \dfrac{3-\sqrt{7}}{3-\sqrt{7}}$

$= \dfrac{4(3-\sqrt{7})}{9-7}$

$= \dfrac{4(3-\sqrt{7})}{2}$

$= 2(3-\sqrt{7})$

$= 6 - 2\sqrt{7}$

$\therefore a = 6$ and $b = -2$. ✓✓ (2 marks)

32 $y = x^2 + 2x - 5$

$y = 3x + 1$

$\therefore\ x^2 + 2x - 5 = 3x + 1$

$x^2 - x - 6 = 0$

$(x - 3)(x + 2) = 0$

$x - 3 = 0$ or $x + 2 = 0$

$x = 3$ or $x = -2$ ✓

When $x = 3$, $y = 3 \times 3 + 1$

$= 10$

When $x = -2$, $y = 3 \times -2 + 1$

$= -5$

$\therefore$ The points of intersection are (3, 10) and (–2, –5). ✓ (2 marks)

33 Surface area = 2124 cm^2

$4\pi r^2 = 2124$

$r^2 = \dfrac{2124}{4\pi}$ ✓

$[= 169.022\,5496\ldots]$

$r = \sqrt{\dfrac{2124}{4\pi}}$ $(r > 0)$

$= 13.000\,867\,26\ldots$

The radius is 13 cm. ✓

$V = \dfrac{4}{3}\pi r^3$

$= \dfrac{4}{3} \times \pi \times (13.000\,867\,26\ldots)^3$

$= 9204.614\,021\ldots$

$= 9205$ (nearest unit)

The volume is 9205 cm^3 to the nearest cubic centimetre. ✓ (3 marks)

34 78, 86, 52, 67, 74, 82, 91, 66, 81, 60, 78, 75

a mean = 74.166 666…
= 74.2 (1 d.p) ✓
(1 mark)

b standard deviation
= 10.691 377 63…
[by calculator]
= 10.7 (1 d.p.) ✓
(1 mark)

c Ryan performed better in the art exam. ✓
In music his score of 81 was less than one standard deviation above the mean, but in art his mark of 73 is two standard deviations above the mean. ✓
(2 marks)

(Total Part B: 25 marks)

SAMPLE EXAM PAPER 3

Part A: Multiple Choice PAGE 103

1 $y = x^2 - 6x + 14$

Axis: $x = \dfrac{-b}{2a}$

$= \dfrac{-(-6)}{2 \times 1}$

$= 3$

When $x = 3$, $y = 3^2 - 6 \times 3 + 14$
$= 5$

So the minimum value, as the parabola is concave up, is 5.
The parabola will not meet the x-axis.

∴ **[A]** ✓ (1 mark)

2 $(16x^{-\frac{1}{2}})^{\frac{1}{2}} = 4x^{-\frac{1}{4}}$

$= \dfrac{4}{\sqrt[4]{x}}$

∴ **[A]** ✓ (1 mark)

3 $y = \dfrac{4}{x+2}$

As $x + 2 \neq 0$, $x = -2$ is an asymptote.
As x gets very large, y gets closer to 0.
$y = 0$ is an asymptote.

∴ **[B]** ✓ (1 mark)

4 If $x = 2at$,
$x^2 - tx = (2at)^2 - t(2at)$
$= 4a^2t^2 - 2at^2$

∴ **[D]** ✓ (1 mark)

5 Find the standard deviation for each:
A $\sigma_n = 3.05$ (2 d.p.)
B $\sigma_n = 3.19$ (2 d.p.)
C $\sigma_n = 3.71$ (2 d.p.)
D $\sigma_n = 3.29$ (2 d.p.)
A has the lowest standard deviation so it is the most consistent.

∴ **[A]** ✓ (1 mark)

6

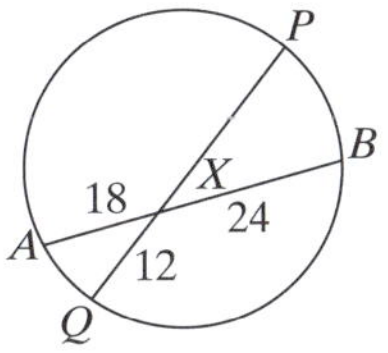

The products of the intercepts of intersecting chords are equal.
$12 \times PX = 18 \times 24$
$PX = 36$ cm
$PQ = 12 + 36$
$= 48$ cm

∴ **[D]** ✓ (1 mark)

7

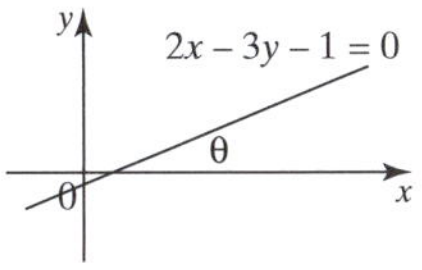

$2x - 3y - 1 = 0$
$3y = 2x - 1$
$y = \dfrac{2}{3}x - \dfrac{1}{3}$

So $m = \dfrac{2}{3}$

$\tan\theta = \dfrac{2}{3}$

$\theta = \tan^{-1}\left(\dfrac{2}{3}\right)$

$= 33.690\,0675\ldots^\circ$
$= 34^\circ$ (nearest degree)

∴ **[C]** ✓ (1 mark)

8

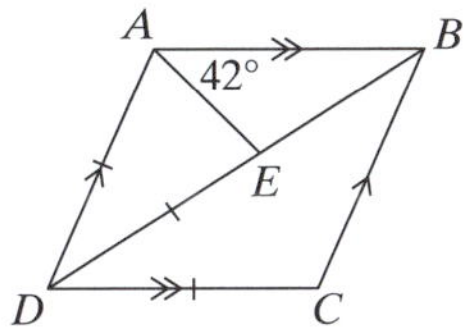

$ABCD$ is a rhombus.
(A parallelogram with adjacent sides equal.)
Let $\angle ADE = x^\circ$
$\angle ADE + \angle DAE + \angle DEA = 180^\circ$
(angle sum of triangle)

So $\angle DAE + \angle DEA = 180^\circ - x^\circ$
but $\angle DAE = \angle DEA$
(angles opposite equal sides isosceles triangle)

∴ $\angle DAE = (90 - \frac{x}{2})^\circ$
$\angle EDC = x^\circ$
(diagonals bisect angles)
$\angle BAD + \angle ADC = 180^\circ$
(co-interior, $AB \parallel DC$)

So $42 + 90 - \dfrac{x}{2} + x + x = 180$

$\dfrac{3x}{2} = 48$

$x = 32$

So $\angle DAE = 74^\circ$
$\angle DAB = 74^\circ + 42^\circ$
$= 116^\circ$
∴ $\angle BCD = 116^\circ$
(opposite angles are equal)

∴ **[C]** ✓ (1 mark)

9 $\sin 40^\circ = \cos(90 - 40)^\circ$
$= \cos 50^\circ$
$\sin 40^\circ = \sin(180 - 40)^\circ$
$= \sin 140^\circ$
Both I and II are correct.

∴ **[D]** ✓ (1 mark)

10 $y = 3x - 6$
Inverse function is
$x = 3y - 6$
$3y = x + 6$
$y = \dfrac{1}{3}x + 2$

∴ **[B]** ✓ (1 mark)

11

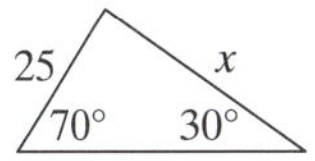

By the sine rule,

$\dfrac{a}{\sin A} = \dfrac{b}{\sin B}$

$\dfrac{x}{\sin 70^\circ} = \dfrac{25}{\sin 30^\circ}$

$x = \dfrac{25 \sin 70^\circ}{\sin 30^\circ}$

$= \dfrac{25 \sin 70^\circ}{½}$

$= 50 \sin 70^\circ$

∴ **[B]** ✓ (1 mark)

12

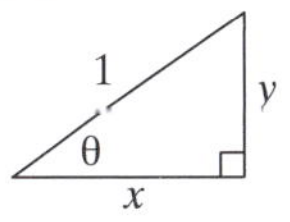

$\tan\theta = \dfrac{y}{x}$

∴ **[B]** ✓ (1 mark)

13 $\dfrac{x^2+7x+10}{x^2-x-6}\times\dfrac{x^2-7x+6}{x^2+4x-5}$

$=\dfrac{(x+2)(x+5)}{(x-3)(x+2)}\times\dfrac{(x-6)(x-1)}{(x+5)(x-1)}$

$=\dfrac{x-6}{x-3}$

∴ **[A]** ✓ (1 mark)

14 $3x^2-6x-1=0$

$a=3, b=-6, c=-1$

$x=\dfrac{-b\pm\sqrt{b^2-4ac}}{2a}$

$=\dfrac{-(-6)\pm\sqrt{(-6)^2-4\times3\times-1}}{2\times3}$

$=\dfrac{6\pm\sqrt{48}}{6}$

$=\dfrac{6\pm4\sqrt{3}}{6}$

$=\dfrac{3\pm2\sqrt{3}}{3}$

∴ **[B]** ✓ (1 mark)

15 $x^2-6x+y^2+4y=36$

x^2-6x+9
$\quad+y^2+4y+4=36+9+4$

$(x-3)^2+(y+2)^2=49$

So $r^2=49$

$r=7\quad(r>0)$

The radius is 7.

∴ **[D]** ✓ (1 mark)

16

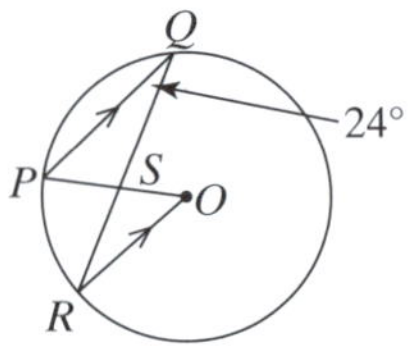

$\angle POR=48°$

(angle at centre is twice angle at circumference standing on the same arc)

$\angle OPQ=\angle POR$

(alternate, $PQ \parallel RO$)

∴ $OPQ=48°$

∴ **[D]** ✓ (1 mark)

17 $\tan A=-0.875$

$A=138.8140748\ldots°$

$=139°$ (nearest degree)

∴ **[B]** ✓ (1 mark)

18 Let the radius of the cylinder be r and the height h.

So $\pi r^2h=108$

The radius of the cone will be $2r$ and the height will be $\dfrac{h}{2}$.

$V=\dfrac{1}{3}\pi(2r)^2\times\dfrac{h}{2}$

$=\dfrac{2}{3}\pi r^2h$

$=\dfrac{2}{3}\times108$

$=72$

The volume of the cone will be 72 cm^3.

∴ **[C]** ✓ (1 mark)

19 Volumes in ratio $1:8=1:2^3$

Areas in ratio $1:2^2=1:4$

The area of the larger figure is four times the area of the smaller figure.

Area of smaller figure

$=(72\div4)\ m^2$

$=18\ m^2$

∴ **[C]** ✓ (1 mark)

20 $P(x)=x^3-4x^2+mx+n$

$P(0)=-5$

∴ $n=-5$

$P(x)=x^3-4x^2+mx-5$

$P(4)=7$

$7=4^3-4\times4^2+m\times4-5$

$7=4m-5$

$4m=12$

$m=3$

∴ $P(x)=x^3-4x^2+3x-5$

$P(-2)=(-2)^3-4\times(-2)^2+3\times(-2)-5$

$=-35$

∴ **[C]** ✓ (1 mark)

(Total Part A: 20 marks)

SAMPLE EXAM PAPER 3

Part B: Written Response PAGE 105

21 Let $u=x^2$

$x^4=(x^2)^2$

$=u^2$

$x^4-10x^2+9=0$

$u^2-10u+9=0$

$(u-9)(u-1)=0$

$u-9=0$ or $u-1=0$

$u=9$ or $u=1$ ✓

∴ $x^2=9$ or $x^2=1$

$x=\pm3$ or $x=\pm1$ ✓

(2 marks)

22 $\dfrac{1}{\sqrt{3}-2}+\dfrac{2}{\sqrt{5}-\sqrt{3}}$

$=\dfrac{1}{\sqrt{3}-2}\times\dfrac{\sqrt{3}+2}{\sqrt{3}+2}$

$+\dfrac{2}{\sqrt{5}-\sqrt{3}}\times\dfrac{\sqrt{5}+\sqrt{3}}{\sqrt{5}+\sqrt{3}}$

$=\dfrac{\sqrt{3}+2}{3-4}+\dfrac{2(\sqrt{5}+\sqrt{3})}{5-3}$ ✓

$=-\sqrt{3}-2+\sqrt{5}+\sqrt{3}$

$=-2+\sqrt{5}$ ✓ (2 marks)

23 $2\sin\theta-\sqrt{2}=0$

$2\sin\theta=\sqrt{2}$

$\sin\theta=\dfrac{\sqrt{2}}{2}$

$=\dfrac{1}{\sqrt{2}}$

Now $\sin 45°=\dfrac{1}{\sqrt{2}}$

$\theta=45°$ or $180°-45°$

$\theta=45°$ or $135°$ ✓✓

(2 marks)

24 $\dfrac{x}{x^2-1}-\dfrac{x+1}{x-1}$

$=\dfrac{x}{(x+1)(x-1)}-\dfrac{x+1}{x-1}$

$=\dfrac{x-(x+1)(x+1)}{(x+1)(x-1)}$

$=\dfrac{x-(x^2+2x+1)}{(x+1)(x-1)}$ ✓

$=\dfrac{x-x^2-2x-1}{(x+1)(x-1)}$

$=\dfrac{-x^2-x-1}{(x+1)(x-1)}$ ✓ (2 marks)

25 $\log_m 4=\log_8 2$

Now $2=\sqrt[3]{8}=8^{\frac{1}{3}}$

So $\log_8 2=\dfrac{1}{3}$

$\log_m 4=\dfrac{1}{3}$

$m^{\frac{1}{3}}=4$

$m=4^3$

$=64$ ✓ (1 mark)

26 $\log_7(x+y)=0$

∴ $x+y=1$ [i]

$\log_7(x-y)=1$

∴ $x-y=7$ [ii] ✓

[i] + [ii]: $2x=8$

$x=4$

Substitute into [i]:

$4 + y = 1$

$y = -3$

$\therefore x = 4$ and $y = -3$ ✓ (2 marks)

27

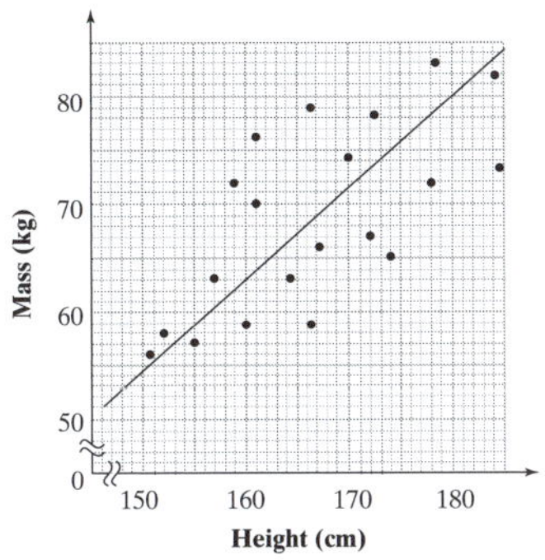

When $h = 160$, $y = 63$

When $h = 180$, $y = 80$

gradient: $m = \dfrac{80 - 63}{180 - 160}$

$= \dfrac{17}{20}$ ✓

Equation is of form $y = mh + b$

$y = \dfrac{17}{20}h + b$

Substitute $h = 180$ and $y = 80$,

$80 = \dfrac{17}{20} \times 180 + b$

$80 = 153 + b$

$b = -73$

$\therefore y = \dfrac{17}{20}h - 73$ ✓ (2 marks)

28 **a** $P(x) = x^3 - 4x^2 + x + 6$

$P(3) = 3^3 - 4 \times 3^2 + 3 + 6$

$= 0$

$\therefore x - 3$ is a factor of

$x^3 - 4x^2 + x + 6$ ✓ (1 mark)

b

$$\begin{array}{r}
x^2 - x - 2 \\
x - 3 \,\big)\, x^3 - 4x^2 + x + 6 \\
\underline{x^3 - 3x^2} \qquad\quad \\
-x^2 + x \qquad \\
\underline{-x^2 + 3x} \qquad \\
-2x + 6 \\
\underline{-2x + 6} \\
0
\end{array}$$
✓

$\therefore P(x) = (x - 3)(x^2 - x - 2)$

$= (x - 3)(x - 2)(x + 1)$

✓ (2 marks)

c y-intercept – 6

x-intercepts are –1, 2 and 3

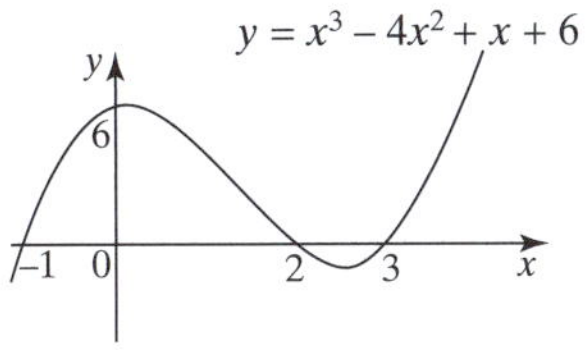

✓✓ (2 marks)

29 **a**

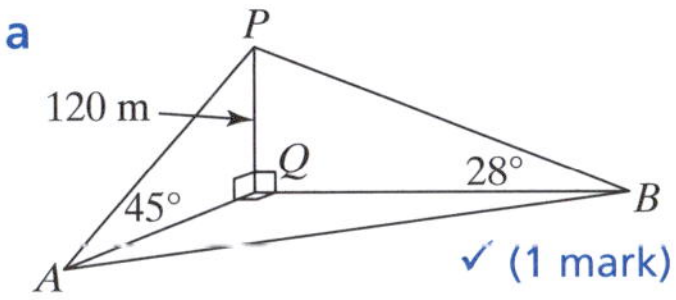

✓ (1 mark)

b ΔAPQ is a right-angled isosceles triangle

$[\angle APQ = \angle PAQ = 45°]$

so $AQ = PQ$

$\therefore AQ = 120$ m ✓ (1 mark)

c In ΔPBQ,

$\tan 28° = \dfrac{120}{BQ}$

$BQ = \dfrac{120}{\tan 28°}$ ✓

$= 225.687\,1758\ldots$

$= 226$ m

(nearest metre) ✓

(2 marks)

d $AB^2 = AQ^2 + BQ^2$

$= 120^2 + (225.687\,1758\ldots)^2$

$= 65\,334.701\,34\ldots$

$AB = 255.606\,5362\ldots$

$(AB > 0)$

$= 256$ m

(nearest metre) ✓

(1 mark)

30 **a**

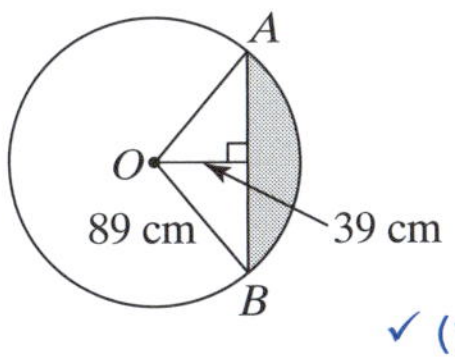

✓ (1 mark)

b Let the length of the chord be $2x$ cm.

$x^2 + 39^2 = 89^2$

$x^2 + 1521 = 7921$

$x^2 = 6400$ ✓

$x = 80 \quad (x > 0)$

So the chord is 2×80 cm or 160 cm long. ✓ (2 marks)

c Let $\angle AOB = \theta$

$\cos\theta = \dfrac{89^2 + 89^2 - 160^2}{2 \times 89 \times 89}$ ✓

$\theta = 128.0215\ldots°$

$= 128°$

(nearest degree) ✓

(2 marks)

d Area ΔAOB:

$A = \dfrac{1}{2}bh$

$= \dfrac{1}{2} \times 160 \times 39$

$= 3120$ cm^2 ✓

Area sector ΔAOB:

$A = \dfrac{128.0215\ldots}{360} \times \pi \times 89^2$

$= 8849.3303\ldots$ cm^2

Shaded area

$= 8849.3303\ldots - 3120$

$= 5729.3303\ldots$

$= 5729$ cm^2

(nearest cm^2) ✓ (2 marks)

(Total Part B: 30 marks)

TEST & EXAM RESULTS

Transfer your **percentage score** that you calculated in the **Your Feedback** box at the end of each **test** and **exam** to the table below. This will help you work out your areas of strength and weakness.

Test Topic	Intermediate Test Score	Advanced Test Score
Number and Algebra: Rational and Irrational Numbers	%	%
Number and Algebra: Surds	%	%
Number and Algebra: Logarithms	%	%
Number and Algebra: Functions	%	%
Number and Algebra: Polynomial Expressions	%	%
Number and Algebra: Further Polynomials	%	%
Number and Algebra: Parabolas	%	%
Number and Algebra: Further Graphs	%	%
Number and Algebra: Curve-sketching Techniques	%	%
Number and Algebra: The Equation of a Straight Line	%	%
Number and Algebra: Equations	%	%
Number and Algebra: Further Factorisation	%	%
Number and Algebra: Quadratic Equations	%	%
Measurement and Geometry: Volume of Pyramids, Cones and Spheres	%	%
Measurement and Geometry: Surface Area of Pyramids, Cones and Spheres	%	%
Measurement and Geometry: Geometry Review	%	%
Measurement and Geometry: Parts of Circles and Chord Properties	%	%
Measurement and Geometry: Angle Properties, Tangents and Secants	%	%
Measurement and Geometry: Problems in Two and Three Dimensions	%	%
Measurement and Geometry: Complementary Ratios, Exact Values and the Unit Circle	%	%
Measurement and Geometry: Sine Rule, Cosine Rule and Area of a Triangle	%	%
Statistics and Probability: Mean and Standard Deviation	%	%
Statistics and Probability: Investigations of Data Sets and Reports	%	%
Sample Exam Papers		
Sample Exam Paper 1		%
Sample Exam Paper 2		%
Sample Exam Paper 3		%

FEEDBACK CHECKLIST TO IMPROVE YOUR TEST & EXAM RESULTS

Do you want to improve your scores in the Sample Exams? Check that:

☑ You are ready.

This is a revision workbook designed specifically for the revision of work already done. It is not a replacement for your textbook or class notes. Cover the topic in class or in your own time by reading your textbook or the ***Excel*** *Advanced Mathematics Study Guide Years 9–10*, before using this book for further practice on the topic.

☑ You are revising in the right order.

Maths is a subject that builds on previous knowledge. Often you will need to have a good grasp of an early topic in order to fully understand a later topic.

☑ Your standards are realistic.

You cannot expect to score 100% all the time. Remember that this book has been designed to help you identify your strengths as well as your weaknesses, so it is OK to make mistakes. The key to success is learning from those mistakes.

☑ You are not rushing through the questions.

Many Maths students fail to read questions properly. This accounts for a large number of the mistakes made in tests and exams. Although your time is limited, you must still take enough time to read each question carefully. If necessary, read the question a second (or even third) time. Don't start your answer until you are sure that you understand the question. For the longer questions, take a moment or two to plan your answer.

☑ You are allowing yourself enough time to study.

Preparation for any test or exam requires a sufficient investment of time to be undertaken successfully. This means that you have to plan ahead. Don't wait until a week or so before an exam, for example, to ask yourself the question: How much time will I need to devote to this subject to revise it thoroughly? Don't forget to allow sufficient time for your other subjects—Maths isn't the only subject you will have to sit an exam for!

☑ You are OK.

You may be having some problems unrelated to your schoolwork. It is wise to sort out such problems as quickly as possible. Speak to your teacher, a parent or your school counsellor if you think you need help.

NOTES

NOTES

NOTES